Handbook of WiMAX Technology

Handbook of
WiMAX Technology

Edited by **Timothy Kolaya**

C LANRYE
I NTERNATIONAL

New Jersey

Published by Clanrye International,
55 Van Reypen Street,
Jersey City, NJ 07306, USA
www.clanryeinternational.com

Handbook of WiMAX Technology
Edited by Timothy Kolaya

International Standard Book Number: 978-1-63240-295-0 (Hardback)

Contents

Preface

WiMAX is a technology standard for long-range wireless networking. This book has been compiled to showcase the novel techniques in WiMAX Technology. The target of the book is the physical layer, and it assembles the contributions of various prominent researchers from across the world. The amount of distinct works on WiMAX depicts the enormous worldwide significance of WiMAX as a wireless broadband access technology. This book is aimed for readers concerned with the transmission process under WiMAX since it includes both theoretical and technical information, which offers an in-depth review of latest advances in the field, for engineers and researchers, and other readers interested in WiMAX.

This book is a comprehensive compilation of works of different researchers from varied parts of the world. It includes valuable experiences of the researchers with the sole objective of providing the readers (learners) with a proper knowledge of the concerned field. This book will be beneficial in evoking inspiration and enhancing the knowledge of the interested readers.

In the end, I would like to extend my heartiest thanks to the authors who worked with great determination on their chapters. I also appreciate the publisher's support in the course of the book. I would also like to deeply acknowledge my family who stood by me as a source of inspiration during the project.

Editor

Part 1

Advanced Transmission Techniques, Antennas and Space-Time Coding

Space-Time Adaptation and MIMO Standardization Status

Ismael Gutiérrez and Faouzi Bader
Centre Tecnològic de Telecomunicacions de Catalunya-CTTC,
Parc Mediterrani de la Tecnologìa, Castelldefels, Barcelona,
Spain

1. Introduction

The use of multiple antennas at the transmitter and/or receiver sides has been used for years in order to increase the signal to noise ratio at the receiver or for beam steering (a.k.a. beamforming) to reduce the amounts of interference at the receiver. However, one of the main benefits of using a Multiple Input Multiple Output (MIMO) channel is the increase of the channel capacity. Nowadays, by using different space-time-frequency coding techniques, orthogonal (or quasi orthogonal) virtual paths between transmitter and receiver can be obtained. These virtual paths can be used in order to increase the spectral efficiency or in order to increase the signal diversity (i.e. space diversity). In fact, a fundamental trade-off between diversity and multiplexing capabilities exists and must be considered when designing a multiple antenna system. In this chapter, the following issues are addressed:

- We start by describing and reviewing the (ergodic) capacity of the MIMO channel in case of perfect channel state information at the transmitter (CSIT). Afterwards, the outage capacity is studied leading to the diversity-multiplexing trade-off.
- Then, we analyze different spatial adaptation and precoding mechanisms that can be applied to increase the performance of the system (either in terms of throughput or robustness). A new spatial adaptation algorithm proposed by the authors and called *Transmit Antenna and space-time Coding Selection* (TACS) is described showing some performance results that illustrate the improve on performance and/or throughput.
- Finally, the well-known Space-Time coding techniques are reviewed, and a summary of the MIMO techniques adopted in WiMAX2 (IEEE 802.16e/m) is provided.

The present analysis is done following the general Linear Dispersion Codes framework, which is of special interest since it allows describing in an elegant way most of the space-time block codes existing in the literature.

2. Characteristics of the MIMO channel

Before going into the details of how MIMO transmission can be carried out, it is important to have a look to the capacity of the MIMO channel. Usually, for capacity evaluation of

MIMO channels it is assumed that the fading coefficients between antenna pairs are i.i.d. Rayleigh distributed. In addition, and without loss of generality, it is also assumed that the channel is constant during the transmission of one MIMO codeword. Under this assumption the channel is referred as a block fading channel.

As shown in [1], the capacity of the MIMO channel can be obtained as follows

$$C(\rho,\mathbf{H}) = \underset{\mathbf{Q}>0:Tr\{\mathbf{Q}\}=M}{\arg\max} \quad \log_2 \det\left[\mathbf{I}_N + \frac{\rho}{M}\mathbf{H}\mathbf{Q}\mathbf{H}^{\mathrm{H}}\right] \qquad (1)$$

where $\mathbf{H}$ is the MIMO channel matrix, M the number of transmit antennas, I_N an identity matrix of size N equal to the number of receive antennas, ρ is the Signal to Noise Ratio (SNR) and $\mathbf{Q}$ is the input covariance matrix whose trace is normalized to be equal to the number of transmit antennas. To gain a further insight on the channel characteristics, a good method is to apply the *Singular Value Decomposition* (SVD) to the MIMO channel matrix, so we express the channel matrix as

$$\mathbf{H} = \mathbf{U}\boldsymbol{\Sigma}\mathbf{V}^{\mathrm{H}}, \qquad (2)$$

where $\boldsymbol{\Sigma}$ is a diagonal matrix, whose entries are the eigenvalues of $\mathbf{H}$, and $\mathbf{U}$ and $\mathbf{V}$ are the lower and upper diagonal matrices respectively. The first important characteristic of the MIMO channel is given by the number of eigenvalues which tells us about the number of the independent virtual channels between transmitter and receiver. In addition, using the SVD of channel matrix, we can rewrite the channel capacity as

$$C(\rho,\mathbf{H}) = \underset{\{p_k\}_{k=1}^n}{\arg\max} \quad \sum_{k=1}^{n}\log_2\left(1 + \rho p_k \sigma_k^2\right) \qquad (3)$$

where σ_k and p_k are the eigenvalues and the power transmitted through each of the said virtual channels respectively.

In case the channel is known at the transmitter, one can use this information to maximize the channel capacity by applying what it is known as multiple eigenmode transmission. This is much less complex that it sounds, and is carried out by multiplying the input vector $\mathbf{x}$ by $\mathbf{V}^{\mathrm{H}}$. At the receiver side, similar operation is needed, therefore the result of multiplying the received signal by $\mathbf{U}^{\mathrm{H}}$ is shown in the following equation

$$\mathbf{y} = \mathbf{U}\boldsymbol{\Sigma}\mathbf{V}^{\mathrm{H}}\mathbf{V}\mathbf{x} + \mathbf{n} = \mathbf{U}\boldsymbol{\Sigma}\mathbf{x} + \mathbf{n} \Rightarrow \hat{\mathbf{x}} = \mathbf{U}^{\mathrm{H}}\mathbf{U}\boldsymbol{\Sigma}\mathbf{x} + \mathbf{U}^{\mathrm{H}}\mathbf{n} = \boldsymbol{\Sigma}\mathbf{x} + \tilde{\mathbf{n}}, \qquad (4)$$

where it is observed that the number of eivenvalues determines the number of independent complex symbols that can be transmitted per each MIMO codeword. It can be also concluded that at low SNR values, the optimum allocation strategy will be to allocate all the available power to the strongest (or dominant) eigenmode, whereas at high SNRs the maximum capacity is obtained by allocating the same power to all the non-zero eigenmodes [1]. Actually, it is also proved that uniform power allocation (UPA) is the optimum strategy for fast fading channels where the transmitter is not able to capture the instantaneous channel state. The (ergodic) channel capacity in case of UPA allocation is given by

$$C_{UPA}(\rho,\mathbf{H}) = E\left\{\sum_{k=1}^{n} \log_2\left(1 + \frac{\rho}{M}\sigma_k^2\right)\right\} = E\left\{\log_2\left(1 + \frac{\rho}{M}\|\mathbf{H}\|_F^2\right)\right\}. \tag{5}$$

where $\|\mathbf{H}\|_F^2$ stands for the Froebenius norm. Similar research studies have been undertaken regarding the effects of antenna correlation on the channel capacity. It is shown in [2] that for low antenna correlation values the optimum strategy is to allocate the same power to all the eigenmodes, whereas for high correlation values the optimum strategy is allocating all the power to the strongest eigenmode.

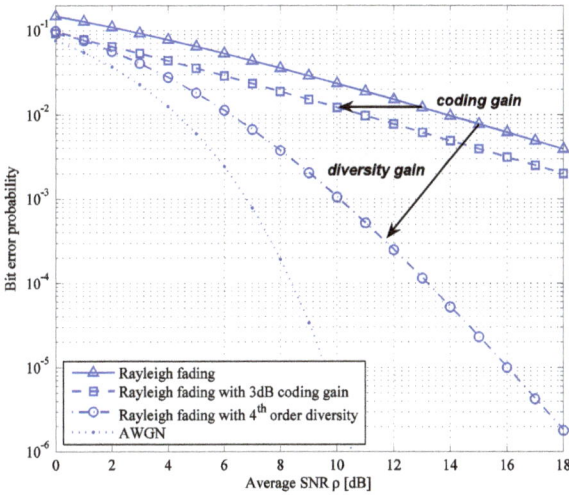

Fig. 1. Array and diversity gain in Rayleigh fading channels.

2.1 The diversity-multiplexing trade-off

When dealing with frequency/time-variant channels, one intrinsic characteristic of the channel is the diversity that can be achieved. In case of single input single output transmission, the general approach is to use coding and interleaving in the frequency and time domains so that one codeword is spread over the highest number possible of channel states. However, frequency and time diversity incur in a loss in bandwidth and/or transmission time delay. Alternatively, in case of multiple input multiple output channels the spatial dimension can be also exploited in order to increase the diversity without neither losing bandwidth nor increasing the transmission delay. Some metrics are defined to characterize the diversity. First, the *diversity gain* (or diversity order) is linked with the number of independent fading branches. Formally, it is defined as the negative asymptotic slope (i.e. for $\rho \to \infty$) of the *log-log* plot of the average error probability P versus the average SNR ρ

$$g_d = -\frac{\log(P)}{\log(\rho)}. \tag{6}$$

Finally, the *coding gain* is defined as the SNR gain (observed as a left shift of the error curve). Then the coding gain g_c is analytically expressed as

$$P_e = \left(\frac{c}{g_c \rho}\right)^{\alpha g_a}, \tag{7}$$

where P_e is error probability, and α and c are scaling constants depending on the modulation level, the coding scheme, and the channel characteristics. The *array gain* g_a represents the decrease of average SNR due to coherent combining (*beamforming*) in case of multiple antennas at both transmitter or receiver sides, and it is formally expressed as

$$g_a = \frac{\rho_{ma}}{\rho_{sa}} \tag{8}$$

where ρ_{sa} is the average SNR for the SISO link, and ρ_{ma} is the average SNR for the MIMO link. The three different concepts are illustrated in Fig. 1 which shows the bit error rate of a QPSK transmission having an AWGN channel and an uncorrelated Rayleigh channel.

In case of multiple antennas at both sides of the link, multiple independent channels exist according to the rank of the channel matrix [1]. The multiplexing gain g_s is defined as the ratio of the transmission rate $R(\rho)$ to the capacity of an AWGN with array gain g_a

$$g_s = \frac{R(\rho)}{\log_2(1 + g_a \rho)} \underset{\rho \to \infty}{\Rightarrow} R(\rho) = g_s \log_2(g_a \rho) \tag{9}$$

where $R(\rho)$ is the transmission rate.

For a slow fading channel (i.e. block fading channel), the maximum achievable rate for each codeword is a time variant quantity that depends on the instantaneous channel realizations. In this case, the outage probability P_{out} metric is preferred and is defined as the probability that a given channel realization cannot support a given rate R

$$P_{out}(R) = \inf_{Q>0:Tr\{Q\}\leq M} P\left(\log_2 \det\left(\mathbf{I}_N + \frac{\rho}{M}\mathbf{HQH}^H\right) < R\right). \tag{10}$$

where *inf* stands for the Q that achieves the lower bound in terms of outage probability. Then, we can gain insights into the channel behaviour by analysing the outage probability as a function of the SNR and ρ for a given transmission rate. Actually, we can establish a relationship between the diversity gain and the multiplexing gain via the outage probability P_{out} as

$$g_d(g_s, \rho) = -\frac{\partial \log(P_{out}(R))}{\partial \log(\rho)} \underset{\rho \to \infty}{\Rightarrow} P_{out}(\rho) = \rho^{-g_d}. \tag{11}$$

Both the multiplexing gain and the diversity gain are upper bounded by $g_s \leq min(N,M)$ and $g_d \leq N \times M$. Intuitively, the multiplexing gain indicates the increase of the transmission rate as a function of the SNR, whereas the diversity gain gives us an idea on how fast the outage probability decreases with the SNR.

The analysis of Eq. (11) at high SNR and uncorrelated Rayleigh channel leads to the diversity-multiplexing trade-off of the channel [3]. It has been shown that $g_d(g_s,\infty)$ is a piece-wise linear function joining the $(g_s, g_d(g_s,\infty))$ points with $g_s=\{0,\ldots, \min(N,M)\}$ and $g_d=(N-g_s)\times(M-g_s)$. This trade-off is illustrated in the following Fig. 2. It is observed that maximum diversity is achieved when there is no spatial multiplexing gain (i.e. the transmission rate is fixed), whereas the maximum spatial multiplexing gain is achieved when the diversity gain is zero (the outage probability is kept fixed).

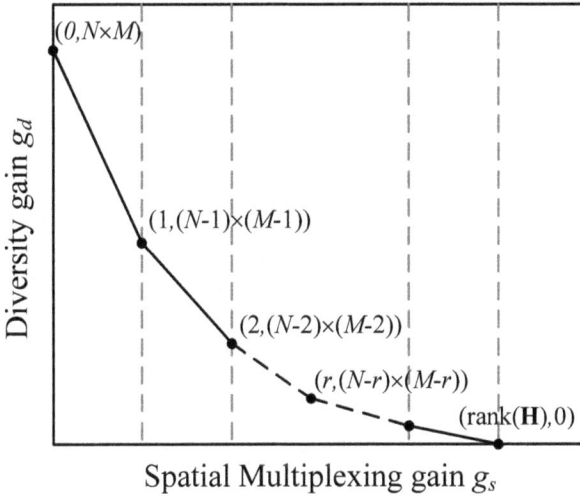

Fig. 2. Asymptotic diversity-multiplexing trade-off in uncorrelated Rayleigh channels.

2.2 Space-time coding over MIMO channels

Based on the above introduction on MIMO channel characteristics, and the very important principle of spatial diversity versus spatial multiplexing tradeoff, we could now start studying the Space-Time MIMO encoding techniques. Analogously to channel coding in SISO links, two types of channel coding have been used for MIMO channels: block coding (referred as Space Time/Frequency Block Coding, STBC/SFBC) and convolutional coding (referred as Space Time Trellis Coding-STTC) [4][5]. For the STBC case, the codeword is only a function of the input bits, whereas the encoder output for the STTC is a function of the input bits and the encoder state. The inherent memory of the STTC provides an additional coding gain compared to the STBC at the expense of higher computational complexity [7][8]. However, since STBC transforms the MIMO channel into an equivalent scalar AWGN channel [6], the concatenation of traditional channel coding with STBC shows good performance and even outperforms STTC for low number of receive antennas $(M,N\leq2)$ [7] and the same number of encoder states. Furthermore, STBC/SFBC are of significantly less complexity than STTC and for this reason they are usually preferred.

2.2.1 Space-time block coding system model

In this section an example of a MIMO system with M, N transmitter and receiver antennas respectively communicating over a frequency flat-fading channel is assumed. A codeword $\mathbf{X}$ is transmitted over T channel accesses (symbols) over the M transmitting antennas, hence $\mathbf{X}=[\mathbf{x}_0 \ldots \mathbf{x}_{T-1}]$ with $\mathbf{x}_i \in C^{M \times 1}$. All the codewords are contained inside a codebook X, and each codeword contains the information from Q complex symbols. The ratio of symbols transmitted per codeword is defined as the spatial multiplexing rate $r_s=Q/T$, where in case of $r_s=M$ the code is referred as *full-rate*. The transmission rate is given by $R=Q \times n_b/T$ [bits/s/Hz] where n_b is the number of bits transmitted by each $\mathbf{x}_i(j)$ complex symbol. Moreover, the spreading of the symbols in the time and spatial domains leads to the increase of diversity, whereas by modifying Q we can modify the spatial multiplexing gain. In consequence, $M \times N \times T$ determines the maximum diversity order, while Q defines the spatial multiplexing rate [10]. During any i^{th} time instant (equivalent to a channel access), the transmitted and received signals are related as

$$\mathbf{y}_i = \sqrt{\frac{\rho}{M}}\mathbf{H}_i\mathbf{x}_i + \mathbf{n}_i, \quad i = \{0,...,T-1\}. \tag{12}$$

In Eq. (12) we have assumed the channel is constant within each channel Access. However, we can go one a step further and assume that the channel is constant during the transmission of one whole codeword transmission ($T_{coh} \gg T$) time, and in this case the channel dependency on the ith time instant subindex can be dropped such that $\mathbf{H}_i=\mathbf{H}$ for any $i=\{0,...,T-1\}$. Under these conditions the quasi-static block fading channel model can be assumed, and we can rewrite Eq. (12) as follows

$$\mathbf{Y}^T = \sqrt{\frac{\rho}{M}}\mathbf{H}\mathbf{X}^T + \boldsymbol{\mathcal{V}}^T \Rightarrow \mathbf{Y} = \sqrt{\frac{\rho}{M}}\mathbf{X}\mathbf{H}^T + \boldsymbol{\mathcal{V}} \tag{13}$$

where $\mathbf{X} \in C^{T \times M}$ means the space-time transmitted codeword, $\mathbf{Y} \in C^{T \times N}$ means the received space-time samples and $\mathbf{V} \in C^{T \times N}$ represents the noise over each receive antenna during each channel access.

2.2.2 Linear Dispersion Codes

The Linear Dispersion Codes (LDC) class belongs to a subclass of the STBC codes where the codeword is given by a linear function of the input data symbols [11][20][21]. When the codeword is a linear function of the data symbols, the transmitted codeword can be expressed as

$$\mathbf{X} = \sum_{q=1}^{Q}\left(\alpha_q\mathbf{A}_q + j\beta_q\mathbf{B}_q\right), \tag{14}$$

where $\mathbf{A}_q \in C^{T \times M}$ determines how the real part of the symbol s_q, a_q, is spread over the space-time domain, and the same for the imaginary part β_q which is spread according to $\mathbf{B}_q \in C^{T \times M}$. For power normalization purposes, it is considered that the transmitted complex symbols s_q have zero mean and unitary energy, this is $E\{s_q^*s_q\}=1$. The matrices $\mathbf{A}_q$, and $\mathbf{B}_q$ are referred as the basis matrices and usually are normalized such that

$$\sum_{q=0}^{Q-1}\left(Tr\left\{\mathbf{A}_q\mathbf{A}_q^H\right\}+Tr\left\{\mathbf{B}_q\mathbf{B}_q^H\right\}\right)=MT. \tag{15}$$

If we impose some conditions on the set of basis matrices $\mathbf{A}_q$, $\mathbf{B}_q$ with $q=\{0,...,Q-1\}$ the mapping between the input symbols and the transmitted codeword $\mathbf{X}$ is unique and the symbols can be perfectly recovered. Substituting Eq. (15) into Eq. (13) and applying the *vec* operator on both sides of the expression, an equivalent real valued system equation can be written as

$$\mathbf{y}=\begin{bmatrix}\Re(\mathbf{y}_0)\\\Im(\mathbf{y}_0)\\\vdots\\\Re(\mathbf{y}_{Q-1})\\\Im(\mathbf{y}_{Q-1})\end{bmatrix}=\sqrt{\frac{\rho}{M}}\mathbf{H}\underbrace{\begin{bmatrix}\alpha_0\\\beta_0\\\vdots\\\alpha_{Q-1}\\\beta_{Q-1}\end{bmatrix}}_{\triangleq s}+\underbrace{\begin{bmatrix}\mathbf{n}_0\\\mathbf{n}_0\\\vdots\\\mathbf{n}_{Q-1}\\\mathbf{n}_{Q-1}\end{bmatrix}}_{\triangleq n}=\sqrt{\frac{\rho}{M}}\mathcal{H}s+n, \tag{16}$$

where $n\in\mathbf{N}\,(0,\,\tfrac{1}{2})$ is the real vector noise with i.i.d. components. The equivalent real valued channel matrix $\mathbf{H}$ is obtained as

$$.\mathcal{H}=\begin{bmatrix}\mathcal{A}_0h_0&\mathcal{B}_0h_0&\cdots&\mathcal{A}_{Q-1}h_0&\mathcal{B}_{Q-1}h_0\\\vdots&\vdots&\ddots&\vdots&\vdots\\\mathcal{A}_0h_{N-1}&\mathcal{B}_1h_{N-1}&\cdots&\mathcal{A}_{Q-1}h_{N-1}&\mathcal{B}_{Q-1}h_{N-1}\end{bmatrix}_{2NT\times2Q}. \tag{17}$$

with:

$$\mathcal{A}_q=\begin{bmatrix}\Re\{\mathbf{A}_q\}&-\Im\{\mathbf{A}_q\}\\\Im\{\mathbf{A}_q\}&\Re\{\mathbf{A}_q\}\end{bmatrix},\mathcal{B}_q=\begin{bmatrix}-\Im\{\mathbf{B}_q\}&-\Re\{\mathbf{B}_q\}\\\Re\{\mathbf{B}_q\}&-\Im\{\mathbf{B}_q\}\end{bmatrix}\in\mathbb{R}^{2T\times2M};h_n=\begin{bmatrix}\Re\{\mathbf{h}_n\}\\\Im\{\mathbf{h}_n\}\end{bmatrix}\in\mathbb{R}^{2M\times1}, \tag{18}$$

where $\mathbf{h}_n$ is the n-th row of the MIMO channel matrix $\mathbf{H}$. Theoretically the maximum number of possible independent streams or channel modes of the effective MIMO channel matrix H is $2NT\times2Q$ (maximum number of singular values different than zero). However, following the LDC design, the number of scalar modes that are excited is equal to the rank of H^TH, which means equal to $2Q$ in the best case [18]. In addition, it is observed that it is possible to use a linear receiver only if $Q\leq NT$, otherwise the system would be undetermined. When $Q<NT$ the system is over-determined, and in consequence more reliability is given to each estimated symbol (i.e. spatial diversity is increased at the expense of spatial multiplexing).

3. Detection techniques: Linear vs non-linear schemes

Recovering the transmitted symbols within each codeword might become a challenging task depending on the set of basis matrices used. Moreover, we have already observed that the ergodic channel is also a function of the basis matrices set. Therefore, a compromise between complexity and performance (in terms of spectral efficiency or decoding errors) exists which motivates the implementation and use of different LDC codes and decoding schemes.

3.1 Maximum Likelihood (ML) decoding

Optimum signal detection requires the maximization of the likelihood function over the discrete set of the code alphabet [13]. Mathematically, this can be expressed as

$$\hat{s} = \arg \min_{\hat{s} \in \Sigma} \left\| y - \sqrt{\frac{\rho}{M}} Hs \right\|_F^2 \tag{19}$$

where Σ is the space of all the transmitted symbol vectors with all the input data combinations having the same likelihood. Regarding the computational complexity of the ML detector, since each vector has a set of $2Q$ symbols each one mapped over $log_2(Z)$ bits, the computational complexity is exponential with $Q \times log_2(Z)$.

The theoretical framework to understand the behavior of a MIMO system using an ML receiver has been extensively studied and analyzed in the scientific literature (e.g. [9]-[13]) leading to important conclusions. The first and more obvious conclusion is that the diversity order is $g_d \leq N \ min(M,T)$ [13]. So, it becomes clear that T should be equal to M to achieve full diversity. However, increasing T requires increasing also the value of Q to maintain the same rate R, which leads to an increase in memory requirements, computational complexity, and delay. It is then apparent that a trade-off exists between achievable diversity and the decoding complexity.

Often the performance of a system is measured in terms of the post-processing SNR or the Effective SNR (ESINR). This ESINR value estimates the SNR required in an AWGN channel to obtain the same performance as in the given system (i.e. our MIMO system). In [14], the author proposed a simple parametrizable expression to estimate the performance of the system under different MIMO transmission schemes and antenna configurations. This model has been used in later sections when the performance evaluation of adaptive MIMO systems with ML receivers is developed and analysed.

3.2 Linear detectors: Zero forcing and minimum mean square error

The high computational cost of the ML receiver $(O(2^{Qlog_2(Z)}))$ makes the use of less computational demanding receiving techniques more appealing, sometimes even despite a degradation on the system performances. Following the expression in Eq. (12), a linear relationship between input and output symbols exists and the system can be solved applying simple algebra as long as $Q \leq NT$, i.e.

$$\hat{s} = Gy = \sqrt{\frac{\rho}{M}} GHs + Gn, \tag{20}$$

where $G \in \mathbb{R}^{2Q \times 2N}$ is the equalizer matrix which compensates the MIMO channel effects. Similar to frequency equalization, the equalizer matrix might be designed to suppress the inter-symbol interference (despite the noise vector might be increased due to the equalization) or to minimize the mean square error (i.e. the MMSE). The first design criterion is known as the Zero-Forcing equalization where

$$\mathbf{G}_{ZF} = \sqrt{\frac{M}{\rho}}\mathbf{H}^{\dagger}. \tag{21}$$

On the other hand, according to the MMSE criterion the following equalizer is obtained

$$\mathbf{G}_{MMSE} = \sqrt{\frac{M}{\rho}}\left(\mathbf{H}^{H}\mathbf{H} + \left(\frac{\rho}{M}\right)^{-1}\mathbf{I}_Q\right)^{-1}\mathbf{H}^{H}. \tag{22}$$

where $\mathbf{I}_Q$ is an identity matrix of size Q. Besides the lower computational cost of the linear receivers, another advantage of using them is that the channel effects can be perfectly estimated on a symbol basis, hence a closed expression for the $ESINR$ per each transmitted symbol can be obtained as

$$ESINR_q = \frac{diag\left[\mathbf{D}\mathbf{D}^{H}\right]_q}{diag\left[\frac{1}{\rho}\mathbf{G}^{H}\mathbf{G} + \mathbf{I}_{self}\mathbf{I}_{self}^{H}\right]_q}. \tag{23}$$

where $\mathbf{D}=diag[\mathbf{GH}]$, and $\mathbf{I}_{self} =\mathbf{GH}-\mathbf{D}$ is the self-interference term. The full expression of ML and ZF receivers can be found in [14].

4. Exploiting the transmit channel knowledge

It has been already stated in previous sections that in case the transmitter has perfect channel information knowledge, the SNR at the receiver is maximized if the transmitter applies all the power over the dominant eigenmode of the channel. Moreover, in order to increase the throughput it may be preferable to transmit over all the non-zero eigenmodes of the channel allocating to each mode a power obtained following the water-filling algorithm [10]. However, both (dominant and multiple eigenmode transmission) beamforming techniques require that the transmitter knows perfectly the channel state information (CSI), and also that the channel doesn't change during a sufficiently large period to allow the CSI estimation and application of the beamforming. In consequence, the beamforming might be only applied for low mobility scenarios and where the channel can be accurately estimated.

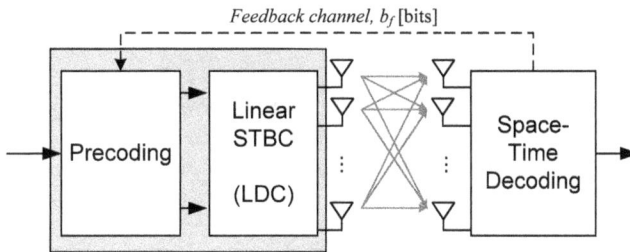

Fig. 3. Linear space-time precoding

In order to obtain the channel information at the transmitter, the most common approach is that the receiver sends some signalling to allow the transmitter to know the status of the downlink channel (in case of FDD system this has to be done explicitly by transmitting the matrix **H**, and for TDD systems the channel reciprocity allows sending some pilots in the reverse link so that the channel is estimated for the forward channel). However, any of these two alternatives will consume bandwidth either in the form of feedback signalling or channel estimation signalling. This triggered lot of work on how to reduce the feedback leading to techniques and metrics such as quantizing the channel information, using the channel condition number, the Demmel condition number, the channel rank, etc [26]. In general, the schemes where the input symbols are adjusted according to the channel status are known as precoders. Actually, precoding and space-time block coding can be considered into the same block where given a set of codes (defined by the codebook) one of them is selected each *Time Transmission Interval* (defined by T or multiples of) according to the b_f feedback bits.

4.1 Transmit and receive antenna selection

Besides the increase of the capacity or the reliability by any of the before mentioned precoding techniques (beamforming, codeword selection based on finite codebooks, etc.), a very simple precoding technique is to select which antenna (or subset of antennas) should be used according to an optimization criterion (e.g. capacity, reliability, etc.). Antenna selection also aids to reduce the hardware cost as well as the signal processing requirements, therefore, it may be good for handheld receivers where, space, power consumption, and cost must be seriously taken into account. Obviously, the reduction of the number of antennas reduces the array gain, however when the channel in any of these antennas is experiencing a deep fade, the capacity loss by not using such antenna is negligible [19]. In consequence, antenna selection at both transmitter and receiver helps in reducing the implementation costs while retaining most of the benefits of MIMO technology.

A MIMO system model considering antenna selection is depicted in Fig. 4, where M and N are the number of transmitter and receiver RF chains respectively, whereas the available antennas are referred by M_a and N_a for transmitter and receiver respectively ($M \le M_a$, $N \le N_a$).

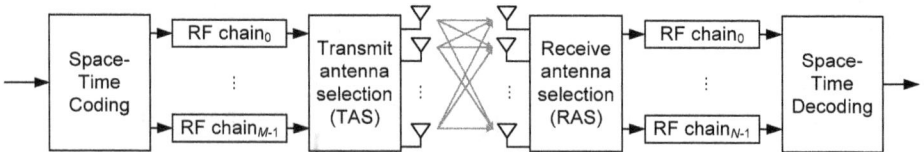

Fig. 4. Antenna selection in MIMO systems with M_a available transmit and N_a receive antennas.

In the SIMO case, it is shown in [10] and [19] that the array gain using a Maximum Ratio Combiner (MRC) without Receive Antenna Selection (RAS) is equal to $g_a = N_a$, whereas when RAS is applied (e.g. the antenna with better channel is selected) the array gain is given by

$$g_a = N\left(1 + \sum_{j=N+1}^{N_a} \frac{1}{j}\right). \tag{24}$$

As a result, we can note that RAS implies a loss in the SNR which becomes larger as the difference between N and N_a is increased. However, the diversity order for both schemes is exactly the same and there is only a coding gain difference [19]. The analysis for Transmit Antenna Selection (TAS) is reciprocal, therefore, the same effects are observed in case of MISO with TAS.

For the MIMO case and if multiple streams are simultaneously transmitted, transmit and/or receive antenna selection has further implications than just a reduction of the array gain. Actually, the inherent spatial multiplexing-diversity trade-off leads to different optimization criteria: diversity optimization (i.e. select the set of antennas that gives a higher Frobenius norm of the channel), improve the link reliability (discard antennas that produces large fadings in any eigenmode), maximize the Shannon capacity, etc. Furthermore, in [19] it is stated that the diversity gain obtained by transmit antenna and receive antenna selection is the same as without selection procedure, hence $g_d=(N-g_s)\times(M-g_s)$ with $g_s=\{0,...,$ min $(N,M)\}$.

4.2 Transmit antenna selection in MIMO systems

Transmit antenna selection techniques were first proposed during the very late 90s in the context of MIMO links in order to improve the array gain. During that period, antenna selection was derived according to the class of ST coding scheme that was involved. Heath et al. in [22] focused on the antenna selection in case of Spatial Multiplexing for linear receivers. The optimization criterion in [22] was to maximize the post-processing SNR in order to minimize the bit error probability. Later, it has been shown that the difference between optimizing the *ESINR* is only 0.5dB better than optimizing the lowest eigenvalue (the ESINR in case of Zero-Forcing is lower bounded by the minimum eigenvalue of H). Similar works have been carried out for Orthogonal STBC (OSTBC) which in this case concluded that maximizing the Frobenius norm of the active channel was the optimum strategy [23]. More recently, Deng et al. extended these transmit antenna selection schemes under the LDC framework concluding that the best selection criteria for minimize the bit error probability is based on maximizing the post-processing (or Effective) SNR [24]. Finally, an interesting application of transmit antenna selection has been proposed by Freitas et al. in [25] where different spatial layers are assumed combining spatial diversity and spatial multiplexing. In [25] the different branches are disposed in parallel hence both spatial diversity and multiplexing gains can be simultaneously achieved. The antennas subsets are then assigned to the spatial layers in order to minimize the bit error probability, where the (more susceptible) SM based layers are assigned the best subset of antennas and the remaining are assigned to the OSTBC layers.

4.3 MIMO precoding based on LDC codes

In the previous section, TAS precoding scheme has been introduced for some of the existing STBC and LDC. Therefore, given a specific code, the number of bits f_b that must be fed-back from the receiver to indicate the optimum transmit antennas set is

$$f_b = \binom{M_a}{M} = \frac{M_a!}{M!(M_a - M)!}. \tag{25}$$

However, if we could afford sending few more bits over the feedback channel, the transmitter/receiver may be able to select which code is more suitable according to the current channel state, or to choose how many spatial streams can be transmitted according to the channel rank [26]. Recent researches have extended the space-time coding selection (i.e. codebook based precoding) into the LDC framework [27]-[32]. An important result was obtained in [29], where it is shown that $f_b=log_2(M)$ feedback bits are enough to achieve full diversity. In [32], the authors showed that the average SNR can be improved up to 2dB compared to the open loop scheme with only 3 feedback bits (i.e. 8 sets of LDC codes).

4.4 Transmit antenna and space-time Code Selection (TACS)

As it has been explained in the previous section, when partial CSIs information is available at the transmitter two common selection techniques could be applied, which are: the space-time code selection, and the transmit antenna selection. One of the first works joining both concepts is that presented by Heath et al. in [33] where the number of the spatial streams (in the SM case) are adapted by selecting the best set of transmitter antennas (i.e. $f_b=M$). Furthermore, it was stated that if the optimum number of streams are transmitted from the optimum selected antenna set, the diversity gain is also maximized ($g_d\leq MN$). Then, given an antenna subset and a fixed rate, the required constellation could be determined as well as the number of spatial streams.

A simplification of this optimization problem is given in [34] where each stream is switched on/off when the post-processing of the SNR value of the stream is above/below a fixed threshold which is related with the rate. Further extensions of space-time code selection with TAS are given by Machado et al. in [35] where the available codes in the codebook are; the Alamouti code, the SM with $M=2$, the Quasi-OSTBC with $M=3$ and single antenna transmission.

In addition, the space-time code selection with transmit antenna selection has been generalized by the authors in [36]-[39] under the LDC framework considering both the linear and the ML receivers and developed within the IEEE 802.16m framework [38]. This generalization allows us to use any type of linear STBC (independently of the optimization criteria) codes and determine which codes are used most of the time and under which channel conditions. Two optimizations criteria have been developed in [14], one following the classical bit error rate optimization (minimizing of the scaled minimum Euclidian distance), and a second one is based on the throughput maximization given a fixed link quality (i.e. fixed packet error rate or bit error rate). This second optimization criterion can be used for resource allocation and scheduling purposes. Nevertheless, it is also shown that for low multiplexing rates the classic STBC codes (i.e. Alamouti, SM and Golden code) with transmit antenna selection are sufficient to explore the Grassmanian subspace [14].

4.4.1 The TACS selection criteria

Given the *ESINR* per stream and the average pairwise error probabilities, two different code and antenna subset optimization scenarios namely *Minimizing the bit error rate* and *Maximizing the throughput* respectively have been evaluated in [14][36]-[39]. In the first scenario, we consider that the same modulation is applied to all the symbols with a fixed rate *R*. In that case, and since transmission power is also fixed, we are interested in selecting the transmit antenna subset and the LDC code that minimizes the error rate probability (i.e. the bit error rate – BER) while the modulation that is required by each LDC is adapted in order to achieve the required rate R. In that case, since the Q-function is monotonically decreasing as a function of the input, the optimization problem is defined as follows

$$\max_{LDC_i, p_i} \min_q \left\{ ESINR_q \left(H, LDC_i, p_i \right) d^2_{min} \left(Z_i \right) \right\}, \tag{26}$$

where i means the LDC index, p_i denotes the transmitting antenna subset, q refers to the spatial stream (i.e. the symbol) index, and d_{min} is the minimum Euclidean distance according to the QAM constellation size used (note that the QAM constellation is a function of the LDC).

In the second scenario, the optimization is performed in order to maximize the system throughput considering a certain quality of service requirement (i.e. a maximum Block Error Rate - BLER). In that case, the problem is formulated as follows

$$\max_{LDC_i, p_i, MCS_j} \min_q R\left(1 - BLER\left(ESINR_q\right)\right) \quad s.t.: BLER \leq \mu \tag{27}$$

where j means the Modulation and Coding Scheme (MCS) index that maximizes the spectral efficiency for the specific channel state subject to a maximum Block Error Rate (BLER). The following Fig. 5 illustrates the scheme of the MIMO system applying the TACS selection algorithm.

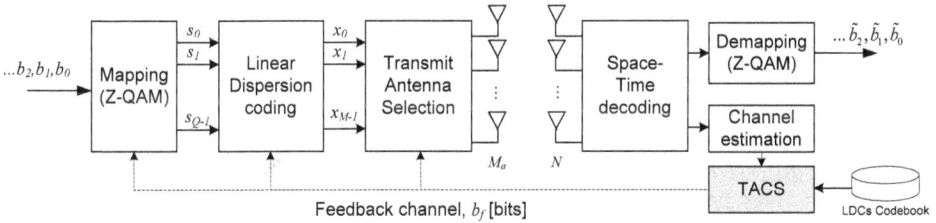

Fig. 5. Proposed TACS spatial adaptation scheme and integration into the transmission scheme.

4.4.2 TACS performance evaluation

In this section we are setting the main parameters to evaluate the performance of the TACS scheme, the IEEE 802.16 standard is here used to carry out the experiment. Some parameters are depicted in Table 1, where perfect synchronization is assumed and inter-cell interference is not considered. The used modulation is a Z-QAM (Z={2,4,16,64}) with Gray mapping. According to the CSI measured, the BS determines: i) the antenna subset, ii) the LDC subset and in case of throughput maximization, iii) the MCS that maximizes the rate for a maximum Block Error Rate - BLER (second optimization criterion). The codebook is composed mainly by the Single Input Multiple Output (Maximum Ratio Combining is used at the receiver) receiver, the Alamouti 's Spatial Diversity (SD) coding scheme (referred as G2 in hereafter plotted the figures) [15], the pure spatial multiplexing (SM) and the Golden code [17]. The performance of the system is evaluated over 100.000 channel realizations,

OFDMA Air Interface and System Level configuration	
Subcarrier Permutation	Distributed (PUSC) and Contiguous (Band AMC)
FFT length, CP	2048, 12.5%
# of used subcarriers	1728
Modulation	{4,16,64}-QAM
Channel coding[1]	Turbo coding with rates: 1/3, 1/2, 2/3, ¾
Channel model	Rayleigh and ITU Pedestrian A
Channel estimation (CQI)	Ideal without any delay
Frame duration, T_{frame}	5ms
DL/UL rate	2:1
OFDM symbols in the DL	30
Number of transmit antennas, M	{1,2,4}
Number of receive antennas, N	{1,2,4}
MIMO detector	MMSE
Rate (spectral efficiency)	{2,4,8} bits per channel use (bpcu)

Table 1. TACS evaluation framework system parameters

where for each realization a tile or a subchannel (specified in each analysis) is transmitted. In case of Partial Usage Subcarrier permutation (PUSC), the tile is formed by 4 subcarriers and 3 symbols, where 4 tones are dedicated to pilots as defined in IEEE 802.16e [17]. For the Band Adaptive Modulation and Coding (AMC) permutation scheme, each bin (equivalent to the tile concept) is comprised by 9 subcarriers where 1 tone is used as pilot. Perfect channel estimation is assumed at the receiver. Every $log_2(Z)$ bits are mapped to one symbol. The channel models used are uncorrelated Rayleigh ($\mathbf{H}\sim CN(0,1)$) and the ITU Pedestrian A [38]. In both cases the channel is considered constant within a tile (*block fading channel model*). In case of uncorrelated Rayleigh the channel between tiles is uncorrelated, whereas in the ITU PedA case the channel is correlated both in frequency and time.

4.4.3 MIMO reference and simulation results

In Fig. 6, the reference performance for a fixed rated is depicted for $N=2$ when no transmit antenna selection neither code selection are used. For uncorrelated Rayleigh channel, we can observe that for low data rate, i.e. $R=\{2,4\}$, the Alamouti code outperforms the rest of the schemes. This is strictly related to the diversity order that G2 achieves equal to $g_d=N\times M=4$, whereas the SM and the Golden code with a linear receiver get a diversity order of

---

[1] Forward error correction is consider only for the throughput maximization case, where the LUT used to predict the BLER as a function of the ESINR, are obtained using the Duo-Binary Turbo code defined for IEEE 802.16e.

$g_d=(N - M + 1)=1$. At higher data rates ($R>8$), all the codes perform similarly in the analysed SNR range despite of the different diversity order between them.

4.4.4 TACS performance under bit error rate minimization criterion

In Fig. 7 and Fig. 8, the bit error rate performance using TACS is shown having a fixed rate $R=4$. Fig. 7 shows the improvement due to the increase in M_a and also the performance achieved when combined with code selection. It can be observed how the TAS increases the diversity order, leading to a large performance increase for the SM and Golden subsets. It is very important to notice that despite the diversity increase for all the LDC subsets, SD and SIMO schemes still perform better when each code is evaluated independently. However, in Fig. 8, we can observe that when the code selection is switched on, SIMO and Golden subsets are selected most times, while the usage of SIMO increases with the SNR and the usage of SM and the Golden code increases with M_a. Furthermore, the achieved improvement by the TACS is clearly appreciated in Fig. 7, where an SNR improvement of approximately 1dB is obtained for $M_a=\{3,4\}$. It is also surprising that the SM code is rarely selected knowing that the Golden code should always outperform SM since it obtains a higher diversity. However, as it is observed in Fig. 8, for less than 5% of the channel realizations the SM may outperform slightly the Golden code. Whether the singular value decomposition of the effective channel H is analysed when SM is selected, it has been observed that when all singular values are very close, both the SM and the Golden code lead to very similar performances, therefore no matter which one is selected.

In Fig. 9 and Fig. 10, the performance using the TACS is again analysed for $R=8$. In Fig. 9 the different diversity orders of SD, SM, and the Golden Code are illustrated. We can appreciate here that the SM and the Golden code show the best performance when $M_a=\{3,4\}$, and also for $M_a=2$ when SNR≤18dB. Furthermore the increase in the diversity order due to TACS can be observed in both Fig. 7 and Fig. 9. The maximum diversity order ($g_d = M_aN$) is achieved since at least one LDC (SIMO and G2) from those in the codebook are able to achieve the maximum diversity order.

Moreover, the BER using the TACS is equivalent to that obtained from the SISO scheme (referred as SISO$_{eq}$ in the plots) over a Rayleigh fading channel with the same rate R, a diversity order $g_d=M_aN$ and a coding gain equal to Δ. The performance of this *equivalent* SISO scheme, in terms of the bit error rate probability P_b, can be obtained directly by close expressions that are found in [41][42] and applying the Craig's formula in [43],

$$P_b = \frac{1}{\log_2 \sqrt{Z}} \sum_{i=1}^{\log_2 \sqrt{Z}} P_b(i) \tag{28}$$

$$P_b(i) = \frac{2}{\sqrt{Z}} \sum_{k=0}^{\left(1-2^{-i}\right)\sqrt{Z}-1} \omega(k,i,Z) \frac{1}{\pi} \int_0^{\pi/2} \left(1+(2k+1)^2 \frac{3\rho_b}{2(Z-1)\sin^2\theta}\right)^{-g_d} d\theta \tag{29}$$

$$\omega(k,i,Z) = (-1)^{\left\lfloor \frac{k\cdot 2^{i-1}}{\sqrt{Z}} \right\rfloor} \left(2^{i-1} - \left\lfloor \frac{k\cdot 2^{i-1}}{\sqrt{Z}} + \frac{1}{2} \right\rfloor\right) \tag{30}$$

where $\rho_b = \Delta \cdot \rho \, / \, log_2(Z)$, $\lfloor x \rfloor$ means the smallest integer of x, and Z is the modulation order of the Z-QAM modulation.

The values of Δ for different combinations of $M_a=\{2,3,4\}$, $N=\{2,3,4\}$ and $R=\{4,8\}$ are depicted in Table 2. These values have been obtained adjusting the BER approximation in Eq. (28) to the empirical BER. As shown in Fig. 7 and Fig. 9 the performance of the TACS schemes is perfectly parameterized under the equivalent SISO model. Notice also that the power gain is constant across the whole SNR range.

Δ		$N=2$	$N=3$	$N=4$
	$M_a=2$	2.66	3.9	6.31
R = 4	$M_a=3$	3.20	5.2	8.41
	$M_a=4$	3.75	6.2	9.44
	$M_a=2$	4.20	9	14
R = 8	$M_a=3$	6.75	14.5	23
	$M_a=4$	9.00	19	28.5

Table 2. Coding gain Δ for the TACS proposal with $M_a=\{2,3,4\}$, $N=\{2,3,4\}$, and $R=\{4,8\}$.

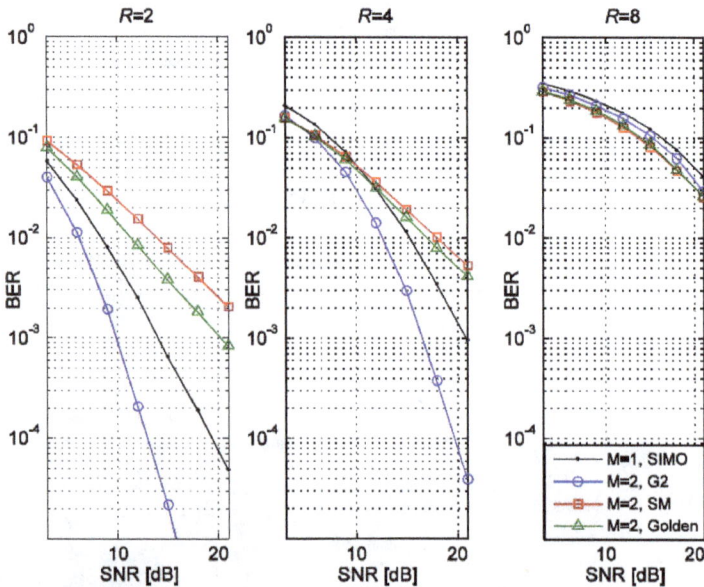

Fig. 6. Uncoded BER for uncorrelated Rayleigh channel with MMSE detector and $N=2$.

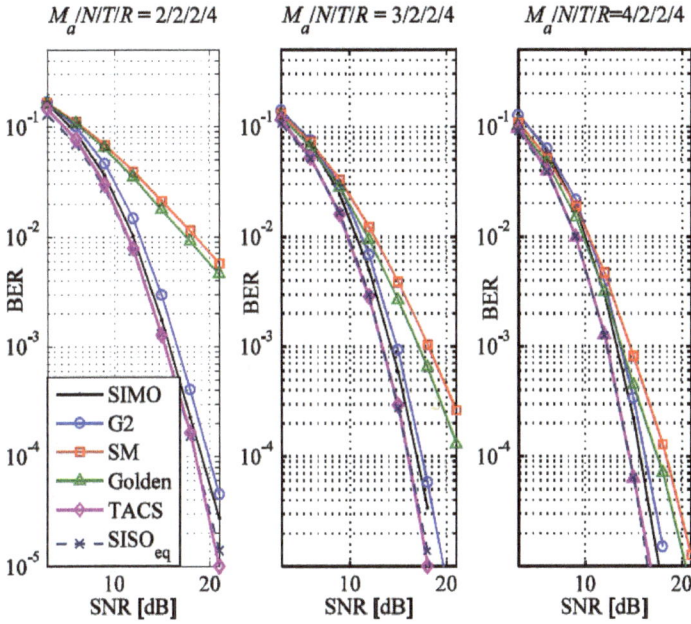

Fig. 7. Uncoded BER performance when $N=2$, $R=4$, $M_a=\{2,3,4\}$ for uncorrelated MIMO Rayleigh channel and MMSE linear receiver.

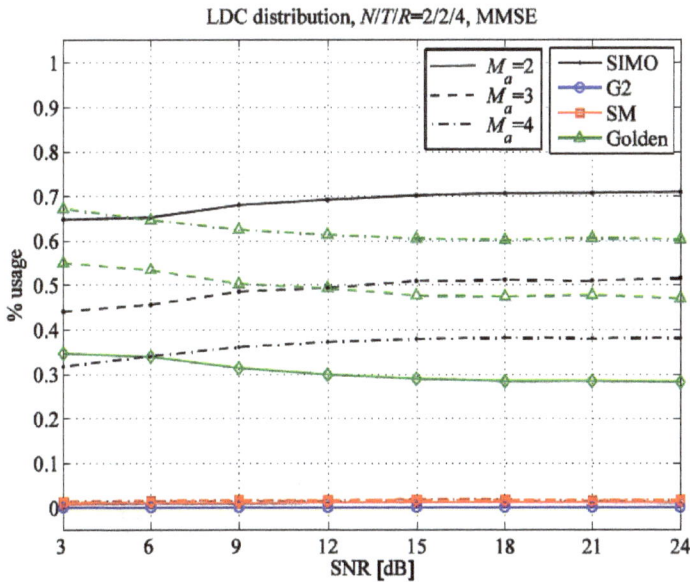

Fig. 8. LDC selection statistics with $N=2$, $R=4$, $M_a=\{2,3,4\}$ for uncorrelated MIMO Rayleigh channel and MMSE linear receiver.

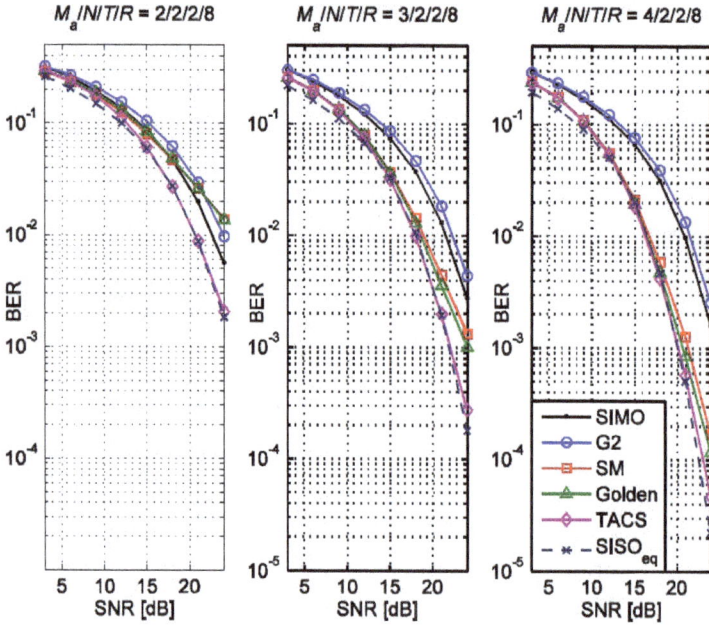

Fig. 9. Uncoded BER performance when $N=2$, $R=8$, $M_a=\{2,3,4\}$ for uncorrelated MIMO Rayleigh channel and MMSE linear receiver.

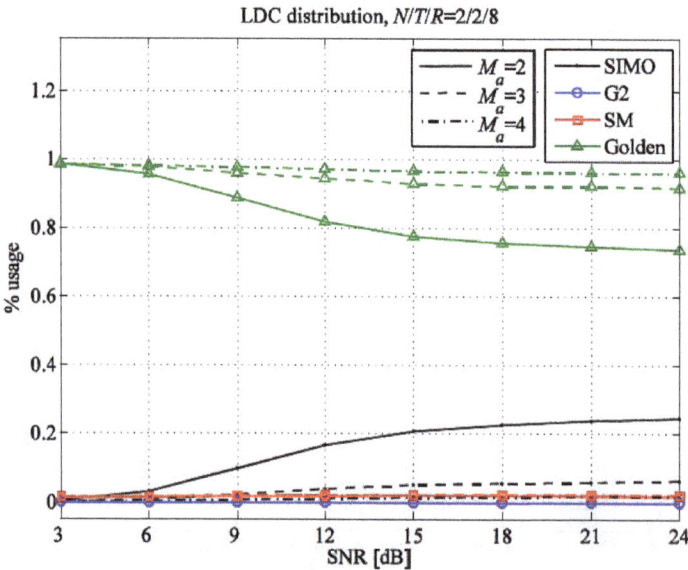

Fig. 10. LDC selection statistics when $N=2$, $R=8$, $M_a=\{2,3,4\}$ for uncorrelated MIMO Rayleigh channel and MMSE linear receiver.

4.4.5 TACS performance under throughput maximization criterion

In this section the performance of the TACS adaptation scheme in case the throughput is maximized (see Eq.(27)) is analysed. Then, for such adaptation scheme, the antenna set and the LDC code that maximizes the throughput is selected. In addition, the highest MCS (in the sense of spectral efficiency) that achieves a BLER<0.01 (1%) is also selected. The look-up-table used for mapping the ESINR to the BLER is shown and described in [14]. In the scenarios considered, the minimum allocable block length according the IEEE 802.16e standard was selected [17] (i.e. the number of sub-channels N_{sch} occupied per block varies between 1 and 4). The number of available antennas is M_a=2 whereas N=2.

In Fig. 11 and Fig. 12, the spectral efficiency achieved by TACS with adaptive Modulation and Coding (AMC) as well as the LDC statistics are shown. For Spatial Multiplexing (SM), two encoding options named *Vertical Encoding* (VE) and *Horizontal Encoding* (HE) are considered. For the first scheme, VE, the symbols within the codeword apply the same MCS format, whereas for the second, HE, each symbol may apply a different MCS. Clearly the first is more restrictive since is limited by the worst stream $(min(ESNR_q))$ whereas the second is able to exploit inter-stream diversity at the expense of higher signalling requirements (at least twice as that required with VE in case of M=2).

Depicted performances shown that at low SNRs (SNR<13dB), the SIMO and Alamouti achieve the highest spectral efficiencies (something that has been already obtained in several previous works [10]). However, as the SNR is increased, the codes with higher multiplexing capacity (e.g. the SM and the Golden code) are preferred. It could be also observed that the SM with VE implies a loss of around 2dB compared to the Golden code, but when HE is used, the Golden code is around 0.5dB worse than the SM-HE.

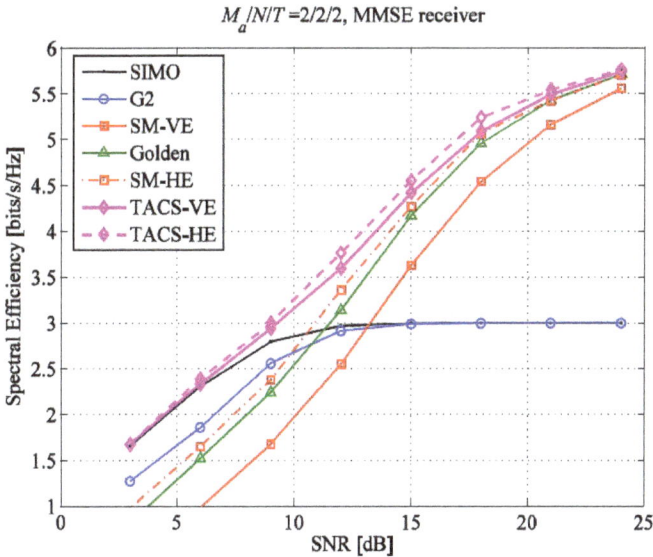

Fig. 11. Spectral efficiency under TACS with throughput maximization criterion with M_a=2, N=2, adaptive MCS and MMSE receiver for an uncorrelated MIMO Rayleigh channel.

$M_a/N/T = 2/2/2$, MMSE receiver

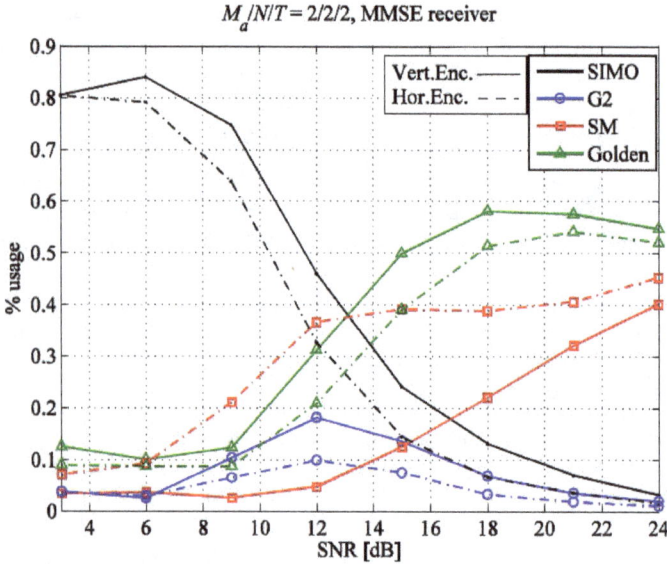

Fig. 12. LDC selection statistics under TACS with throughput maximization criterion with $M_a=2$, $N=2$, adaptive MCS and MMSE receiver for an uncorrelated MIMO Rayleigh channel.

To gain further insights of the TACS behaviour, the statistics of LDC selection as a function of the average SNR are plotted in Fig. 12. We can clearly appreciate that at low SNR the preferred scheme is SIMO where all the power is concentrated in the best antenna, while as the SNR is increased full rate codes ($Q=M$) are more selected since they permit to use lower size constellations. Moreover, comparing SM-VE with SM-HE, we can observe that SM-HE is able to exploit the stream's diversity and hence achieves a higher spectral efficiency than if the Golden code is used. Actually, at average SNR=12, the SM with HE is the scheme selected for most frames, even more than SIMO. These results show that in case of linear receivers (e.g. MMSE) the TACS scheme with AMC gives a noticeable SNR gain (up to 3dB) in a large SNR margin (SNR from 6 to 18dB) and also is a good technique to achieve a smooth transition between diversity and multiplexing.

5. MIMO in IEEE 802.16e/m

The use of MIMO may improve the performance of the system both in terms of link reliability and throughput. As it was discussed in previous sections, both concepts pull in different directions, and in most cases a trade-off between both is meet by each specific space-time code. From a system point of view, and due to the inherent time/freq variability of the wireless channel, no code is optimal for all channel conditions, and at most, the codes can be optimized according to the ergodic properties of the channel. In fact, this is the reason why the TACS scheme is able to bring significant gain compared to a scheme where the same space-time code is always used. This situation is well-known and it is the reason why in most of the Broadband Wireless Access (BWA) systems, the number of space-time codes is increasing.

In IEEE 802.16e/m, two types of MIMO are defined, Single User MIMO and Multiuser MIMO, the first corresponding to the case where one resource unit (the minimum block of frequency-time allocable subcarriers) is assigned to a single user, and the second when this one is shared among multiple users.

In case of two transmit antennas, IEEE 802.16e/m defines two possible encoding schemes referred as Matrix A and Matrix B. Matrix A corresponds to the Alamouti scheme, while Matrix B corresponds to the Spatial Multiplexing (SM) case. In case of using SM, WiMAX allows both Vertical Encoding (VE) and Horizontal Encoding (HE). In the first case, VE, all the symbols are encoded together and belong to the same *layer*. In addition to Matrix A and Matrix B, IEEE 802.16 also defines a Matrix C which corresponds to the Golden Code. This code is characterized for providing the highest spatial diversity for the spatial rate $R=2$. In case of 3 and 4 transmit antennas, WiMAX also defines the encoding schemes of Matrix A, Matrix B, and Matrix C, all of them providing different trade-offs between diversity and spatial multiplexing.

The list of combinations is even longer since WiMAX allows antenna selection and antenna grouping, therefore, the list of encoding matrices also includes the possibility that not all antennas are used, and only a subset are selected (the list of matrices in Table 3 do not show this possibility). In case not all the antennas are used, the power is normalized so that the same power is transmitted disregard of the number of active antennas.

Besides the possibility to select among any of the previous coding matrices, IEEE 802.16e/m also allows the use of precoding. In this case, the space-time coding output is weighted by a matrix before mapping onto transmitter antennas

$$z = Wx \qquad (31)$$

where x is $M_t \times 1$ vector obtained after ST encoding, where M_t is the number of streams at the output of the space time coding scheme. The matrix W is a $M \times M_t$ weighting matrix where M is the number of transmit antennas. The weighting matrix accepts two types of adaptation depending on the rate of update, named short term closed-loop precoding and long term closed-loop precoding.

In the later IEEE 802.16m, the degrees of flexibility has been broadened, allowing several kinds of adaptation [44]. On top of this, IEEE 802.16m includes also ST codes for up to 8 transmitter antennas, enabling the transmission at spectral efficiencies as high as 30bits/sec/Hz which become necessary to achieve the very high throughputs demanded for IMT-Advanced systems [45].

M	N_{min}	T	Q	R	MIMO Encoding Matrix	Name
2	1	2	2	1	$\begin{bmatrix} s_0 & -s_1^* \\ s_1 & s_0^* \end{bmatrix}$	Alamouti (a.k.a. Matrix A)
2	2	1	2	2	$\begin{bmatrix} s_0 & s_1 \end{bmatrix}^T$	Spatial Multiplexing (a.k.a. Matrix B)
2	2	2	4	2	$\dfrac{1}{\sqrt{1+r^2}}\begin{bmatrix} s_0 + jrs_3 & rs_1 + s_2 \\ s_1 - rs_2 & s_3 + jrs_0 \end{bmatrix}, r = \dfrac{-1+\sqrt{5}}{2}$	Golden Code (a.k.a. Matrix C)
3	2	4	4	1	$\begin{bmatrix} s_0 & -s_1^* & 0 & 0 \\ s_1 & s_0^* & s_2 & -s_3^* \\ 0 & 0 & s_3 & s_2^* \end{bmatrix}$	Matrix A²
3	2	4	4	1	$\begin{bmatrix} \sqrt{\dfrac{3}{4}} & 0 & 0 \\ 0 & \sqrt{\dfrac{3}{4}} & 0 \\ 0 & 0 & \sqrt{\dfrac{3}{2}} \end{bmatrix}\begin{bmatrix} s_0 & -s_1^* & s_4 & -s_5^* \\ s_1 & s_0^* & s_5 & s_4^* \\ s_6 & -s_7^* & s_2 & -s_3^* \end{bmatrix}$	Matrix B
3	2	4	4	1	$\begin{bmatrix} s_0 & s_1 & s_2 \end{bmatrix}^T$	Matrix C
4	1	4	4	1	$\begin{bmatrix} s_0 & -s_1^* & 0 & 0 \\ s_1 & s_0^* & 0 & 0 \\ 0 & 0 & s_2 & -s_3^* \\ 0 & 0 & s_3 & s_2^* \end{bmatrix}$	Matrix A
4	2	4	8	2	$\begin{bmatrix} s_0 & -s_1^* & s_4 & -s_5^* \\ s_1 & s_0^* & s_5 & s_4^* \\ s_2 & -s_3^* & s_6 & -s_7^* \\ s_3 & s_2^* & s_7 & s_6^* \end{bmatrix}$	Matrix B
4	4	1	4	4	$\begin{bmatrix} s_0 & s_1 & s_2 & s_3 \end{bmatrix}^T$	Matrix C

Table 3. WiMAX IEEE 802.16e MIMO encoding matrices.

[2] In case of 3 and 4 transmit antennas, Matrix A, B and C accept different antenna grouping and selection schemes. This antenna grouping does similar effects as TACS, indicating which antennas and Space-time codes are preferred.

6. Summary

The use of multiple antenna techniques at transmitter and receiver sides is still considered a hot research topic where the channel capacity can be increased if multiple streams are multiplexed in the spatial domain. The study on the trade-off between diversity and multiplexing has motivated the emergence of many different space-time coding architectures where most of the proposed schemes lie in the form of Linear Dispersion Codes. Furthermore, as it was shown by the authors in previous sections, when the transmitter disposes of partial channel state information, robustness and throughput can be very significantly improved. One of the simplest adaptation techniques is the use of antenna selection, which increases the diversity of the system up to the maximum available $(g_d=M_a{\times}N_a)$. On the other hand, when transmit antenna selection is combined with code selection a coding gain is achieved. In this chapter, a joint Transmit Antenna and space-time Coding Selection (TACS) scheme previously proposed by the authors has been described. The TACS algorithm allows two kind of optimization: i) bit error rate minimization, and ii) throughput maximization. One important result obtained from these studies is that the number of required space-time coding schemes is quite low. In fact, previous studies by the author have shown that in case of spectral efficiencies of 8bits/second/Hertz or lower, using SIMO, Alamouti, SM, and the Golden code is enough to maximize the performance (for higher rates, codes with higher spatial rate would be required). Furthermore, the worse performance achieved by linear receivers (e.g. ZF, MMSE) is compensated by the TACS scheme, which allows to achieve performances close to those obtained with the non-linear receivers (e.g. the Maximum Likelihood) with much lower computational requirements. As a final conclusion, it can be considered that transmit antenna selection with linear dispersion code selection can be an efficient spatial adaptation technique whose low feedback requirements make it feasible for most of the Broadband Wireless Access systems, especially in case of low mobility.

7. Acronyms

3GPP	3rd Generation Partnership Project
AWGN	Additive White Gaussian Noise
BLER	Block Error Rate
BS	Base Station
CSI	Channel State Information
FDD	Frequency Division Duplexing
LDC	Linear Dispersion Codes
LTE	Long Term Evolution
MCS	Modulation and Coding Scheme
MIMO	Multiple Input Multiple Output
MMSE	Minimum Mean Square Error
OSTBC	Orthogonal Space-Time Block Code
QAM	Quadrature Amplitude Modulation
SIMO	Single Input Multiple Output
SISO	Single Input Single Output
SM	Spatial Multiplexing
SNR	Signal To Noise Ratio

STBC   Space-Time Block Code
TACS   Transmit Antenna and (space-time) Code Selection
TDD   Time Division Duplexing
UPA   Uniform Power Allocation
ZF   Zero Forcing

8. References

[1] E. Telatar, "Capacity of multi-antenna Gaussian channels", *European Transactions on Telecommunications*, Nov. 1999.

[2] A. Saad, M. Ismail, N. Misran, "Correlated MIMO Rayleigh Channels: Eigenmodes and Capacity Analyses", *International Journal of Computer Science and Network Security*, Vol. 8 No. 12 pp. 75-81, Dec. 2008.

[3] L. Zheng, D. Tse, "Diversity and multiplexing: a fundamental tradeoff in multiple antenna channels", *IEEE Trans. On Information Theory*, May, 2003.

[4] V. Tarokh, N. Seshadri, A. R. Calderbank, "Space-Time codes for high data rates wireless communication: Performance criterion and code construction", IEEE. Trans. on Information Theory, vol.44, pp.744-765, March, 1998.

[5] V. Tarokh, H. Jafarkhani, A.R. Calderbank, "Space-time block codes from orthogonal designs," *IEEE Transactions on Information Theory*, vol.45, no.5, pp.1456-1467, Jul 1999.

[6] G. Ganesan, P. Stoica, "Space-time diversity scheme for wireless communications", in Proc. ICASSP, 2000.

[7] S. Sandhu, R. Heath, A. Paulraj, "Space-time block codes versus space-time trellis codes," *IEEE International Conference on Communications*, 2001 (ICC-2001), vol.4, pp.1132-1136 vol.4, 2001.

[8] J. Cheng; H. Wang; M. Chen; S. Cheng, "Performance comparison and analysis between STTC and STBC," the 54th *IEEE Vehicular Technology Conference*, (VTC 2001 Fall), vol.4, no., pp.2487-2491 vol.4, 2001.

[9] L. Yu, P.H.W. Fung, W. Yan, S. Sumei, "Performance analysis of MIMO system with serial concatenated bit-interleaved coded modulation and linear dispersion code," *IEEE International Conference on Communications*, 2004, vol.2, no., pp. 692-696 Vol.2, 20-24 June 2004.

[10] C. Oestges, B. Clerckx, "MIMO Wireless Communications", Academic Press, Elsevier, USA, 2007.

[11] B. Hassibi, B.M. Hochwald, "High Rates Codes that are Linear in Space and Time", April 2001.

[12] W. Zhang, X. Ma, B. Gestner, D. V. Andreson, "Designing Low Complexity Equalizers for Wireless Systems", *IEEE Communications Magazine*, January, 2009, pp. 56-62.

[13] G. J. Foschini, "Layered space-time architecture for wireless communications in a fading environment when using multiple antennas", Bell Lab Tech. J. v.1., n.2, 1996.

[14] I. Gutierrez, "Adaptive Communications for Next Generation Broadband Wireless Access Systems", Ph.D. Thesis, June 2009.

[15] S. M. Alamouti, "A simple transmit diversity technique for wireless communications", IEEE J. Selected Areas in Communications, vol. 17, pp. 1451-1458, Oct. 1998.

[16] J.C. Belfiore, G. Rekaya, E. Viterbo: "The Golden Code: A 2 x 2 Full-Rate Space-Time Code with Non-Vanishing Determinants," IEEE Transactions on Information Theory, vol. 51, n. 4, pp. 1432-1436, Apr. 2005.

[17] IEEE Standard for Local and metropolitan area networks, Part 16: Air Interface for Fixed and Mobile Broadband Wireless Access Systems, Amendment 2: Physical and Medium Access Control Layers for Combined Fixed and Mobile Operation in Licensed Bands and Corrigendum 1, IEEE Std 802.16e™-2005, Feb.2006.

[18] M. Vu, A. Paulraj, "MIMO Wireless Linear Precoding", IEEE Signal Processing Magazine, Vol. 4, no.5, pp.87-105, Sept. 2007.

[19] A.B. Gershman, N.D. Sidiropoulos, "Space Time Processing for MIMO Communications", John Wiley & Sons, UK, 2005.

[20] R. W. Heath, A.J. Paulraj, "Linear Dispersion Codes for MIMO Systems Based in Frame Theory", IEEE Transactions on Signal Processing, Vol.50, n.10, Oct.2002

[21] R. Gohary, T. Davidson, "Design of Linear Dispersion Codes: Asymptotic Guidelines and Their Implementation", IEEE Transactions on Wireless Communications, Vol.4, No.6, Nov.2005.

[22] R. Heath, S. Sandhu, A. Paulraj, "Antenna Selection for Spatial Multiplexing Systems with Linear Receivers", IEEE Communications Letters, Vol.5, no.4, April 2001.

[23] D.A. Gore, A.J. Paulraj, "MIMO antenna subset selection with space-time coding," *IEEE Transactions on Signal Processing*, vol.50, no.10, pp. 2580-2588, Oct 2002.

[24] D. Deng, M. Zhao, J. Zhu, "Transmit Antenna Selection for Linear Dispersion Codes Based on Linear Receiver", Proc. Vehicular Technology Conference, 2006. VTC 2006-Spring.

[25] W.C. Freitas, F.R.P. Cavalcanti, R.R. Lopes, "Hybrid MIMO Transceiver Scheme with Antenna Allocation and Partial CSI at Transmitter Side", *IEEE 17th International Symposium on Personal, Indoor and Mobile Radio Communications, 2006*, pp.1-5, 11-14 Sept. 2006.

[26] R.W. Heath, A.J. Paulraj, "Switching between diversity and multiplexing in MIMO systems," *IEEE Transactions on Communications*, vol.53, no.6, pp. 962-968, June 2005.

[27] L. Che, V.V. Veeravalli, "A Limited Feedback Scheme for Linear Dispersion Codes Over Correlated MIMO Channels," *IEEE International Conference on Acoustics, Speech and Signal Processing, 2007*, (ICASSP-2007), vol.3, pp. 41-44, 15-20 April 2007.

[28] A. Osseiran, V. Stankovic, E. Jorswieck, T. Wild, M. Fuchs, M. Olsson, "A MIMO framework for 4G systems: WINNER concept and results", *IEEE 8th Workshop on Signal Processing Advances in Wireless Communications, 2007*, (SPAWC-2007), pp.1-5, 17-20 June 2007.

[29] D.J. Love, R.W. Heath, T. Strohmer, "Grassmannian beamforming for multiple-input multiple-output wireless systems," *IEEE Transactions on Information Theory*, vol.49, no.10, pp. 2735-2747, Oct. 2003.

[30] D. Deng, J. Zhu, "Linear Dispersion Codes Selection Based on Grassmannian Subspace Packing",submitted to IEEE Journal of Selected Topics in Signal Processing, 2007.

[31] R. Machado, B.F. Uchoa-Filho, T.M. Duman, "Linear Dispersion Codes for MIMO Channels with Limited Feedback," *IEEE Wireless Communications and Networking Conference*, 2008 (WCNC-2008), pp.199-204, March 31 2008-April 3 2008.

[32] D. Yang, N. Wu, L.L. Yang, L. Hanzo, "Closed-loop linear dispersion coded eigen-beam transmission and its capacity," *Electronics Letters*, vol.44, no.19, pp.1144-1146, September 11 2008.

[33] R.W. Heath, D. J. Love, "Multimode Antenna Selection for Spatial Multiplexing Systems with Linear Receivers", *IEEE Transactions on Signal Processing*, Vol.53, no.8, Aug.2005.

[34] K. Youngwook, C.Tepedelenlioglu, "Threshold-Based Substream Selection for Closed-Loop Spatial Multiplexing", *IEEE Transactions on Vehicular Technology*, vol.57, no.1, pp.215-226, Jan. 2008.

[35] R. Machado, B.F. Uchôa-Filho, "Extended Techniques for Transmit Antenna Selection with STBCs", Journal of Communications and Information Systems, vol.21, n.3, pp.118-195, 2006.

[36] T. Lestable, M. Jiang, A. Mourad, D. Mazzarese, Uplink MIMO Schemes for IEEE 802.16m, IEEE S80216m-08_534r2, July, 2008.

[37] T. Lestable, M. Jiang, A. Mourad, D. Mazzarese, S. Han, H. Choi, H. Kang , I. Gutierrez, "Linear Dispersion Codes for Uplink MIMO schemes in IEEE 802.16m", IEEE C802.16m-08/535, July, 2008.

[38] I. Gutierrez, F. Bader, A. Mourad, Spectral Efficiency Under Transmit Antenna and STC Selection with Throughput Maximization Using WiMAX, in Proceedings of the 17th International Conference on Telecommunications (ICT2010), 4-7 April 2010, Doha (Qatar).

[39] I. Gutiérrez, F. Bader, A. Mourad, Joint Transmit Antenna and Space-Time Coding Selection for WiMAX MIMO System, in Proceedings of the 20th IEEE Personal, Indoor and Mobile Radio Communications Symposium 2009 (PIMRC 2009), 13-16 September 2009, Tokyo (Japan).

[40] R. Srinivasan et al., "Evaluation Methodology for P802.16m-Advanced Air Interface", IEEE 802.16m-07/037r2.

[41] W. Lopes, W. Queiroz, F. Madeiro, M. Alencar, "Exact Bit Error Probability of M-QAM modulation Over Flat Rayleigh Fading Channels", Microwave and Optoelectronics Conference, 2007. Oct. 2007.

[42] J. Lu, T. Tjhung, C. Chai, "Error probability Performance of *L*-Branch Diversity Reception of MQAM in Rayleigh Fading", IEEE Transactions on Communications, vol. 46, No.2, Feb. 1998.

[43] M. K. Simon, "Evaluation of Average Bit Error Probability for Space-Time Coding based on a Simpler Exact Evaluation of Pairwise Error Probability", Journal of Communications and Networking, 3 (3): 257-267, Sept. 2001.

[44] "IEEE Standard for Local and metropolitan area networks Part 16: Air Interface for Broadband Wireless Access Systems Amendment 3: Advanced Air Interface", IEEE Std 802.16m-2011, 06-May-2011

[45] Report ITU-R M.2134, Requirements related to technical system performance for IMT-Advanced radio interface(s), November 2008 <http://www.itu.int/publ/R-REP-M.2134-2008/en>.

Hexa-Band Multi-Standard Planar Antenna Design for Wireless Mobile Terminal

Yu-Jen Chi[1] and Chien-Wen Chiu[2]

[1]Department of Electrical Engineering, National Chiao Tung University,
[2]Department of Electric Engineering, National Ilan University,
Taiwan

1. Introduction

Electronic devices such as mobile phones and laptop computers are parts of modern life. Users of portable wireless devices always desire such devices to be of small volume, light weight, and low cost. Thanks to the rapid advances in very large scale integration (VLSI) technology, this dream has become a reality in the past two decades. As technology grows rapidly, a mobile is not just a phone recently. The highly integration of circuits makes the mobile phone and the PDA (personal digital assistant) been combined into a single handset, which is called a smart phone. Also, the Internet carries various information resources and services, such as electronic mail, online chat, file transfer and file sharing, these attractive proprieties make wireless internet service becomes an important function that should be integrated into mobile devices. There are many ways for the user to connect to the internet. The traditional wireless local area network (WLAN) is a popular communication system for accessing the Internet. However, the reach of WiFi is very limited. WLAN connectivity is primarily constrained to hotspots, users need to find the access points and can only use it in certain rooms or areas. As the user get out of range of the hotspot, the signal will become very weak and the user may lose the connection. This disadvantage limits the mobility of wireless communication. Except for the widely used wireless local area network, third generation (3G) mobile telephony based on the High Speed Downlink Packet Access (HSDPA), which is part of the UMTS standards in 3G communications protocol, is another high speed wireless internet access service. It has become popular nowadays that people can get to the internet via cellular communication system. This technology gives the users the ability to access to the Internet wherever the signal is available from the cellular base station. However, the quality sometimes depends on the number of users simultaneously connected per cellular site. In addition to utilizing WLAN/3G dual-mode terminals to enhance efficiency of mobile number portability service, WiMAX (the Worldwide Interoperability for Microwave Access) is an emerging telecommunications technology that provides wireless data transmission in a variety of ways, ranging from point-to-point links to full mobile cellular-type access. WiMAX is similar to Wi-Fi but it can also permit usage at much greater distances. The bandwidth and range of WiMAX make it suitable for the applications like VoIP (Voice over Internet Protocol) or IPTV (Internet Protocol Television). Many people expect WiMAX to emerge as another technology that may be adopted for handset devices in the near future.

The rapid progress in mobile communication requires that many functions and wireless communication systems be integrated into a mobile phone. When portability is taken into account, antenna that can be built in the phone device is desirable. This has led to a great demand for designing multiband antennas for handset devices. Among existing built-in or internal type scheme, the inverted-F (IFA) or planar inverted-F antenna (PIFA) are the most promising candidates. The linear inverted-F antenna, which is the original version of the PIFA, has been described by R. King in 1960 as a shunt-driven inverted-L antenna-transmission line with open-end (king et al., 1960). The PIFA, which is constructed by replacing the linear radiator element of IFA with a planar radiator element, can also be evolved from a microstrip antenna. Taga first investigated PIFA's performance for 800MHz band portable unit radio in 1987 (Taga & Tsunekawa, 1987). He also wrote a chapter in his textbook to teach how to design a single band PIFA (Hirasawa & Haneishi, 1922). The PIFA or IFA are not only small in size but also have a broadband bandwidth. Since it is cheap and easy to fabricate, it has become very popular with mobile phone manufacturers. Many references concerning PIFA and its relatives were published in the decade.

In the past decade, researches for variation of the PIFA and multiband antenna grow rapidly like mushroom. Tri-band, quad-band, penta-band or hexa-band antenna can be found in many journals (Chiu & Lin, 2002; Guo et al., 2003, 2004; Ciais et al., 2004; Chen, 2007; Bancroft, 2005; Ali & Hayes, 2000; Soras et al., 2002; Nepa et al., 2005; Wong et al., 2005; Liu & Gaucher, 2004, 2007; Wang et al., 2007). For example, Chiu presented a tri-band PIFA for GSM800/DCS1800/PCS1900 in 2002 (Chiu & Lin, 2002) . Using two folded arms between the two plates, Guo at el. proposed a compact internal quad-band for covering GSM900/DCS1800/PCS1900 and ISM2450 bands (Guo, et al., 2003). By adding three quarter-wavelength parasitic elements to create new resonances, Ciais et al. presented a design of a compact quad-band PIFA for mobile phones (Ciais et al., 2004). In 2004, Guo & Tan proposed a new compact six-band but complicated internal antenna. His antenna is comprised of a main plate, a ground plane, a parasitic plate and a folded stub perpendicular to the two main plates (Guo & Tan, 2004).

In order to integrate all the wireless services into a mobile terminal and have an effective usage of the precious board space in the mobile device, multiband antenna that is designed to operate on several bands is necessary. However, designing a multiband antenna in a narrow space is a great challenge; a method that decrease the complexity of the antenna structure is also necessary to be investigated. Guo et. al. have recently designed quad-band antennas for mobile phones (Chiu & Lin, 2002; Nashaat et al., 2005; Karkkainen, 2005) and dual-band antennas for WLAN operations (Su & Chou, 2008). However, few of these antennas simultaneously cover the following communication standards: GSM (880-960 MHz), DCS (1710-1880 MHz), PCS (1850-1990 MHz), UMTS2100 (1920-2170 MHz), WLAN + Bluetooth (2400-2480 MHz), WiMAX (2500-2690 MHz), HiperLAN/2 in Europe (5150-5350 / 5470-5725 MHz) and IEEE 802.11a in the U.S. (5150-5350 / 5725-5825 MHz) (Liu & Gaucher, 2004, 2007; Wang et al., 2007; Rao & Geyi, 2009; Nguyen et al., 2009; Anguera et al., 2010; Kumar et al., 2010; Liu et al., 2010; Hsieh et al., 2009; Yu & Tarng, 2009; Hong et at., 2008; Guo et al., 2004; Li et al., 2010). This chapter proposes a planar multiband antenna that comprises a dual-band inverted-F resonator and two parasitic elements to cover all the communication standards mentioned above. One element is devoted to generating a dipole mode and another is helpful to excite a loop mode so as to broaden the impedance bandwidth. This hepta-band antenna is designed for a mobile device and the parasitic element broadens the impedance bandwidth to about 45.5%. This antenna is extended to simultaneously operate in WLAN, WiMAX, and WWAN systems. It covers all cellular bands world-wide and all wireless network bands, such as the following communication standards: GSM/DCS/PCS/UMTS/WLAN/WiMAX/HIPERLAN2/IEEE 802.11. The antenna structure that measures only 50 mm x 12 mm x 0.5 mm can be easily fabricated by stamping from a metal plate. The following describes the details of the proposed antenna as well as the experimental results.

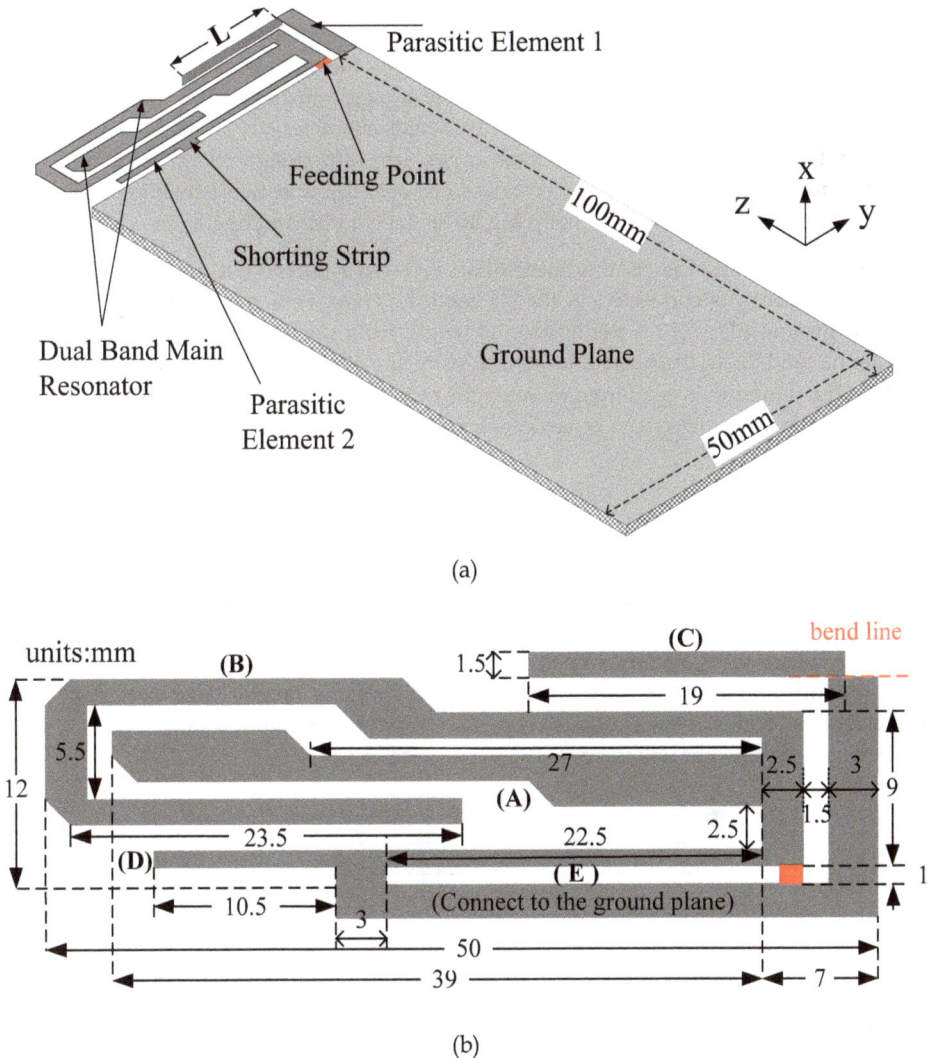

(a)

(b)

Fig. 1. The proposed antenna (a) Three-dimensional configuration of the proposed antenna (b) Plane view of the antenna structure.

2 Antenna design

2.1 Design of a dual-band antenna

Modern mobile terminals require small and thin design, therefore, planar inverted-F antenna, which requires a spacing of about 7 mm ~ 12 mm between the antenna and the substrate to achieve the sufficient operating bandwidth, is not suitable to be integrated with the present thin mobile terminals although it is popular and widely used. Fig. 1(a) shows a three dimensional view of the proposed design. The antenna, which is mounted on the top edge of the printed circuit board (PCB), is fed by a 50 Ω coaxial cable. The antenna is coplanar with the system ground of the PCB. The dielectric constant of the PCB used here is 4.4 and the thickness is 1.58 mm. As shown in Fig. 1(b), this radiating structure measures 50 mm × 12 mm × 1.5 mm and can be extended to a single metallic plate. It is basically an inverted-F antenna in which the quarter-wavelength characteristic is obtained thanks to a short-circuited metallic strip. As indicated in Fig. 1(b), this design comprises a direct-feed dual band main resonator with two branches (A) and (B), and two parasitic elements (C) and (D) excited by electromagnetic coupling, to achieve multiband operation.

Shown in Fig. 2 is a typical configuration of an inverted-F antenna. It can be fed by a mini-coaxial cable which is connected to the RF module. Here, H is the height of the radiator above the ground plane, L_F is the horizontal length from the feed point to the open end of the antenna, and L_B is the horizontal length from the feed point to the closed end of the antenna. This antenna is a quarter-wavelength radiator with one short end and one open end. The resonant frequency can be easily calculated by the formula:

$$f = \frac{c}{4(H + L_B + L_F)}$$

the where c is the speed of light. The resonant frequency can be adjusted by changing the value L_F, and the distance L_B between the feed point and shorting strip can be used to adjust the input impedance. The height H of the antenna is closely related to the impedance bandwidth where the Q factor can be reduced by increasing the antenna height to broaden the bandwidth and vice versa. Variations of IFA Antenna height cause some effects on bandwidth. Fig. 3 shows the simulation results with different antenna height H. It is found that increasing the height will increase the impedance bandwidth.

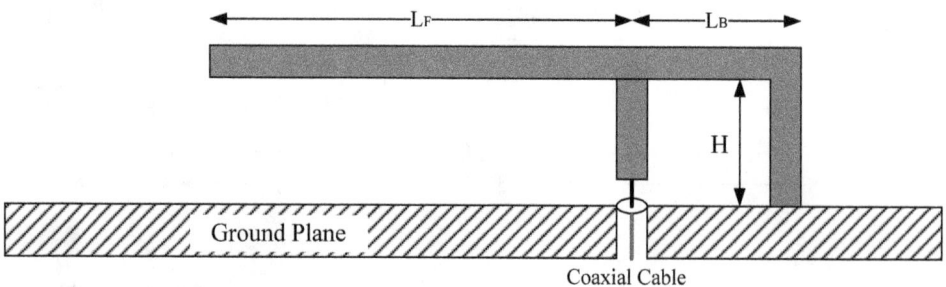

Fig. 2. A typical inverted-F Antenna.

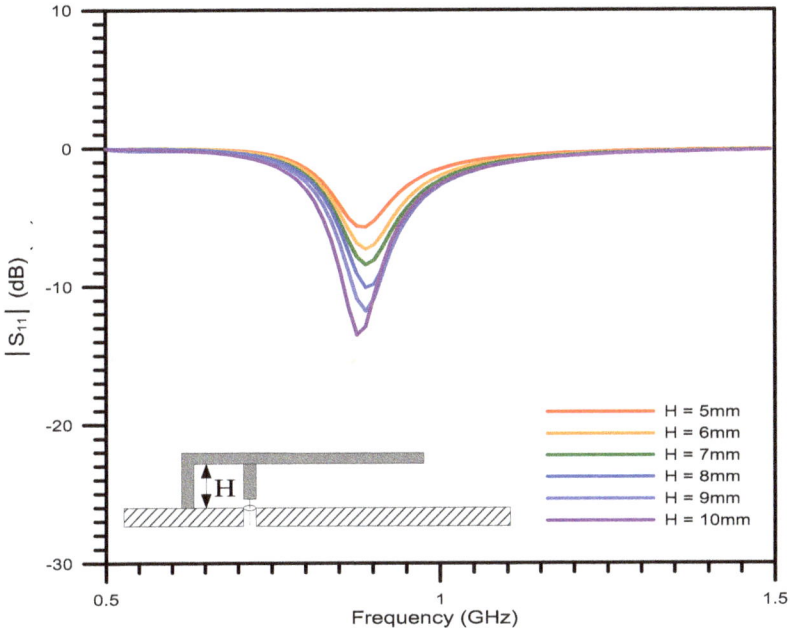

Fig. 3. Antenna height influences on the impedance bandwidth for a simple IFA.

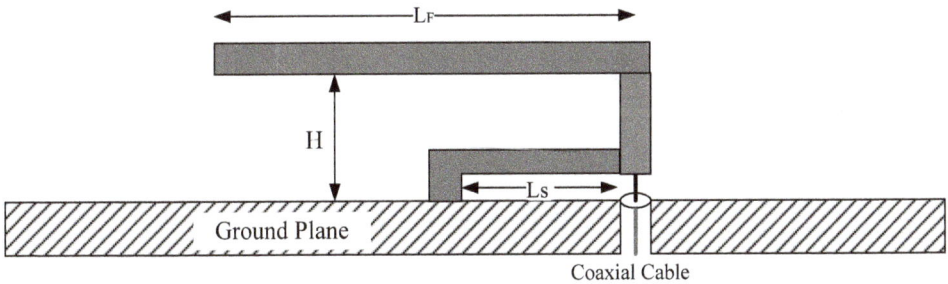

Fig. 4. A variation of typical inverted-F antenna.

Fig. 4 shows another kind of inverted-F antenna while the shorting pin is moved to the bottom for size reduction. The mechanism of this alternative is the same as the previous one, but the input impedance is matched by adjusting the length of the shorting strip L_S.

The dual band inverted-F antenna can be simply accomplished by creating two resonant paths of the antenna element. As shown in Fig. 5, the dual-band main resonator consists of two branches (A and B). The length of the longer branch (B) is about 83 mm (9 + 44.5 + 6 + 23.5 mm) which is one-quarter of the wavelength at 900MHz. The lower resonant mode for GSM operation can be excited on this resonator. On the other hand, branch (A) in the middle creates a shorter path of 42 mm, which is about a quarter of wavelength at 1800 MHz. As a result, the resonant mode for DCS operation can be excited. Simulation result of the dual band antenna is shown in Fig. 6. The input impedance can be adjusted by changing the length of the shorting strip L_s. In this case, L_s is selected to be 22.5 mm to have the widest bandwidth at both lower and upper band.

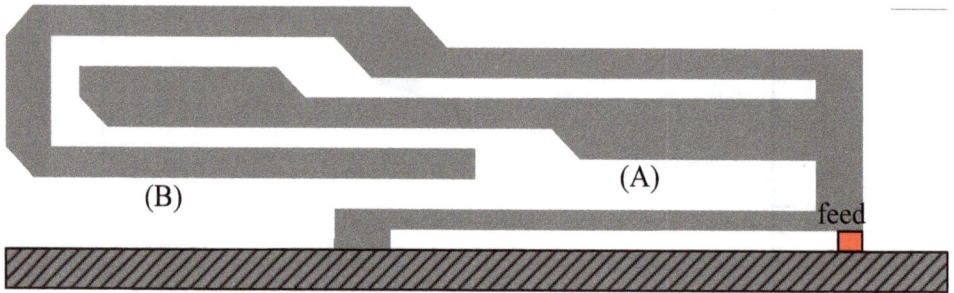

Fig. 5. A dual band inverted-F main resonator.

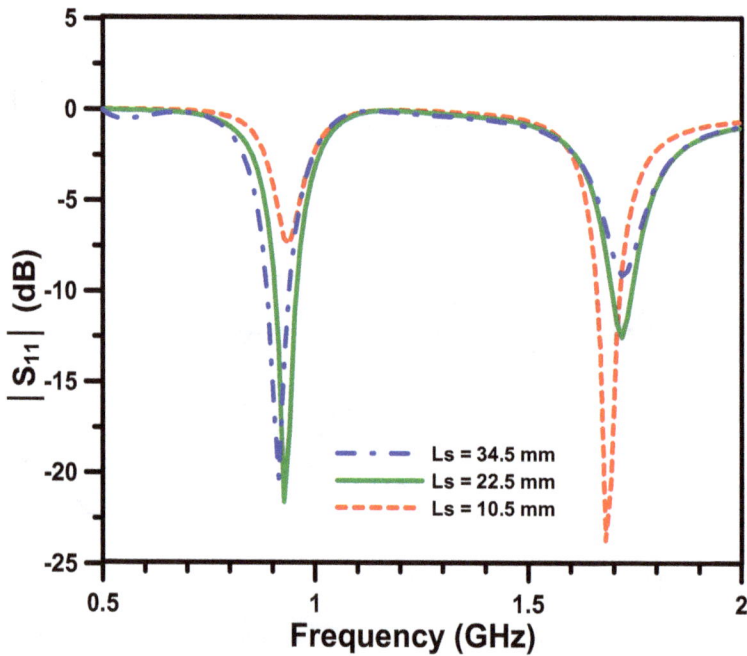

Fig. 6. Parameter study with different value of Ls.

2.2 Bandwidth enhanced by a parasitic element

Creating multiple resonant paths of the inverted-F antenna is helpful to generate multiple resonances. However, the coupling between each resonant path makes it difficult to match the antenna at each frequency band. To cover the wide bandwidth from 1900 MHz to 2700 MHz, this work introduces a parasitic resonator C near the main driven resonator. This parasitic element is excited by electromagnetic coupling from the main dual band resonator. Thus, a dipole-like antenna that resonates at 2250 MHz is formed by both the introduced resonator C, and the main resonators A and B. Fig. 7 shows the surface current distributions on the resonators and the ground plane. Finding show that part of the dual band resonator and the parasitic element form a dipole antenna. From point a, through point b, c, and d, then to point f in Fig. 7, the total length (39 mm + 3 mm + 9 mm + 19 mm = 70 mm) is closed to 0.5 wavelength at 2250MHz (67 mm). This allows the antenna to generate an additional 0.5-wavelength resonant mode at 2250 MHz to cover the desired operation bands.

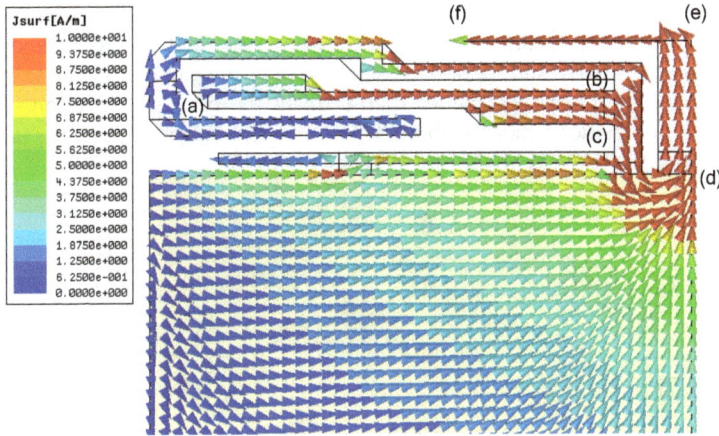

Fig. 7. Victor surface current distribution at 2.25 GHz.

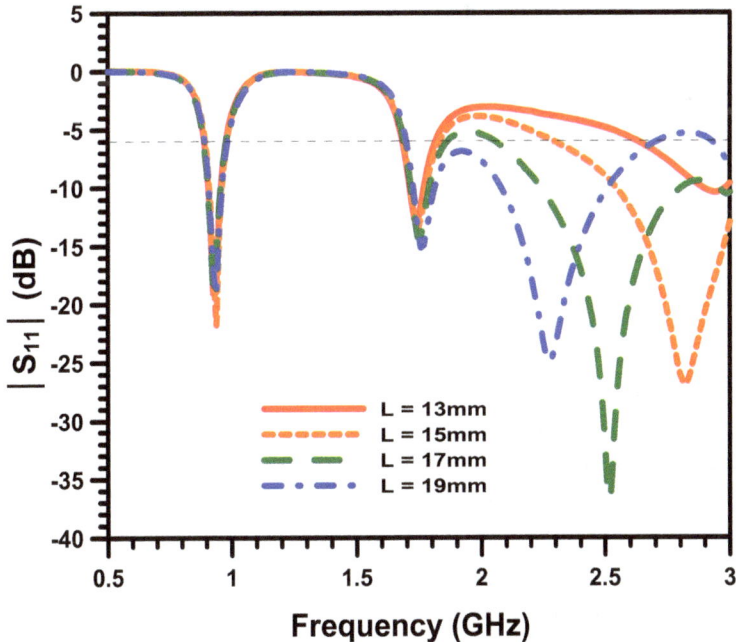

Fig. 8. Parameter study with different length of the parasitic resonator.

To demonstrate the effect of the parasitic element covering from 1900 MHz to 2700 MHz, Fig. 8 shows the parameter study of the proposed antennas with different length of the parasitic element. By Investigating the Smith chart shown in Fig. 9, it is evident that the input impedance is closer to 50 Ω as length L increases, because the longer the parasitic element, the more the loaded capacitance (Chi, 2009). The narrow gap between the main resonator and the parasitic element C introduces a proper capacitance to compensate for possible inductance contributed from the dual-band main resonator. Increasing capacitance neutralizes the effect due to inductance of the strip. Therefore, the capacitive coupled parasitic element creates a new resonant mode but does not change the original two resonant modes at 900 MHz and 1800 MHz. The length of the parasitic element is selected to be 19 mm to have the return loss better than 6 dB in the band of operation. The achieve bandwidth of the parasitic element is about 34.78 %, covering from 1900 MHz to 2700 MHz, which is enough for WLAN, WMAN, and WWAN operations.

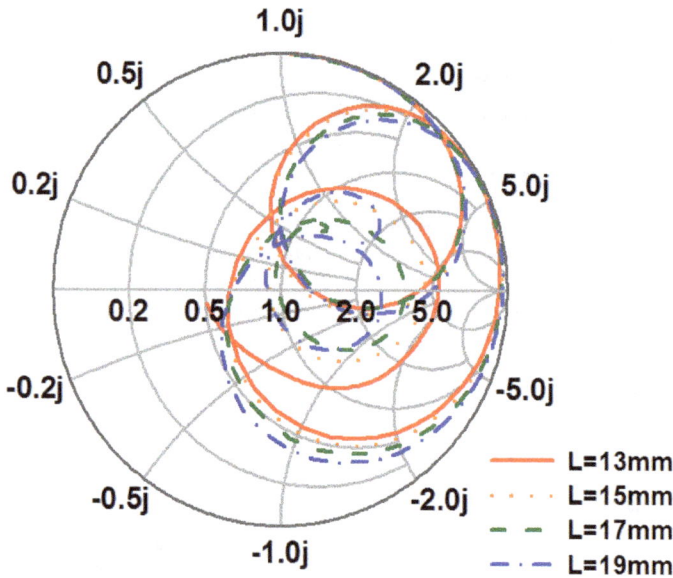

Fig. 9. Parametric study – Smith Chart.

2.3 Create resonances at the U-NII band

So far, a hexa-band Inverted-F antenna has been designed, except IEEE 802.11a or HYPERLAN/2. The current research will include the U-NII (Unlicensed National Information Infrastructure) band in this design by a tuning parasitic resonator D, as Fig. 1(b) shows. First, the third harmonics of the resonating frequency in the second band (1.72 GHz) is about 5.20 GHz. This mode which contributes to the U-NII band is also excited. The surface current distribution on the resonator A in Fig. 10(a) demonstrates that the 1.5 wavelength mode generates at the resonating frequency. The vector current distribution is shown in Fig. 11(a). Second, the loop resonator E in Fig. 1(b) is designed as a one-wavelength rectangular loop antenna. The perimeter of the loop antenna (25.5 mm + 1 mm +

25.5 mm + 1 mm) is roughly equal to a wavelength of the resonant frequency 5.59 GHz (53.67 mm). Fig. 10 (b) shows surface current distributions at the resonating frequency 5.59 GHz, The vector current distribution shown in Fig. 11(b) demonstrates that one-wavelength loop mode is excited on the resonator E.

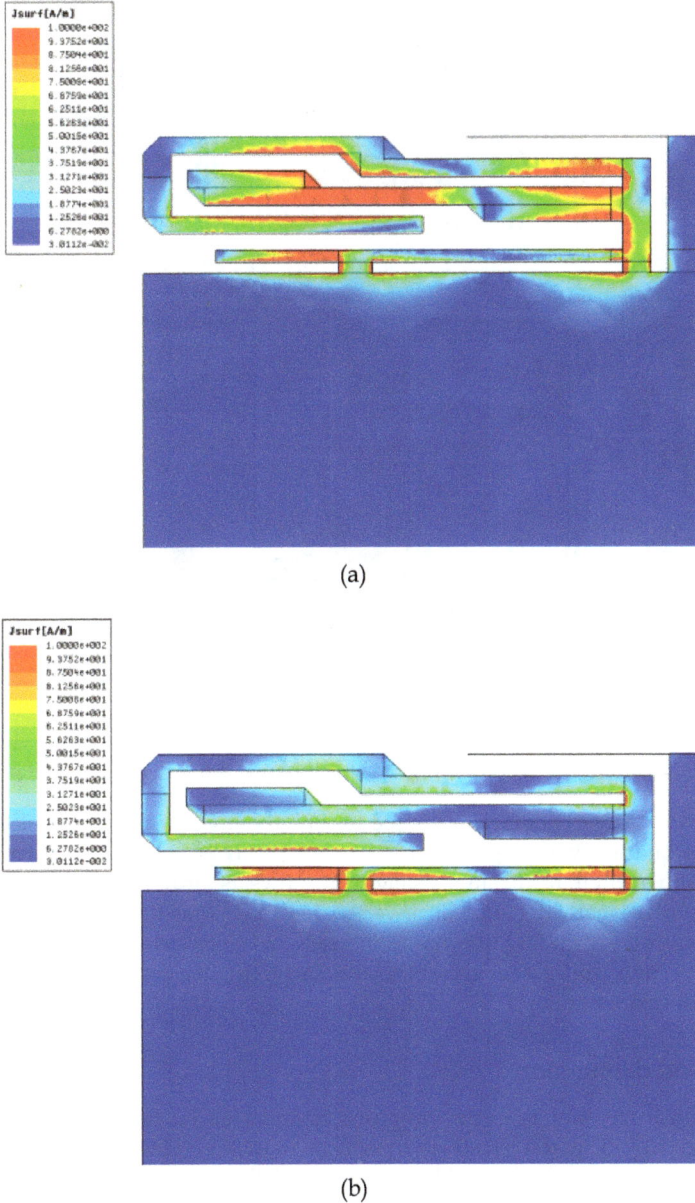

(a)

(b)

Fig. 10. Surface current distribution at (a) 5.20 and (b) 5.59 GHz.

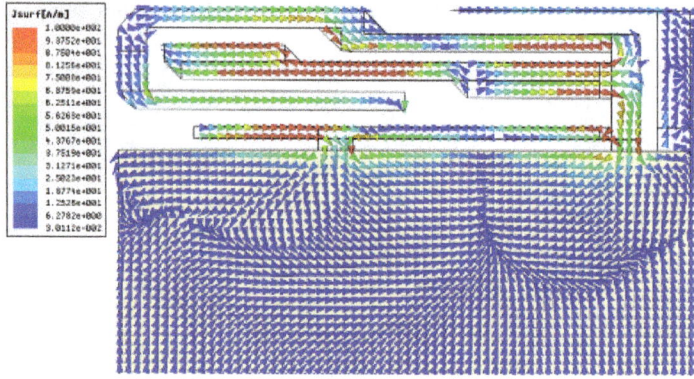

(a)

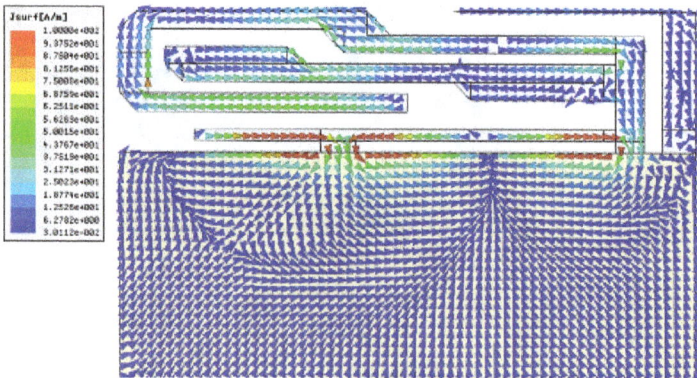

(b)

Fig. 11. Victor current distributions at higher U-NII bands: (a) 5.20 and (b) 5.59 GHz.

Finally, this work applies another technique to tune the higher order resonances for the U-NII band. The quarter wavelength resonating at 6.0 GHz is only about 12.5 mm. A short resonator D with a length of 10.5 mm, as Fig. 1(b) shows, is introduced to the short-circuited pin of the main resonator to form an inverted L-shape parasitic element. The capacitive coupling between the strip and the chassis increases its electrical length since the radiating strip is only 1 mm above the ground plane. Adding this parasitic element improves resonance performance at the U-NII band.

3. Results and discussion

This study constructs and tests the proposed antenna based on the design dimensions shown in Fig. 1(b). The test structure was shown in Fig. 12 and the measurement of scattering parameters was performed by an Agilent E5071B network analyzer. Fig. 13 shows the measured and simulated return loss where the solid red line is the measured result and the dotted blue line is the simulated one. Findings show good agreement between the measured data and simulated results. The antenna covers all cellular bands used world-wide is evident. The achieved bandwidths with return loss better than 6 dB are 80 MHz (880–960 MHz) in the GSM band, 1000 MHz (1700–2700 MHz) in the DCS/PCS/UMTS/WiFi /WiMAX band and 1270 MHz (4820–6090 MHz) in the 5 GHz U-NII band. When ground plane length varies from 80 mm to 120 mm, frequency shifting is slight (Chi, 2009).

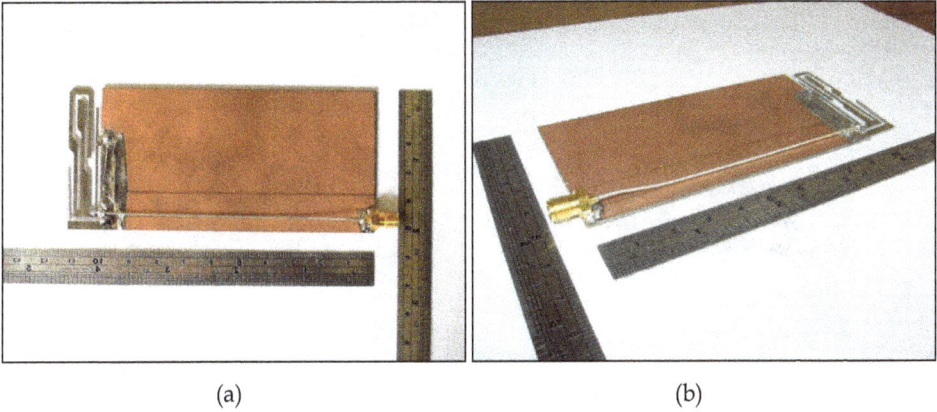

(a)   (b)

Fig. 12. Photography of the fabricated antenna (a) top view, (b) side view.

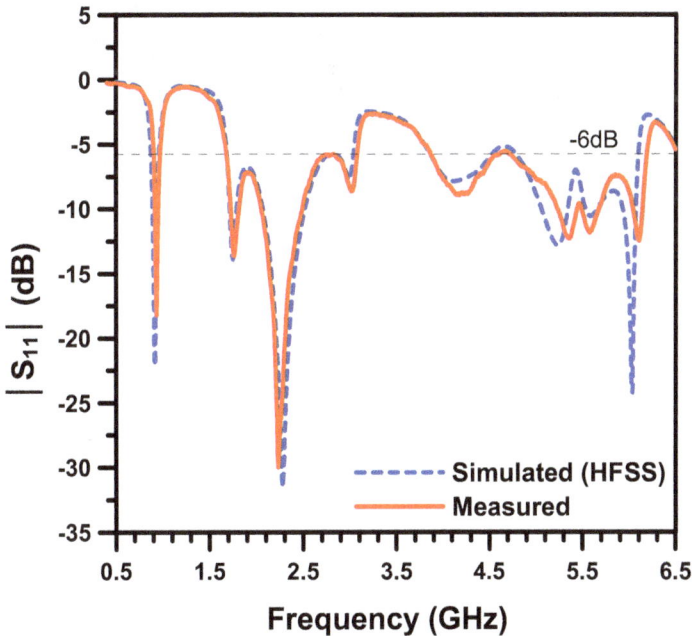

Fig. 13. Measured and simulated results of the proposed antenna.

This study performed radiation-pattern and gain measurement in the anechoic chambers of SGS Ltd. Taiwan, as shown in Fig. 14. Fig. 15 shows the measured and simulated radiation patterns at the xy-cut, xz-cut, and yz-cut. The measured radiation patterns show a good match to the simulation results except at 925MHz. In the small antenna measurement, the patterns are easily affected by the feeding RF cable in the GSM band (Chen et al., 2005). This work finds that the dual-polarization radiation-patterns have very suitable characteristics for portable devices. For the radiation shown in Fig. 14(a), more energy for E_θ is radiated in the lower band as compared to $E\varphi$. The $E\varphi$ field has some dips at 900 MHz on the xz-plane or 1800 MHz on the xy-plane. This is probably due to current cancellation on the strips and the ground plane.

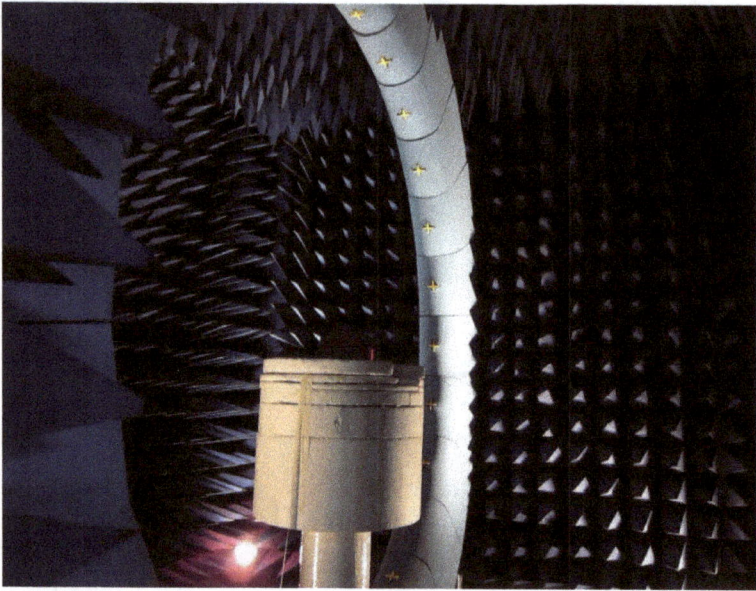

Fig. 14. Radiation Pattern measurement in a 3D anechoic chamber.

Findings also show a dipole-like pattern at the frequency 2170 MHz. Radiation patterns shown in Fig. 15(b) confirm this deduction. The radiation pattern of this mode is similar to a small dipole oriented in the y–axis leading to a directional pattern in the E-plane (xy-plane, blue line) and omni-directional pattern in the H-plane (xz-plane, blue line), as Fig. 15(b), shows respectively. The resonators C and B at 2170 MHz have strong current distributions along the z-direction which also contribute to radiation fields. The radiation pattern of this current distribution is due to a small dipole oriented in the z–axis leading to a bidirectional pattern in the E-plane (xz-plane, red line) and omni-directional pattern in the H-plane (xy-plane, red line), as Fig. 15(b), shows respectively. Findings also show an asymmetric radiation pattern at the U-NII band (5-6 GHz) and some variation and nulls, since different modes are excited in this U-NII band.

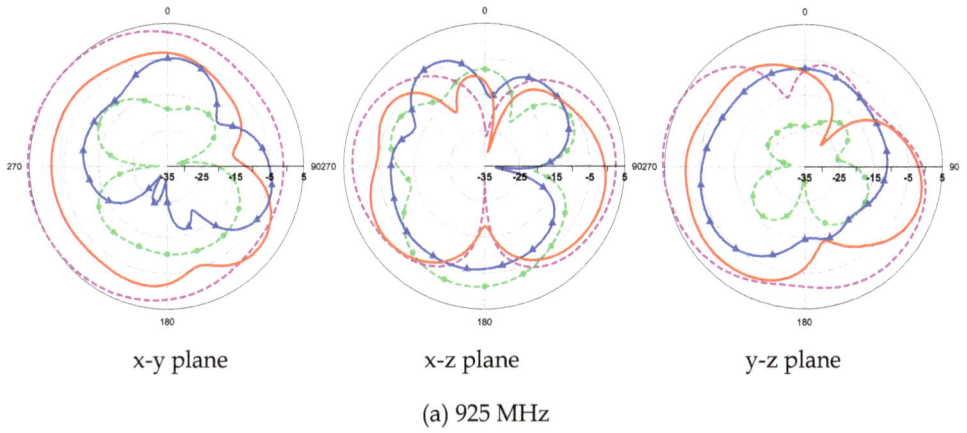

x-y plane   x-z plane   y-z plane

(a) 925 MHz

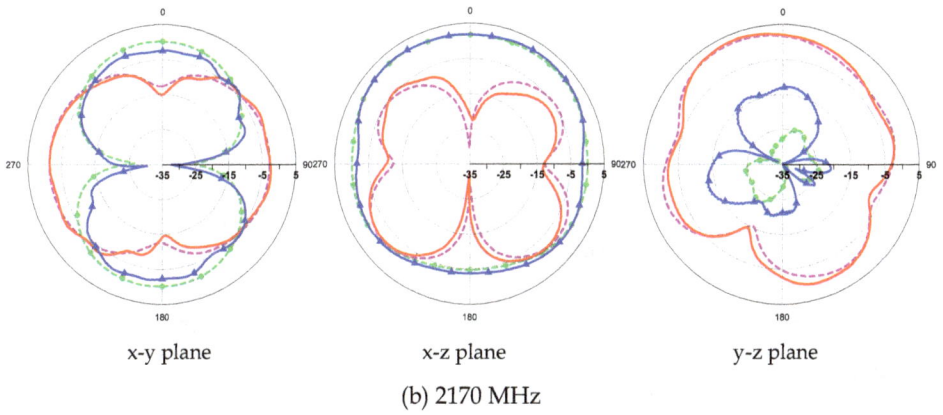

x-y plane   x-z plane   y-z plane

(b) 2170 MHz

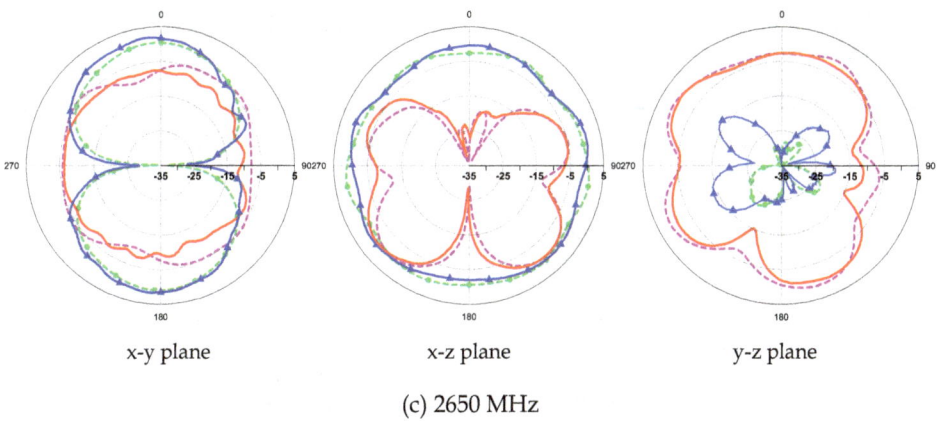

x-y plane   x-z plane   y-z plane

(c) 2650 MHz

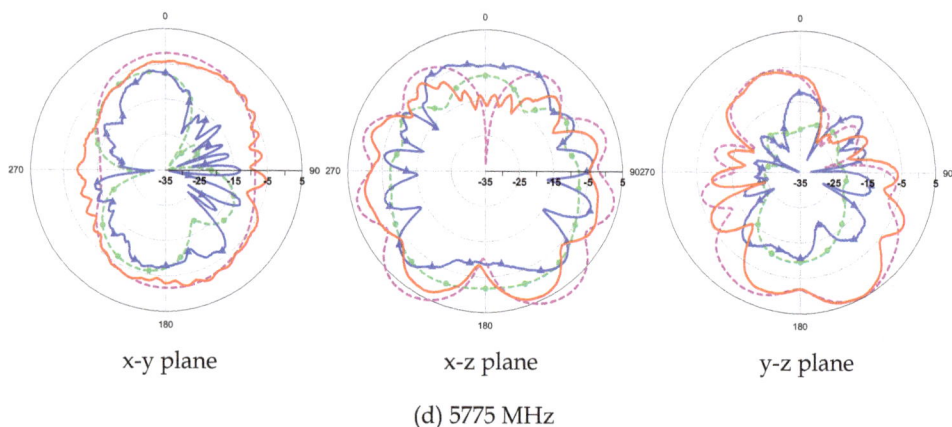

x-y plane   x-z plane   y-z plane

(d) 5775 MHz

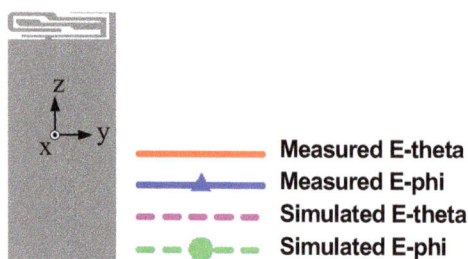

Measured E-theta
Measured E-phi
Simulated E-theta
Simulated E-phi

Fig. 15. Measured and simulated radiation patterns in three cuts (a) 925 MHz (b) 2170 MHz (c) 2650 MHz (d) 5775 MHz.

Frequency (MHz)	925	1710	1795	1920	1990
Peak Gain (dBi)	-0.25	2.4	2.05	1.39	1.63
Average Gain (dBi)	-1.96	1.10	-0.63	-0.01	-0.51
Efficiency	51.42%	61.94%	64.85%	70.35%	78.80%
Frequency	2170	2420	2650	5250	5800
Peak Gain	2.95	2.5	2.48	6.91	8.35
Average Gain	1.10	1.15	0.58	-0.31	-1.99
Efficiency	90.11%	86.83%	71.42%	70.24%	71.80%

Table 1. Measured three-dimensional peak gain, average gain, and radiation efficiency.

By using the commercial electromagnetic simulation software HFSS, this research carries out simulations for the theoretical gains to investigate antenna performance and compare it with the measured results (Chi, 2009). Good agreement confirms that the measured data are accurate. The two-dimensional average gain is determined from pattern measurements made in the horizontal (azimuth) plane for both polarizations of the electric field. The results are then averaged over azimuth angles and normalized with respect to an ideal isotropic radiator (Chen, 2007). Finally, Table 1 lists the measured peak gain, two-dimensional average gain and radiation efficiency for all the operation bands, showing that all radiation efficiencies are over 50 percent, meeting the specification requirement.

4. Summary

This chapter reported a down-sized multiband inverted-F antenna to integrate the 3.5G and WLAN/WiMAX antenna systems. It is comprised of a dual-band antenna with one feed point and two parasitic elements to cover many mobile communication systems including GSM900 /DCS /PCS /UMTS /WLAN/ WiMAX /HiperLAN2 /IEEE802.11a. Measured parameters including return loss, radiation patterns, three-dimensional peak gain and average gain as well as radiation efficiency were presented to validate the proposed design. Since this antenna can be formed by a single plate, it is both low cost and easy to fabricate, making it suitable for any palm-sized mobile device applications.

5. References

C. Soras, M. Karaboikis, and G. T. V. Makios, "Analysis and design of an inverted-F antenna printed on a PCMCIA card for the 2.4 GHz ISM band," IEEE Antenna's and propagation magazine, vol. 44, no. 1, February 2002.

C. W. Chiu and F. L. Lin, "Compact dual-band PIFA with multi-resonators," Electronics Letters, vol. 38, pp. 538-540, June 2002.

C.-L. Liu, Y.-F. Lin, C.-M. Liang, S.-C. Pan, and H.-M. Chen, "Miniature Internal Penta-Band Monopole Antenna for Mobile Phones," IEEE Trans. Antennas Propag., vol. 58, no. 3, March 2010.

D. Liu and B. Gaucher, "A new multiband antenna for WLAN/Cellular application," Vehicular Technology Conference, vol. 1, 60th, pp. 243 - 246, Sept. 2004.

D. Liu and B. Gaucher, "A quadband antenna for laptop application," International Workshop on Antenna Technology, pp. 128-131, March 2007.

D.M. Nashaat, H. A. Elsadek, and H. Ghali, "Single feed compact quad -band PIFA antenna for wireless communication applications," IEEE Trans. Antennas Propagat., vol. 53, No. 8, pp. 2631-2635, Aug. 2005.

H.-W. Hsieh, Y.-C. Lee, K.-K. Tiong, and J.-S. Sun, "Design of A Multiband Antenna for Mobile Handset Operations," IEEE Antennas Wireless Propag. Lett., vol. 8, 2009.

J. Anguera, I. Sanz, J. Mumbrú, and C. Puente, "Multiband Handset Antenna with A Parallel Excitation of PIFA and Slot Radiators," IEEE Trans. Antennas Propag., vol. 58, no. 2, February 2010.

K. Hirasawa and M. Haneishi, "Analysis, design and measurement of small and low profile antennas," ch.5, Norwood, MA, Artech House, 1922.

K.-L. Wong, L.-C. Chou, and C.-M. Su, "Dual-band flat-plate antenna with a shorted parasitic element for laptop applications," IEEE Transactions on Antennas and Propagation, vol. 53, no. 1, pp. 539-544, January 2005.

M. Ali and G. J. Hayes, "Analysis of intergated inverted-F antennas for bluetooth applications," IEEE International symposium on antenna and propagation, 2000.

M. K. Karkkainen, "Meandered multiband PIFA with coplanar parasitic patches," IEEE Microw. Wireless Compon. Lett., vol.15, pp. 630-632, Oct. 2005.

P. Ciais, R. Staraj, G. Kossiavas, and C. Luxey, "Design of an internal quad-band antenna for mobile phones," IEEE Microwave and wireless components letters, vol. 14, no. 4, April 2004.

P. Kumar.m, S. Kumar, R. Jyoti, V. Reddy, and P. Rao1, "Novel Structural Design for Compact and Broadband Patch Antenna," 2010 International Workshop on Antenna Technology (iWAT), 1-3 March 2010.

P.Nepa, G. Manara, A. A. Serra, and G. Nenna, "Multiband PIFA for WLAN mobile terminals," IEEE antenna and wireless propagation letters, vol. 4, 2005.

Q. Rao and W. Geyi, "Compact Multiband Antenna for Handheld Devices," IEEE Trans. Antennas Propag., vol. 57, no. 10, October 2009.

R. Bancroft, "Development and integration of a commercially viable 802.11a/b/g HiperLan/ WLAN antenna into laptop computers," Antennas and Propagation Society International Symposium, vol. 4A, pp. 231- 234, July 2005.

R. King, C. W. Harisson, and D. H. Denton, "Transmission-line missile antenna," IRE Trans. Antenna Propagation, vol. 8, no. 1, pp. 88-90, 1960.

S. Hong, W. Kim, H. Park, S. Kahng, and J. Choi, "Design of An Internal Multiresonant Monopole Antenna for GSM900/DCS1800/US-PCS/S-DMB Operation," IEEE Trans. Antennas Propag., vol. 56, no. 5, May 2008.

S.W. Su and J.H. Chou, "Internal 3G and WLAN/WiMAX antennas integrated in palm-sized mobile devices," Microw. Opt. Technol. Lett., vol. 50, no. 1, pp. 29-31, Jan. 2008.

T. K. Nguyen, B. Kim, H. Choo, and I. Park, "Multiband dual Spiral Stripline-Loaded Monopole Antenna," IEEE Antennas Wireless Propag. Lett., vol. 8, 2009.

T. Taga and K. Tsunekawa, "Performance analysis pf a built-in planar inverted-F antenna for 800MHz and portable radio units," IEEE Trans. on selected areas in communications, vol. SAC-5, no. 5, June 1987.

W. X. Li, X. Liu, and S. Li, "Design of A Broadband and Multiband Planar Inverted-F Antenna," 2010 International Conference on Communications and Mobile Computing, vol. 2, 12-14 April 2010.

X. Wang, W. Chen, and Z. Feng, "Multiband antenna with parasitic branches for laptop applications," Electronics letters, vol. 43, no. 19, 13th, September 2007.

Y. J., Chi, "Design of internal multiband antennas for portable devices," Master Thesis, National Ilan University, June 2009

Y.-C. Yu and J.-H. Tarng, "A Novel Modified Multiband Planar Inverted-F Antenna," IEEE Antennas Wireless Propag. Lett., vol. 8, 2009.

Y.-X. Guo and H. S. Tan, "New compact six-band internal antenna," IEEE antenna and wireless propagation letters, vol. 3, 2004.

Y.-X. Guo, I. Ang, and M. Y. W. Chia, "Compact internal multiband antennas for mobile handsets," IEEE antenna and wireless propogation letters, vol. 2, 2003.

Y.-X. Guo, M. Y. W. Chia, and Z. N. Chen, "Miniature Built-In Multiband Antennas for Mobile Handsets," IEEE Trans. Antennas Propag., vol. 52, no. 8, August 2004.

Z. N. Chen, Antennas for Portable Devices, pp.125-126, John Wiley & Sons, Inc. 2007.

Z. N. Chen, N. Yang, Y. X. Guo, and M. Y. W. Chia, "An investigation into measurement of handset antennas," IEEE. Trans. Instrum. Meas., vol. 54, no.3, pp. 1100–1110, June 2005.

Zhi Ning Chen, "Antennas for Portable Devices," John Wiley & Sons, Inc. 2007, ch.4, pp.115-116.

3

A Reconfigurable Radial Line Slot Array Antenna for WiMAX Application

Mohd Faizal Jamlos
School of Computer and Communication Engineering,
University of Malaysia Perlis (UniMAP) to University Malaysia Perlis,
Kangar, Perlis,
Malaysia

1. Introduction

WiMAX refers to interoperable deployments of IEEE 802.16 protocol, in similarity with wireless fidelity (Wi-Fi) of IEEE 802.11 protocol but providing a larger radius of coverage. WiMAX is a potential replacement for current mobile technologies such as Global System Mobile (GSM) and High Speed Downlink Packet Access (HSDPA) and can be also applied as overlay in order to enlarge the capacity and speed.

WiMAX is a broadband platform and needs larger bandwidth compared to existing cellular bandwidth. Fixed WiMAX used fiber optic networks instead of copper wire which is deployed in other technology. WiMAX has been successfully provided three up to four times performance of current 3G technology, and ten times performance is expected in the future. Currently, the operating frequencies of WiMAX are at 2.3 GHz, 2.5 GHz, and 3.5 GHz whereas the chip of WiMAX that operated in those frequencies is already integrated into the laptops and netbooks. As transmitter, TELCO Company requires to prepare a better transmitting communication tower in providing better WiMAX's coverage and data rates. Hence, the need of superior reconfigurable WiMAX's antenna is extremely crucial to sustain the signal strength at the highest level (dB).

Traditional transmission line microstrip antenna has been widely used as a reconfigurable antenna due to its less complexity and easiness to fabricate. However, the reconfigurable beam shape application especially point-to-point communication required an antenna that can provide a better gain since incorporating a PIN diode switches has been known to deteriorate the gain characteristic of an antenna [1, 7]. A lot of efforts have been allocated to enhance the gain of the conventional microstrip antenna [2-3, 5, 9]. For high gain purpose, a radial line slot array (RLSA) antenna design is more beneficial [5]. An RLSA antenna has as much as 50% higher gain than the conventional microstrip antenna [6]. Conventionally, the RLSA antenna has no reconfigurable ability due to its feeding structure which is via coaxial-to-waveguide transition probe. However, it is made realizable by using feed line, PIN diodes and an aperture coupled feeding structure [7-8, 10-12].

Another significant problem of conventional microstrip antenna is the narrowing of half-power beamwidth (HPBW) which could only cover forward radiated beam from −50° to 50° [9]. This antenna also has another salient advantage where it can generate a broadside radiation pattern with a wider HPBW covering from −85° to 85°. Such wide HPBW is deemed as an interesting characteristic in which the antenna can function as WiMAX application.

As the proposed antenna is etched from FR4 substrate, it is inexpensive in terms of fabrication. Dimension wise, the proposed antenna length and width are 150 mm and 150 mm respectively, which is smaller than conventional microstrip antenna that could achieve the same function and performance [10]. In [3, 8, 9-13], switching mechanisms are utilized to alter the radiation pattern efficiently. The antenna, proposed in this paper, can dynamically be used in a beam shaping and broadside radiation pattern for WiMAX application.

This chapter is organized as follows: In Section 2, the RLSA radiating surface, aperture slots and feed line designs incorporates with PIN diode switches are explained and the effects of different configuration of the switches are investigated. The measurement and simulation of beam shaping and broadside radiation pattern using PIN diodes switching results will be shown in Section 3. Finally, conclusion will be drawn in Section 4.

2. Antenna structure

The proposed antenna structure, as shown in figure 1, has the ability to exhibit two major types of radiation patterns; the beam shape and the broadside radiation pattern. The 'circular' and a 'bridge' feed line are interconnected by switches, which consists of end-fire beam-shaped reconfigurable switches (EBRS) and broadside reconfigurable switches (BRS). The EBRS are referring to the first up to the fifth switches while the BRS are the first, fifth, sixth and seventh switches as shown in figure 1(a).

Four aperture slots are used to couple the feeding line to the radiating surface as shown in figure 1(b). Inaccuracy of alignment between the layer of feed line and aperture slots to the radiating surface can significantly deteriorate the antenna's performance especially on the gain characteristic. The aperture slots determine the amount of coupling to the RLSA radiating surface from the feed line of the proposed antenna. Hence, the feed line must be aligned beneath the aperture slots accurately as shown in figure 1(c). The length of the four aperture slots are 40 mm while their width are 3 mm.

The RLSA pattern that is used as the radiating surface in the proposed antenna has the arrangement as shown in figure 1(d) in order to provide a linear polarization along the beam direction. There are 96 slots, with 16 slots in the inner-most ring, and 32 slots in the outer-most ring. The width and length of the RLSA slots are 1.5 mm and 15 mm respectively. The gaps between the slots are mostly 8 mm. The diameter of the circular radiating surface is 150 mm.

Generally, by turning the EBRS ON and the sixth and seventh of the BRS OFF, it will result in a beam shape radiation pattern. The pattern will becomes narrower with an increasing number of EBRS switches turned ON. While by turning ON the BRS and the second up to fourth of EBRS turned OFF, a broadside radiation pattern will be obtained.

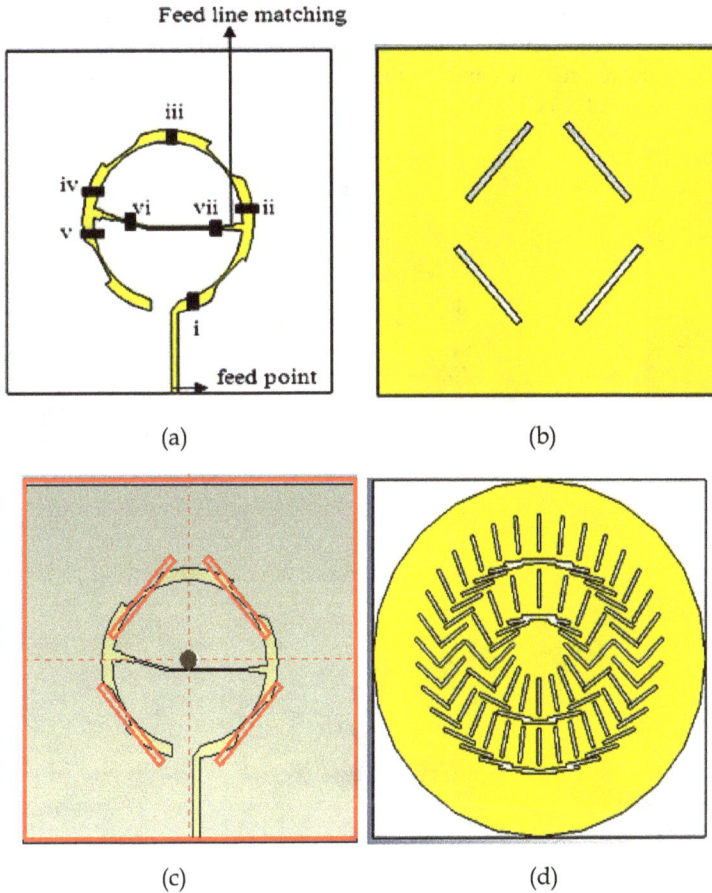

Fig. 1. Simulation structure of the proposed antenna (a) feed line (b) Aperture slots (c) Alignment of aperture slots and feed line (d) RLSA radiating surface

Figure 2 shows the photographs of the proposed antenna. Each of the PIN diodes is surrounded by two inductors and two capacitors forming the switching circuit as shown in figure 2(a). The inductors intend to choke off the alternating current (AC) and radio frequency (RF) signals from flowing into the feeding line while the capacitors allow the flow of the AC and block the direct current (DC) simultaneously.

The proposed antenna is developed using an aperture coupled configuration where the upper and bottom substrate are made of FR4 dielectric substrates (relative permittivity = 4.7, loss tangent = 0.019). The sizes of the substrates are 150 mm x 150 mm. The feed probe's radius is 0.5 mm while the heights of the substrates are 1.6 mm. The back plane reflector is a made up of copper foil with 0.035 mm thickness. The foil is attached on a piece of 2 mm thickness wood. The reflector is placed under the proposed antenna by using PCB stands of 5 mm height, as shown in figure 2(d). The height between the reflector and the feed line is influential in determining the operating frequency of the antenna. If the height is larger than its optimized height, which is 5 mm for this antenna, the operating frequency will be shifted to a lower centre frequency, and vice versa. The reflector width and length are both 150 mm, thus making its surface area the same as the size of the antenna. The proposed antenna is operating at frequency of 2.3 GHz.

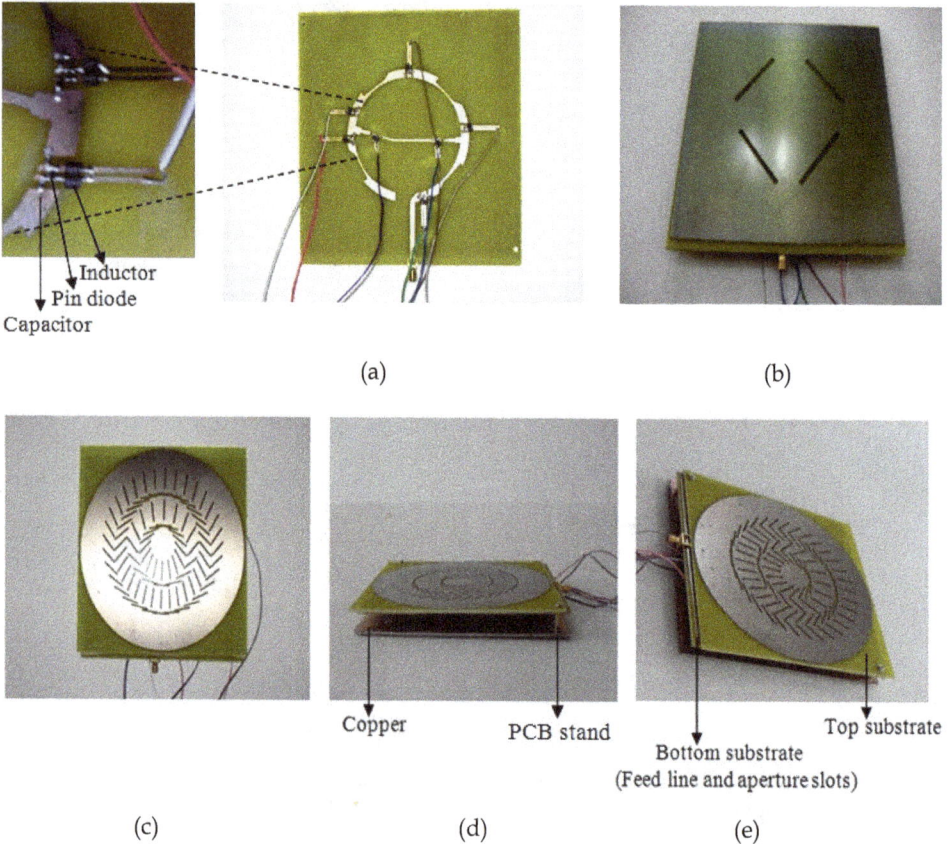

(a)   (b)

(c)   (d)   (e)

Fig. 2. Photograph of the proposed antenna (a) Feed line with PIN diodes switches (b) Aperture slots (c) RLSA radiating surface (d) side view (e) Layout view

3. Result and discussion

Measurement shows that four different types of beam shape radiation pattern can be well reconfigured with the configuration of the EBRS. Different activation of EBRS will results in different gain and HPBW. By turning ON the first switch of the EBRS, gain and HPBW of 4.85 dB and -65° to 70° are obtained respectively, as shown in figure 3(a). While in figure 3(b), turning ON the first and second switches of the EBRS will narrow the HPBW from -40° to 45° with a gain of 7.2 dB. Figure 3(c) demonstrates the beam shape of the radiation pattern with the HPBW from -15° to 20° and a gain of 9.9 dB by turning ON the first, second and third switch of the EBRS simultaneously.

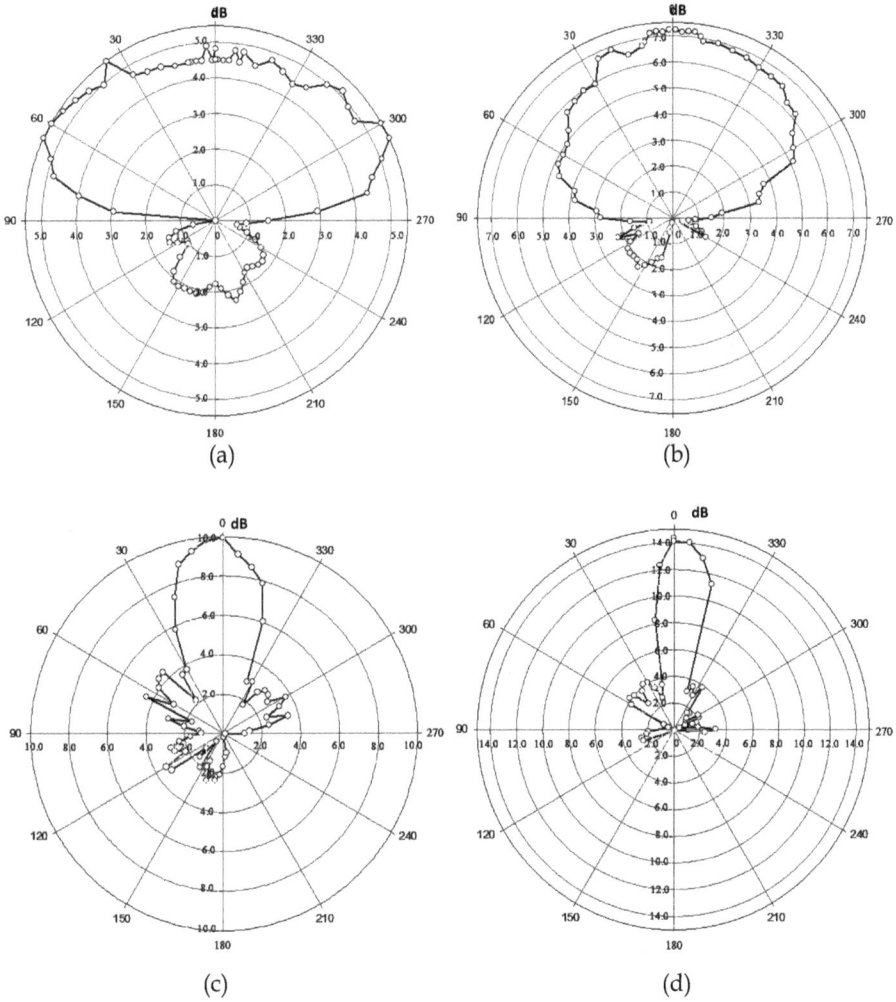

Fig. 3. The measurement of beam shape radiation patterns by turning ON the EBRS (a) switch i (b) switches i and ii (c) switches i, ii, and iii (d) switches i, ii, iii, iv and v

In figure 3(d), the HPBW of the radiation pattern is from -10° to 15° and antenna gain of 14.64 dB are obtained when the first until the fifth switches of the EBRS are turned ON. It is obvious that the proposed antenna can be tuned to have a wide HPBW which covered from -65° to 70° beam angle range, compared to conventional microstrip antenna that can only cover from -50° to 50° beam angle range. The maximum gain of the proposed antenna is 14.64 dB, which can be considered high for an antenna of such size. Computer Simulation Technology (CST) Studio Suite 2009 is used as a platform to design and simulate the radiation pattern of the proposed antenna. It is clearly shown that the simulations have the same behaviour with the measurements where the higher the produced gain, the antenna's

HPBW will becomes narrower as shown in figure 4(a) and figure 4(b). 3D representation of the far field radiation patterns are shown in figure 5. The measurement has comparable results with the simulations. Nevertheless, the measured antenna's gain is slightly higher than simulation results but smaller in terms of HPBW.

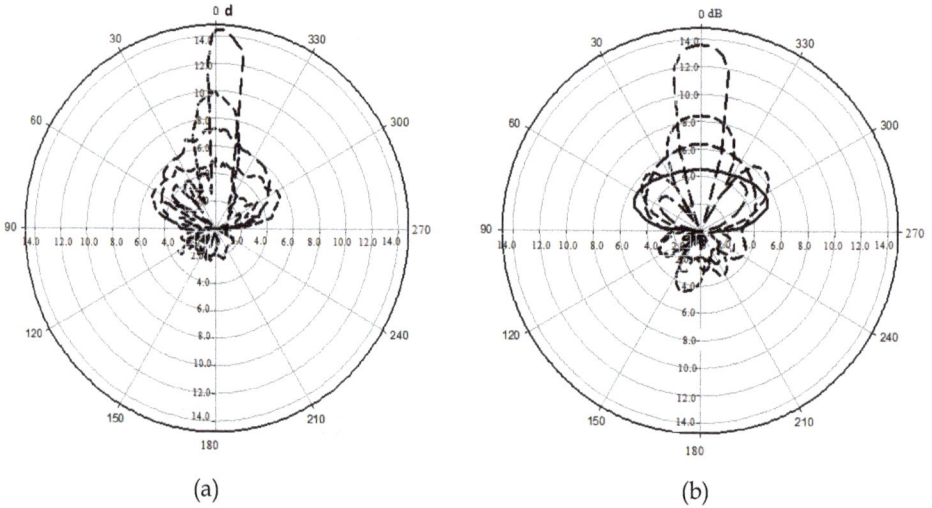

(a)   (b)

Fig. 4. The complete beam shape radiation patterns (a) Measured (b) Simulation

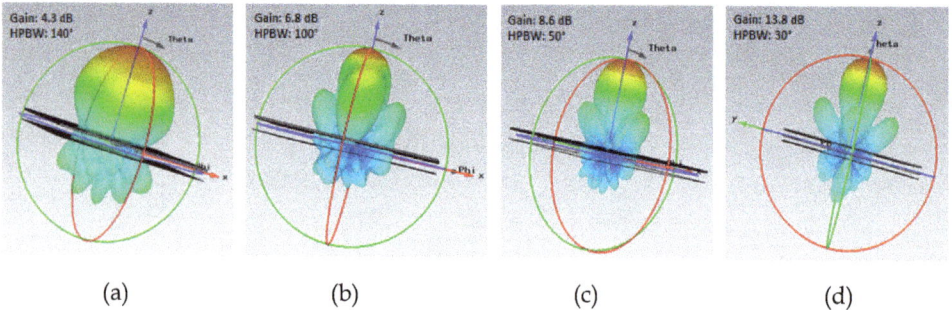

(a)   (b)   (c)   (d)

Fig. 5. 3D-polar plot of beam shape radiation patterns by turning ON the EBRS (a) switch i (b) switches i and ii (c) switches i, ii, and iii (d) switches i, ii, iii, iv and v

The BRS configuration has the ability to turn the radiation pattern from the beam shape to broadside radiation pattern perfectly as shown in figure 6 and figure 7. Figure 6 shows a divisive broadside radiation pattern with a maximum gain of 10.8 dB and a wider HPBW of -85°- 85° when turning all of the BRS ON simultaneously. Certain combination configuration between the BRS and EBRS are able to generate another radiation pattern which is a broadside single-sided radiation pattern. This kind of pattern is lean to the right with HPBW of -80°- 80° when turning ON the sixth and seventh of BRS and the first up to fourth of EBRS concurrently with a maximum gain increased up to 12.8 dB as shown in Figure 7. Since in this switching configuration the direction of radiation pattern is focused on one side, it achieves a higher gain in comparison to the divisive broadside pattern. Figure 8 depicts the 3D simulation of the far field radiation patterns of the proposed antenna which is aligned with the measured radiation patterns as shown in figure 6 and figure 7. However, the measured gain and HPBW are slightly less compared to the simulations due to CST simulation's ideal and free loss environment.

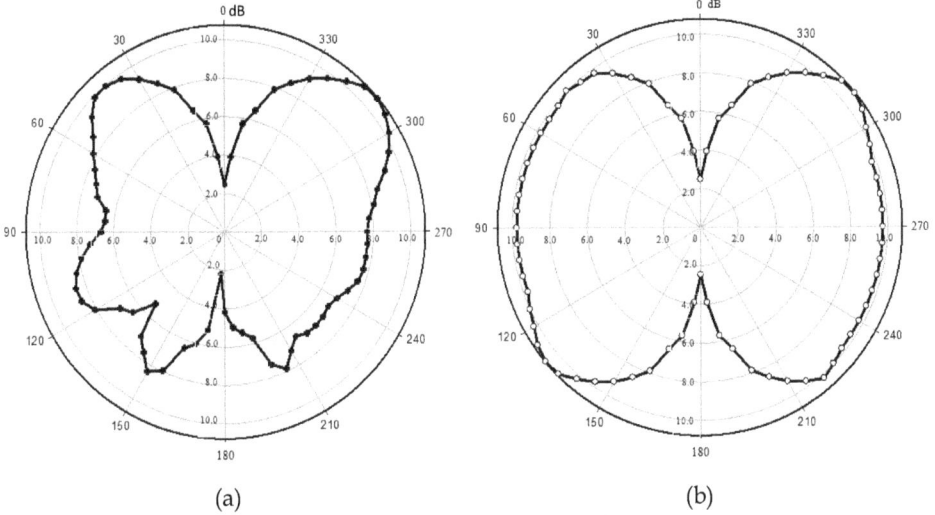

(a)   (b)

Fig. 6. Broadside divisive radiation patterns (a) Measured (b) Simulation

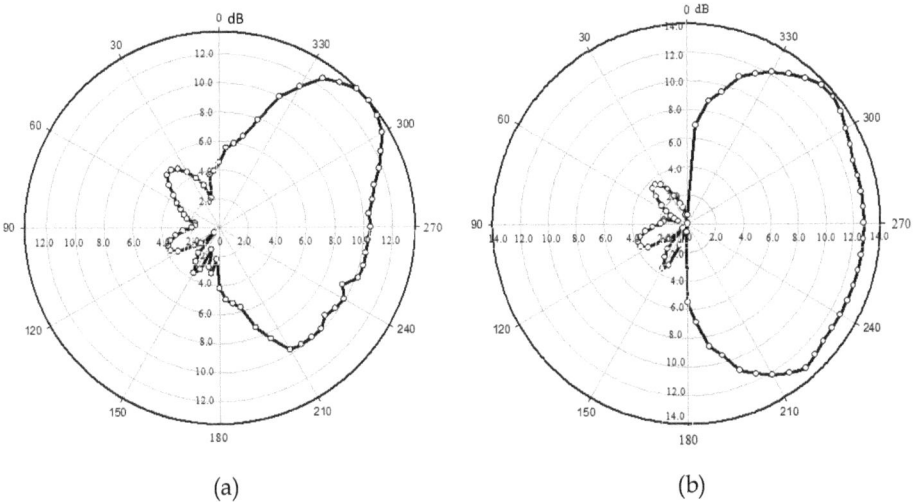

(a)   (b)

Fig. 7. Broadside single-sided radiation patterns (a) Measured (b) Simulation

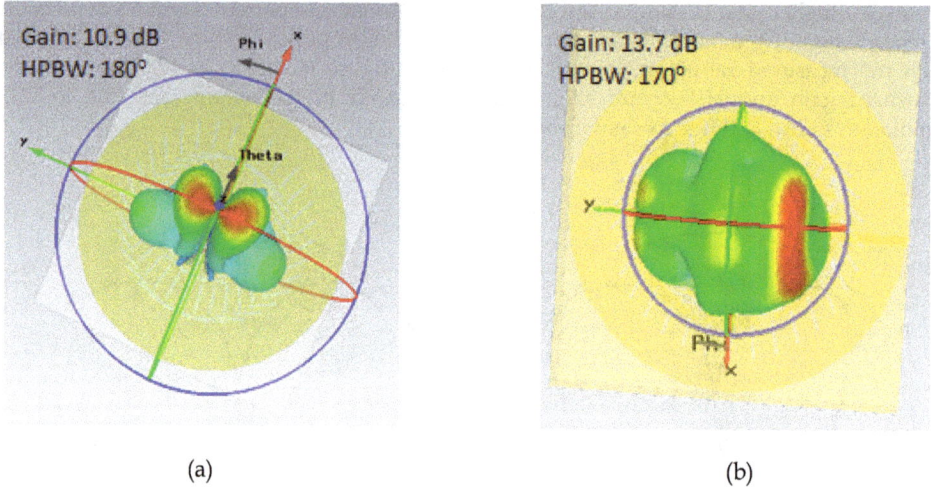

(a)   (b)

Fig. 8. 3D-polar plot of broadside radiation patterns (a) Divisive (b) Single-sided

The measurements show a very good agreement with simulations where the radiation patterns are formed successfully with respect to the beam shaping and broadside characteristics. The high gain measurements and simulations of the proposed antenna can be attributed to the good coupling from the feed line to the RLSA radiating surface through the appropriate sizing, positioning and shape of aperture slots. The outputs of the PIN diodes switching scheme involving the EBRS and BRS are summarized in table 1. All the radiation patterns of the proposed antenna are relatively at frequency 2.3 GHz as depicted by figure 9.

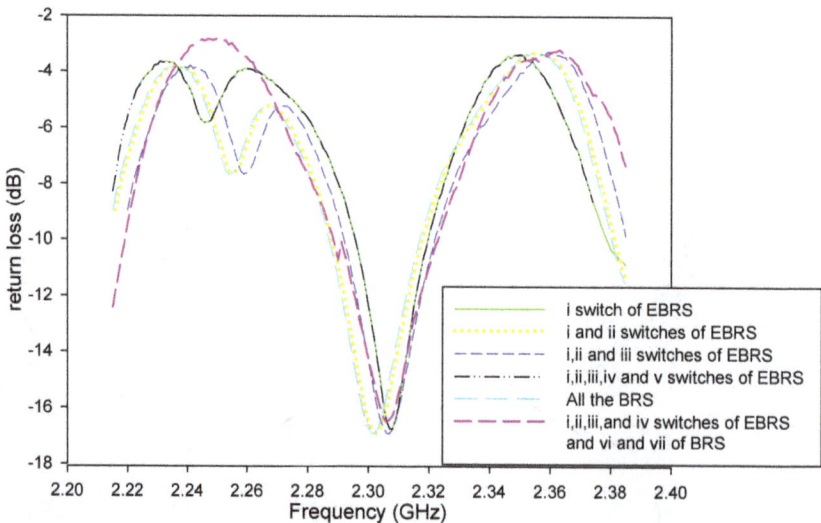

Fig. 9. The measurement of return loss by variation activation of the EBRS and BRS

Type of switch	Number of PIN diode switch	PIN diode status					
End-fire beam-shaped reconfigurable switches (EBRS)	i	ON	ON	ON	ON	ON	ON
	ii	OFF	ON	ON	ON	OFF	ON
	iii	OFF	OFF	ON	ON	OFF	ON
	iv	OFF	OFF	OFF	ON	OFF	ON
	v	OFF	OFF	OFF	ON	ON	OFF
Broadside reconfigurable switches (BRS)	vi	OFF	OFF	OFF	OFF	ON	ON
	vii	OFF	OFF	OFF	OFF	ON	ON
Gain (dB)		4.85	7.2	9.9	14.64	10.8	12.8
HPBW(°)		-65°- 70°	-40° to 45°	-15° to 20°	-10° to 15°	-85°- 85°	-80°- 80°
Type of radiation pattern		Beam shaping				Divisive radiation pattern	Single sided radiation pattern

Table 1. Configuration of PIN diode switches of the measured proposed antenna at 2.3 GHz

4. Conclusion

A novel reconfigurable radiation pattern microstrip antenna using RLSA is introduced in this paper. This chapter has taken advantages of the high performances of RLSA in terms of gain and less signals reflection, to make the proposed antenna becomes more efficient. This antenna is designed based on aperture coupled structure. The ability of the beam shape and broadside radiation pattern is attributed with the usage of PIN diode switches that integrated in the feed line of the proposed antenna. It is shown through the measurements that the radiation patterns can be well reconfigured through the assist of the orientation and geometry of the RLSA slots. The proposed antenna which has a dimension size of 150 mm X 150 mm, can be tuned to reach a high gain of 14.64 dB. The antenna can also provide wider value of HPBW that covered from -85° to 85° which is far better than -50° to 50° HPBW of a conventional microstrip antenna. The broadside patterns are achieved by turning ON selected configuration of the combination between the BRS and EBRS. The structure of the proposed antenna which is not bulky compared to the conventional microstrip antenna would be greatly suitable for beam shape and broadside radiation pattern application such as WiMAX.

5. References

[1] G. Monti, R. De Paolis, and L. Tarricone, "Design of a 3-state reconfigurable crlh transmission line based on mems switches," *Progress In Electromagnetics Research*, PIER 95, 283-297, 2009.

[2] M. T. Islam, M. N. Shakib and N. Misran, "Design Analysis Of High Gain Wideband L-Probe Fed Microstrip Patch Antenna" *Progress In Electromagnetics Research*, PIER 95, 397-407, 2009.

[3] X. Li, L. Yang, S.-X. Gong, and Y.-J. Yang, "Bidirectional High Gain Antenna for Wlan Applications," *Progress In Electromagnetics Research Letters*, Vol. 6, 99-106, 2009.

[4] N. Romano, G. Prisco and F. Soldovieri, "Design Of A Reconfigurable Antenna For Ground Penetrating Radar Applications" *Progress In Electromagnetics Research*, PIER 94, 1-18, 2009.

[5] O. Beheshti-Zavareh and M. Hakak, "A Stable Design Of Coaxial Adaptor For Radial Line Slot Antenna", *Progress In Electromagnetics Research*, PIER 90, 51–62, 2009.

[6] Paul W. Davis and Marek E. Bialkowski, "Experimental Investigations into a Linearly Polarized   Radial Slot Antenna for DBS TV in Australia", *IEEE Transactions on Antennas and Propagation*, vol. 45, No. 7, July1997.

[7] J. L. Masa-Campos and F. Gonzalez-Fernandez, "Dual linear/circular polarized plannar antenna with low profile double-layer polarizer of 45° tilted metallic strips for wimax applications," *Progress In Electromagnetics Research*, PIER 98, 221-231, 2009.

[8] G. M. Rebeiz, RF MEMS: Theory, Design, and Technology. Hoboken, NJ: Wiley-Interscience, 2003.

[9] J. Ouyang, F. Yang, S. W. Yang, Z. P. Nie and Z. Q. Zhao., "A novel radiation pattern and frequency reconfigurable microstrip antenna on a thin substrate for wide-band and wide-angle scanning application", *Progress In Electromagnetics Research*, PIER 4, 167–172, 2008.

[10] M. F. Jamlos, O. A. Aziz, T. A. Rahman and M. R. Kamarudin "a beam steering radial line slot array (rlsa) antenna with reconfigurable operating frequency" J. of Electromagn. Waves and Appl., Vol. 24, 1079–1088, 2010

[11] M. F. Jamlos, T. A. Rahman, and M. R. Kamarudin "a novel adaptive wi-fi system with rfid Technology" *Progress In Electromagnetics Research, Vol. 108, 417-432, 2010*

[12] M. F. Jamlos, T. A. Rahman, and M. R. Kamarudin "adaptive beam steering of rlsa antenna with rfid technology," *Progress In Electromagnetics Research, Vol. 108, 65-80, 2010*

[13] Huff, G. H., J. Feng, S. Zhang, and J. T. Bernhard, "A novel radiation pattern and frequency reconfigurable single turn square spiral microstrip antenna," *IEEE Microwave Wireless Components Letter*, Vol. 13, 57–59, February 2003.

4

CPW-Fed Antennas for WiFi and WiMAX

Sarawuth Chaimool and Prayoot Akkaraekthalin
Wireless Communication Research Group (WCRG), Electrical Engineering,
Faculty of Engineering, King Mongkut's University of Technology North Bangkok,
Thailand

1. Introduction

Recently, several researchers have devoted large efforts to develop antennas that satisfy the demands of the wireless communication industry for improving performances, especially in term of multiband operations and miniaturization. As a matter of fact, the design and development of a single antenna working in two or more frequency bands, such as in wireless local area network (WLAN) or WiFi and worldwide interoperability for microwave access (WiMAX) is generally not an easy task. The IEEE 802.11 WLAN standard allocates the license-free spectrum of 2.4 GHz (2.40-2.48 GHz), 5.2 GHz (5.15-5.35 GHz) and 5.8 GHz (5.725-5.825 GHz). WiMAX, based on the IEEE 802.16 standard, has been evaluated by companies for last mile connectivity, which can reach a theoretical up to 30 mile radius coverage. The WiMAX forum has published three licenses spectrum profiles, namely the 2.3 (2.3-2.4 GHz), 2.5 GHz (2.495-2.69 GHz) and 3.5 GHz (3.5-3.6 GHz) varying country to country. Many people expect WiMAX to emerge as another technology especially WiFi that may be adopted for handset devices and base station in the near future. The eleven standardized WiFi and WiMAX operating bands are listed in Table I.

Consequently, the research and manufacturing of both indoor and outdoor transmission equipment and devices fulfilling the requirements of these WiFi and WiMAX standards have increased since the idea took place in the technical and industrial community. An antenna serves as one of the critical component in any wireless communication system. As mentioned above, the design and development of a single antenna working in wideband or more frequency bands, called multiband antenna, is generally not an easy task. To answer these challenges, many antennas with wideband and/or multiband performances have been published in open literatures. The popular antenna for such applications is microstrip antenna (MSA) where several designs of multiband MSAs have been reported. Another important candidate, which may complete favorably with microstrip, is coplanar waveguide (CPW). Antennas using CPW-fed line also have many attractive features including low-radiation loss, less dispersion, easy integration for monolithic microwave circuits (MMICs) and a simple configuration with single metallic layer, since no backside processing is required for integration of devices. Therefore, the designs of CPW-fed antennas have recently become more and more attractive. One of the main issues with CPW-fed antennas is to provide an easy impedance matching to the CPW-fed line. In order to obtain multiband and broadband operations, several techniques have been reported in the literatures based on CPW-fed slot antennas (Chaimool et al., 2004, 2005, 2008; Sari-Kha et al., 2006; Jirasakulporn,

2008), CPW-fed printed monopole (Chaimool et al., 2009; Moekham et al., 2011) and fractal techniques (Mahatthanajatuphat et al., 2009; Honghara et al., 2011).

In this chapter, a variety of advanced CPW-fed antenna designs suitable for WiFi and WiMAX operations is presented. Some promising CPW-fed slot antennas and CPW-fed monopole antenna to achieve bidirectional and/or omnidirectional with multiband operation are first shown. These antennas are suitable for practical portable devices. Then, in order to obtain the unidirectional radiation for base station antennas, CPW-fed slot antennas with modified shape reflectors have been proposed. By shaping the reflector, noticeable enhancements in both bandwidth and radiation pattern, which provides unidirectional radiation, can be achieved while maintaining the simple structure. This chapter is organized as follows. Section 2 provides the coplanar waveguide structure and characteristics. In section 3, the CPW-fed slot antennas with wideband operations are presented. The possibility of covering the standardized WiFi and WiMAX by using multiband CPW-fed slot antennas is explored in section 4. In order to obtain unidirectional radiation patterns, CPW-fed slot antennas with modified reflectors and metasurface are designed and discussed in section 5. Finally, section 6 provides the concluding remarks.

System	Designed Operating Bands		Frequency Range (GHz)
WiFi IEEE 802.11	2.4 GHz		2.4-2.485
	5 GHz	5.2 GHz	5.15-5.35
		5.5 GHz	5.47-5.725
		5.8 GHz	5.725-5.875
Mobile WiMAX IEEE 802.16 2005	2.3 GHz		2.3-2.4
	2.5 GHz		2.5-2.69
	3.3 GHz		3.3-3.4
	3.5 GHz		3.4-3.6
	3.7 GHz		3.6-3.8
Fixed WiMAX IEEE 802.16 2004	3.7 GHz		3.6-3.8
	5.8 GHz		5.725-5.850

Table 1. Designed operating bands and corresponding frequency ranges of WiFi and WiMAX

2. Coplanar waveguide structure

A coplanar waveguide (CPW) is a one type of strip transmission line defined as a planar transmission structure for transmitting microwave signals. It comprises of at least one flat conductive strip of small thickness, and conductive ground plates. A CPW structure consists of a median metallic strip of deposited on the surface of a dielectric substrate slab with two narrow slits ground electrodes running adjacent and parallel to the strip on the same surface

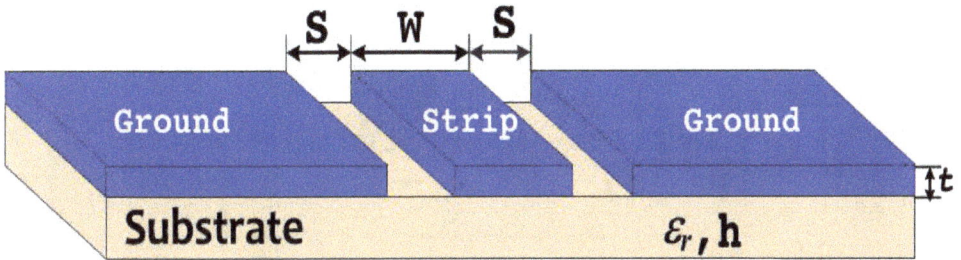

Fig. 1. Coplanar waveguide structure (CPW)

as shown in Fig 1. Beside the microstrip line, the CPW is the most frequent use as planar transmission line in RF/microwave integrated circuits. It can be regarded as two coupled slot lines. Therefore, similar properties of a slot line may be expected. The CPW consists of three conductors with the exterior ones used as ground plates. These need not necessarily have same potential. As known from transmission line theory of a three-wire system, even and odd mode solutions exist as illustrated in Fig. 2. The desired even mode, also termed coplanar mode [Fig. 2 (a)] has ground electrodes at both sides of the centered strip, whereas the parasitic odd mode [Fig. 2 (b)], also termed slot line mode, has opposite electrode potentials. When the substrate is also metallized on its bottom side, an additional parasitic parallel plate mode with zero cutoff frequency can exist [Fig. 2(c)]. When a coplanar wave impinges on an asymmetric discontinuity such as a bend, parasitic slot line mode can be exited. To avoid these modes, bond wires or air bridges are connected to the ground places to force equal potential. Fig. 3 shows the electromagnetic field distribution of the even mode at low frequencies, which is TEM-like. At higher frequencies, the fundamental mode evolves itself approximately as a TE mode (H mode) with elliptical polarization of the magnetic field in the slots.

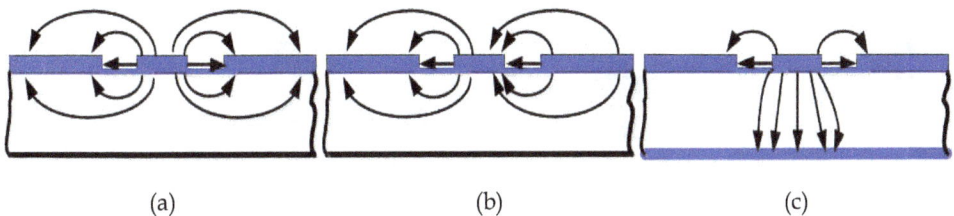

(a)   (b)   (c)

Fig. 2. Schematic electrical field distribution in coplanar waveguide: (a) desired even mode, (b) parasitic odd mode, and (c) parasitic parallel plate mode

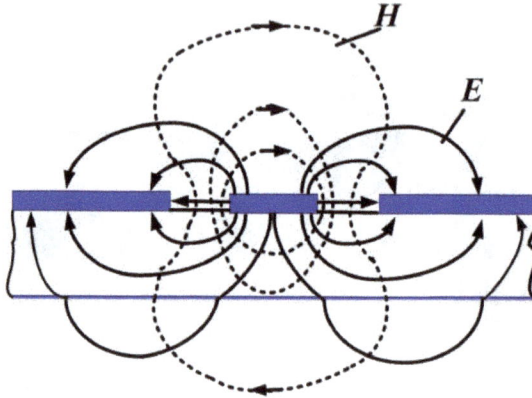

Fig. 3. Transversal electromagnetic field of even coplanar mode at low frequency

3. Wideband CPW-fed slot antennas

To realize and cover WiFi and WiMAX operation bands, there are three ways to design antennas including (i) using broadband/wideband or ultrawideband techniques, (ii) using multiband techniques, and (iii) combining wideband and multiband techniques. For wideband operation, planar slot antennas are more promising because of their simple structure, easy to fabricate and wide impedance bandwidth characteristics. In general, the wideband CPW-fed slot antennas can be developed by tuning their impedance values. Several impedance tuning techniques are studied in literatures by varying the slot geometries and/or tuning stubs as shown in Fig. 4 and Fig. 5. Various slot geometries have been carried out such as wide rectangular slot, circular slot, elliptical slot, bow-tie slot, and hexagonal slot. Moreover, the impedance tuning can be done by using coupling mechanisms, namely inductive and capacitive couplings as shown Fig. 5. For capacitively coupled slots, several tuning stubs have been used such as circular, triangular, rectangular, and fractal shapes. In this section, we present the wideband slot antennas using CPW feed line. There are three antennas for wideband operations: CPW-fed square slot antenna using loading metallic strips and a widened tuning stub, CPW-fed equilateral hexagonal slot antennas, and CPW-fed slot antennas with fractal stubs.

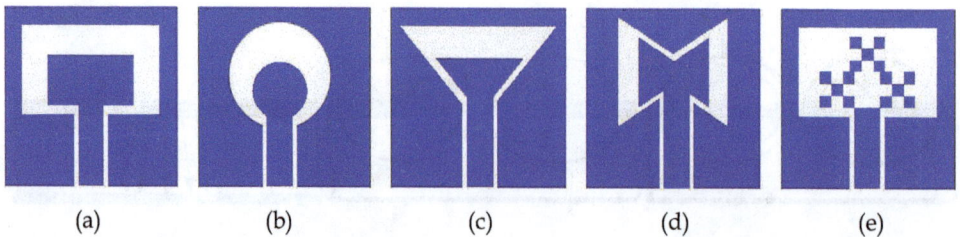

Fig. 4. CPW-fed slots with various slot geometries and tuning stubs (a) wide rectangular slot, (b) circular slot, (c) triangular slot, (d) bow-tie slot, and (e) rectangular slot with fractal tuning stub

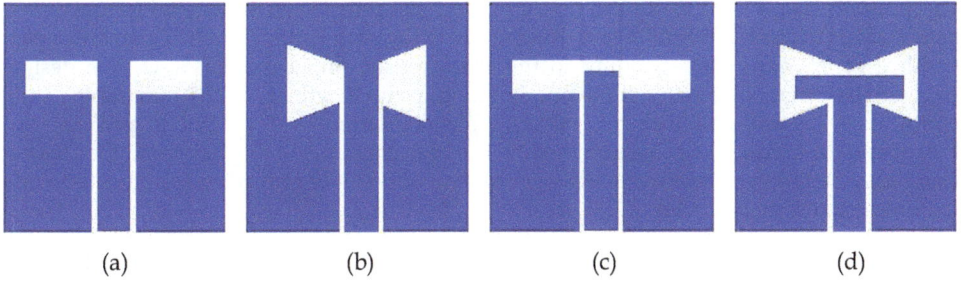

Fig. 5. CPW-fed slots with (a)-(b) inductive coupling and (c)–(d) capacitive coupling

3.1 CPW-fed square slot antenna using loading metallic strips and a widened tuning stub

The geometry and prototype of the proposed CPW-fed slot antenna with loading metallic strips and widen tuning stub is shown in Fig. 6(a) and Fig. 6(b), respectively. The proposed antenna is fabricated on an inexpensive FR4 substrate with thickness (h) of 1.6 mm and relatively permittivity (ε_r) of 4.4. The printed square radiating slot has a side length of L_{out} and a width of G. A 50-Ω CPW has a signal strip of width W_f, and a gap of spacing g between the signal strip and the coplanar ground plane. The widened tuning stub with a length of L and a width of W is connected to the end of the CPW feed line. Two loading metallic strips of the same dimensions (length of L_1 and width of 2 mm) are designed to protrude from the top corners into the slot center. The spacing between the tuning stub and edge of the ground plane is S. In this design, the dimensions are chosen to be G =72 mm, and L_{out} = 44 mm. Two parameters of the tuning stub including L and W and the length of loading metallic strip (L_1) will affect the broadband operation. The parametric study was presented from our previous work (Chaimool, et. al., 2004, 2005).

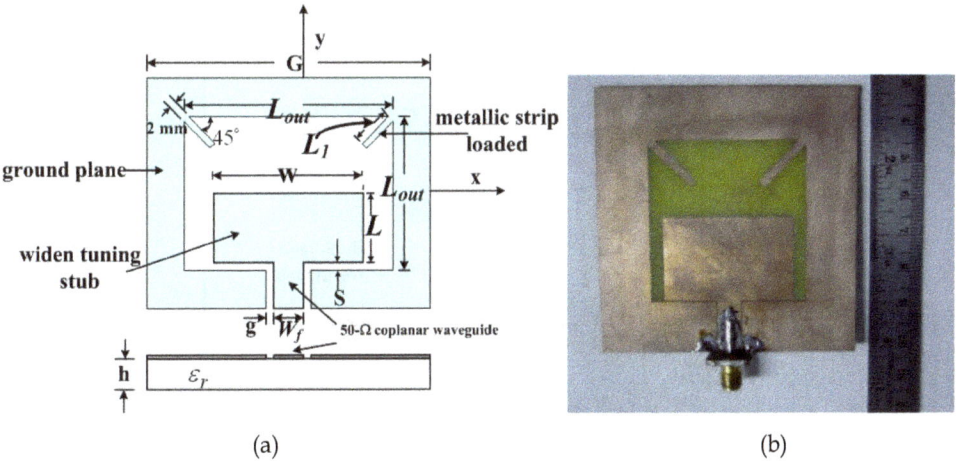

(a)   (b)

Fig. 6. (a) geometry of the proposed CPW-fed slot antenna using loading metallic strips and a widened tuning stub and (b) photograph of the prototype

The present design is to make the first CPW-fed slot antenna to form a wider operating bandwidth. Firstly, a CPW-fed line is designed with the strip width W_f of 6.37 mm and a gap width g of 0.5 mm, corresponding to the characteristic impedance of 50-Ω. The design structure has been obtained with the optimal tuning stub length of L =22.5 mm, tuning stub width W = 36 mm, and length of loading metallic strips L_1 = 16 mm to perform the broadband operation. The proposed antenna has been constructed (Fig. 6(b)) and then tested using a calibrated vector network analyzer. Measured result of return losses compared with the simulation is shown in Fig. 7.

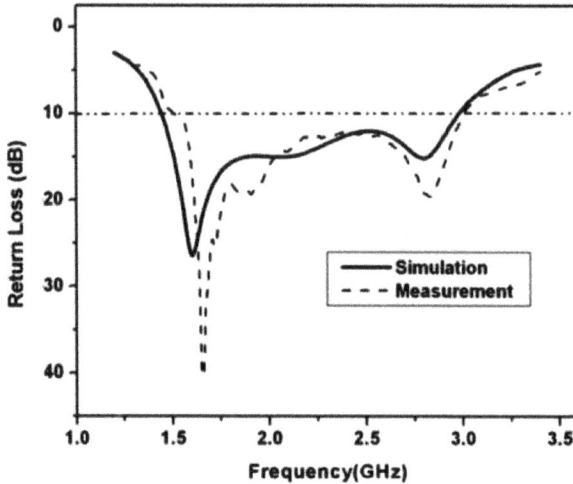

(a)

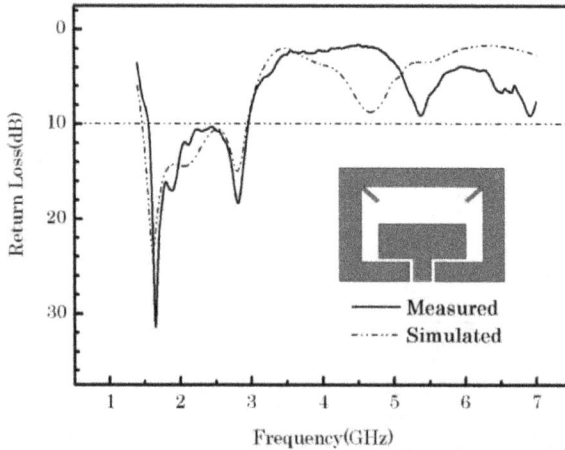

(b)

Fig. 7. Measured and simulated return losses for tuning stub width W = 36 mm, L = 22.5 mm, L_{out} = 44 mm, G=72 mm, L_1=16 mm, W_f=6.37 mm, and g = 0.5 mm, and (a) narrow band, (b) wideband views

The far-field radiation patterns of the proposed antenna with the largest operating bandwidth using the design parameters of L_1 =16 mm, W = 36 mm, L =22.5 mm, and S = 0.5 mm have been then measured. Fig. 8 shows the plots of the radiation patterns measured in y-z and x-z planes at the frequencies of 1660 and 2800 MHz. It has been found that we can obtain acceptable broadside radiation patterns.

This section introduces a new CPW-fed square slot antenna with loading metallic strips and a widened tuning stub for broadband operation. The simulation and experimental results of the proposed antenna show the impedance bandwidth, determined by 10-dB return loss, larger than 67% of the center frequency. The proposed antenna can be applied for WiFi (2.4 GHz) and WiMAX (2.3 and 2.5 GHz bands) operations.

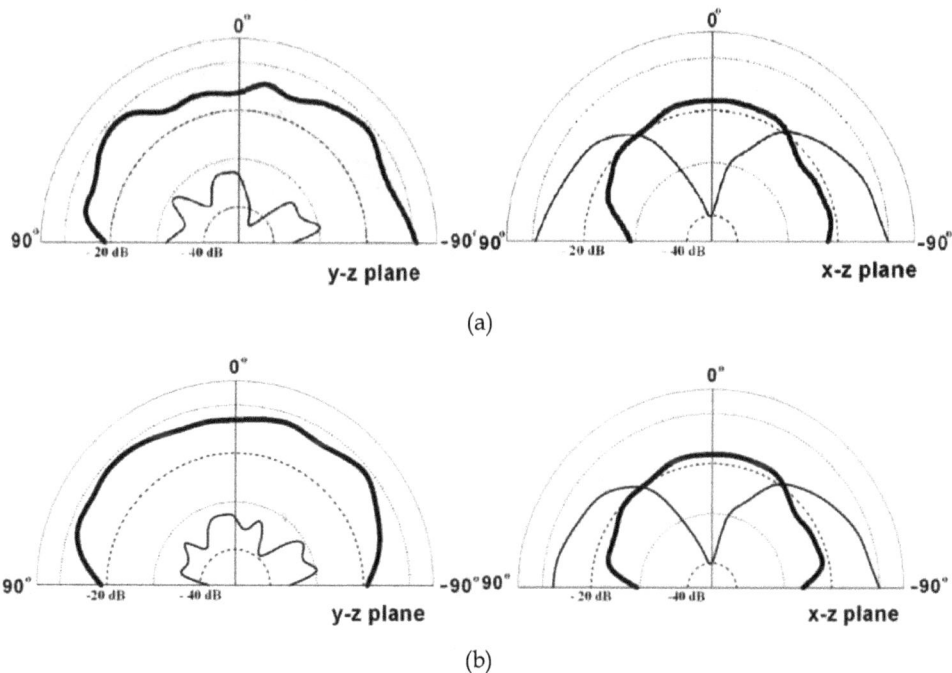

(a)

(b)

Fig. 8. Measured radiation patterns in the y-z and x-z planes for the proposed (a) f = 1660 MHz and (b) f = 2800 MHz

3.2 CPW-fed equilateral hexagonal slot antenna

Fig. 9 shows the geometry and the prototype of the CPW-fed hexagonal slot antenna. It is designed and built on an FR4 substrate with thickness (h) of 1.6 mm and relatively permittivity (ε_r) of 4.4. The ground plane is chosen to be an equilateral hexagonal structure with outer radius (R_o) and inner radius (R_i). A 50-Ω CPW feed line consists of a metal strip of width (W_f) and a gap (g). This feed line is used to excite the proposed antenna. The tuning stub has a length of L_f and a width of W_f. For our design, the key dimensions of the proposed antenna are initially chosen to be R_o = 55 mm, R_i = 33 mm, W_f = 6.37 mm, and g =

0.5 mm, then we have adjusted three parameters including R_o, R_i, and L_f to obtain a broadband operation.

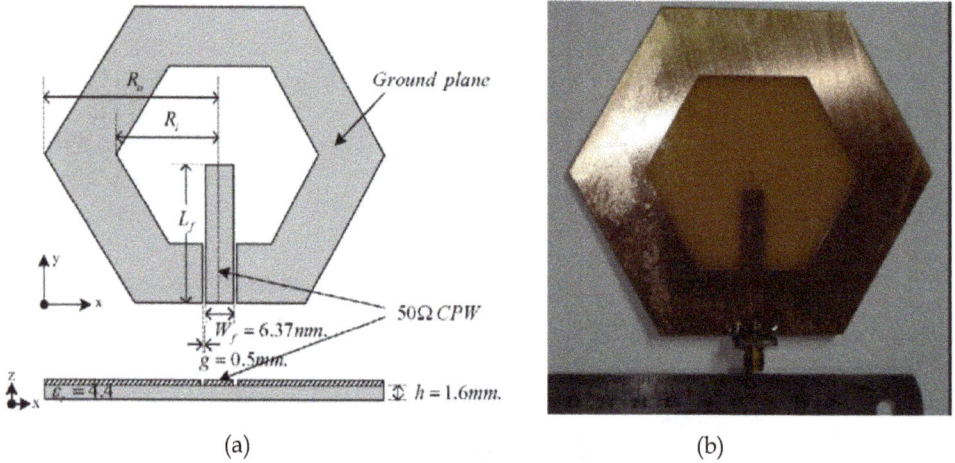

(a)   (b)

Fig. 9. (a) geometry of the proposed CPW-fed equilateral hexagonal slot antenna and (b) the prototype of the proposed antenna (Sari-Kha et al., 2005)

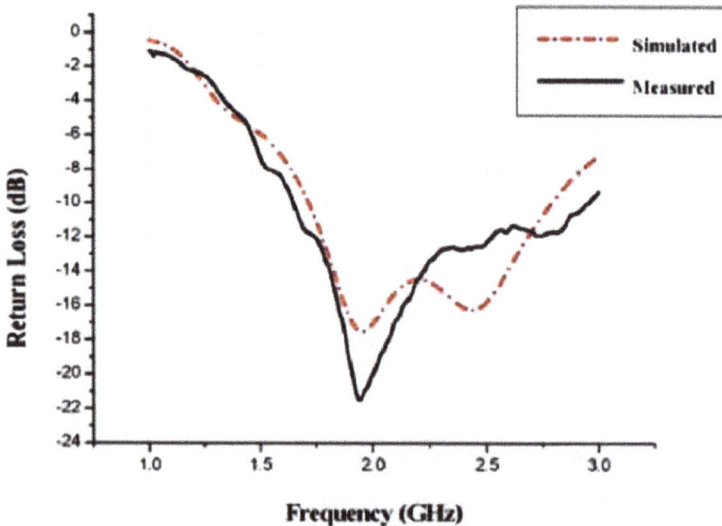

Fig. 10. Simulated and measured return losses of the CPW-fed equilateral hexagonal slot antenna with R_o = 55 mm, R_i = 33 mm, and L_f = 42.625 mm

The optimal dimensions have been used for building up the proposed antenna. Measured return loss using a vector network analyzer is now shown in Fig.10. As we can see that the measured return loss agrees well with simulation expectation. It is also seen that the proposed antenna has an operational frequency range from 1.657 to 2.956 GHz or bandwidth about 55% of the center frequency measured at higher 10 dB return loss.

This section presents design and implementation of the CPW-fed equilateral hexagonal slot antenna. The transmission line and ground-plane have been designed to be on the same plane with the antenna slot to be applicable for wideband operation. It is found that the proposed antenna is accessible to bandwidth about 55.39%, a very large bandwidth comparing with conventional microstrip antennas, which mostly provide 1-5 % bandwidth. The proposed antenna can be used for many wireless systems such as WiFi , WiMAX, GSM1800, GSM1900, and IMT-2000.

3.3 CPW-fed slot antennas with fractal stubs

In this section, the CPW-fed slot antenna with tuning stub of fractal geometry will be investigated. The Minkowski fractal structure will be modified to create the fractal stub of the proposed antenna. The proposed antennas have been designed and fabricated on an inexpensive FR4 substrate of thickness h = 0.8 mm and relative permittivity ε_r = 4.2. The first antenna consists of a rectangular stub or zero iteration of fractal model (0 iteration), which has dimension of 10 mm × 25 mm. It is fed by 50–Ω CPW-fed line with the strip width and distance gap of 7.2 mm and 0.48 mm, respectively. In the process of studying the fractal geometry on stub, it is begun by using a fractal model to repeat on a rectangular patch stub for creating the first and second iterations of fractal geometry on the stub, as shown in Fig. 11. Then, the fractal stub is connected by 50–Ω CPW-fed line. On the second iteration fractal stub of the antenna, the fraction of size between the center element and four around elements is 1.35 because this value is suitable for completely fitting to connect between the center element and four around elements. As shown in Fig. 12(a), the dimensions of the second iteration antenna are following: W_T= 48 mm, L_T= 50 mm, W_{S1} = 39.84 mm, L_{S1} = 20.6 mm, W_{S2} = 15.84 mm, L_{S2} = 19.28 mm, W_{S3} = 7.42 mm, L_{S3} = 7.72 mm, W_A = 25 mm, L_B = 10 mm, W_{TR}= 7.2 mm, and h = 0.8 mm.

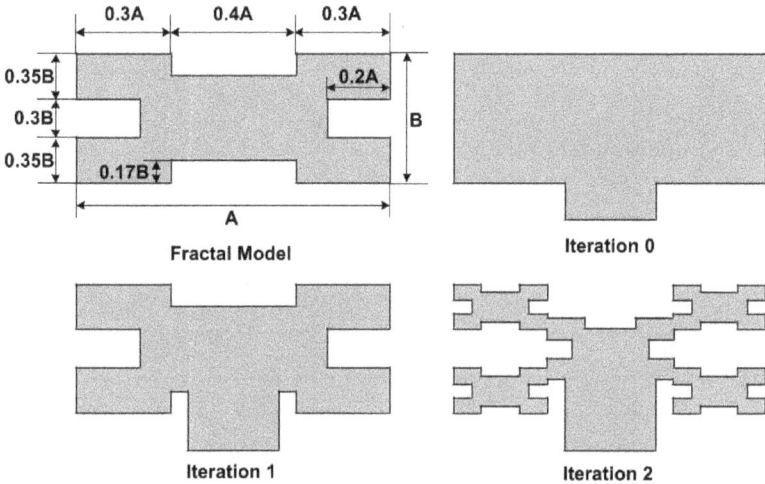

Fig. 11. The fractal model for stubs with different geometry iterations

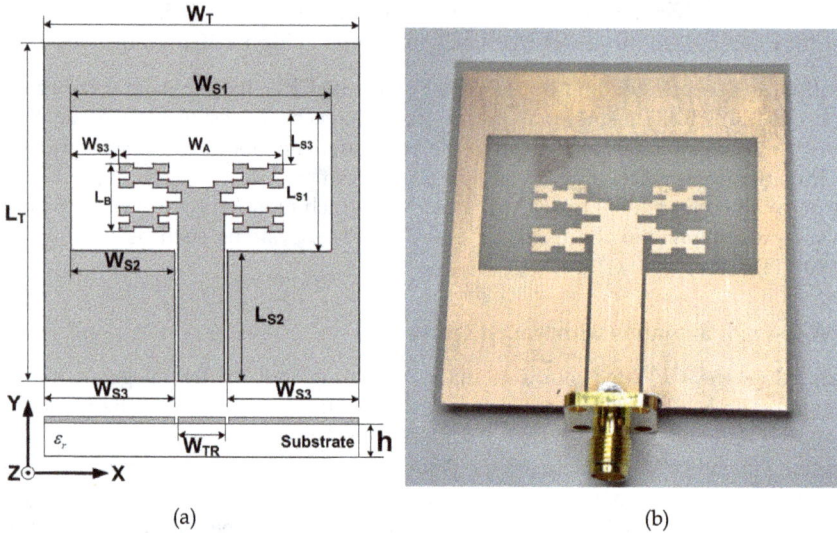

(a)   (b)

Fig. 12. (a) Geometry of the proposed CPW-fed slot antenna with the 2nd iteration fractal stub and (b) photograph of the fabricated antenna

In order to study the effects of fractal geometry on the stub of the slot antenna, IE3D program is used to simulate the characteristics and frequency responses of the antennas. The simulated return loss results of the 1st and 2nd iterations are shown in Fig. 13 and expanded in Table 2. The results show that all of return loss bandwidth tendencies and center

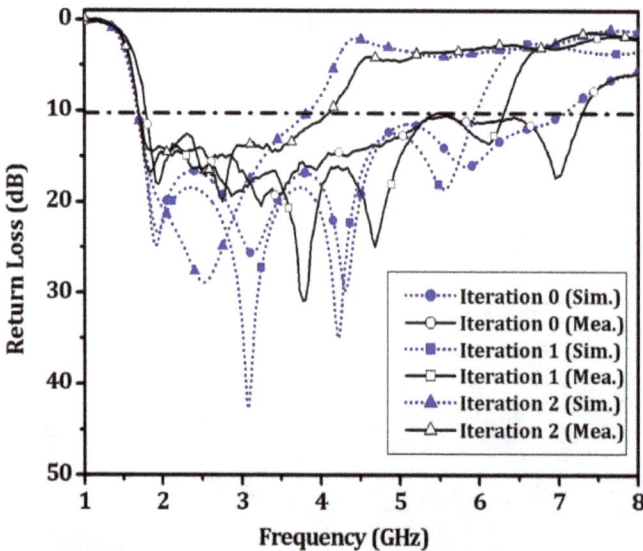

Fig. 13. Simulated and measured return losses of the proposed antenna with different iterations of fractal stubs

Antenna type	Center Frequency (GHz)		Return Loss Bandwidth (RL ≥10 dB)			
			BW (GHz)		BW (%)	
	Sim.	Mea.	Sim.	Mea.	Sim.	Mea.
Iteration 0	4.3	4.5	1.6 - 7.1	1.7 – 7.1	123	121
Iteration 1	3.8	4.0	1.6 – 5.9	1.7 – 6.3	112	115
Iteration 2	2.7	2.8	1.6 – 3.8	1.7 – 4.0	78	82

Table 2. Comparison of characteristic results with different iterations of fractal stubs.

frequencies decrease as increasing the iteration for fractal stub. Typically, the increasing iteration in the conventional fractal structure affects to the widely bandwidth. However, these results have inverted because the electrical length on the edge of stub, which the stub in the general CPW-fed slot antenna was used to control the higher frequency band, is increased and produced by the fractal geometry. In Table 3, simulation results show the antenna gains at operating frequency of 1.8 GHz, 2.1 GHz, 2.45 GHz, and 3.5 GHz above 3dBi. As the higher operating frequency, the average antenna gains are about 2 dBi. The overall dimension of CPW-fed fabricated slot antennas with fractal stub is 48× 50 × 0.8 mm³, as illustrated in Fig. 12(b). The simulated and measured results of the proposed antennas are compared as shown in Fig. 13. It can be clearly found that the simulated and measured results are similarity. However, the measured results of the return loss bandwidth slightly shift to higher frequency band. The error results are occurred due to the problem in fabrication because the fractal geometry stubs need the accuracy shapes. Moreover, the radiation patterns of 0, 1st and 2nd iteration stubs of the antennas are similar, which are the bidirectional radiation patterns at two frequencies, 2.45 and 3.5 GHz, as depicted in Fig. 14.

Operating Frequency		Antenna Gain (dBi)		
		Iteration 0	Iteration 1	Iteration 2
1.8 GHz	Sim.	3.1	3.1	3.1
	Mea.	2.1	2.5	2.7
2.1 GHz	Sim.	3.3	3.3	3.3
	Mea.	2.3	2.1	2.3
2.45 GHz	Sim.	3.3	3.3	3.3
	Mea.	2.9	2.8	2.6
3.5 GHz	Sim.	3.5	3.5	3.3
	Mea.	1.6	1.5	1.3
5.2 GHz	Sim.	1.8	2.2	N/A
	Mea.	1.1	1.7	N/A
5.8 GHz	Sim.	1.8	2.4	N/A
	Mea.	1.3	2.2	N/A
6.9 GHz	Sim.	2.2	N/A	N/A
	Mea.	2.1	N/A	N/A

Table 3. Summarized results of the antenna gains

(a)

(b)

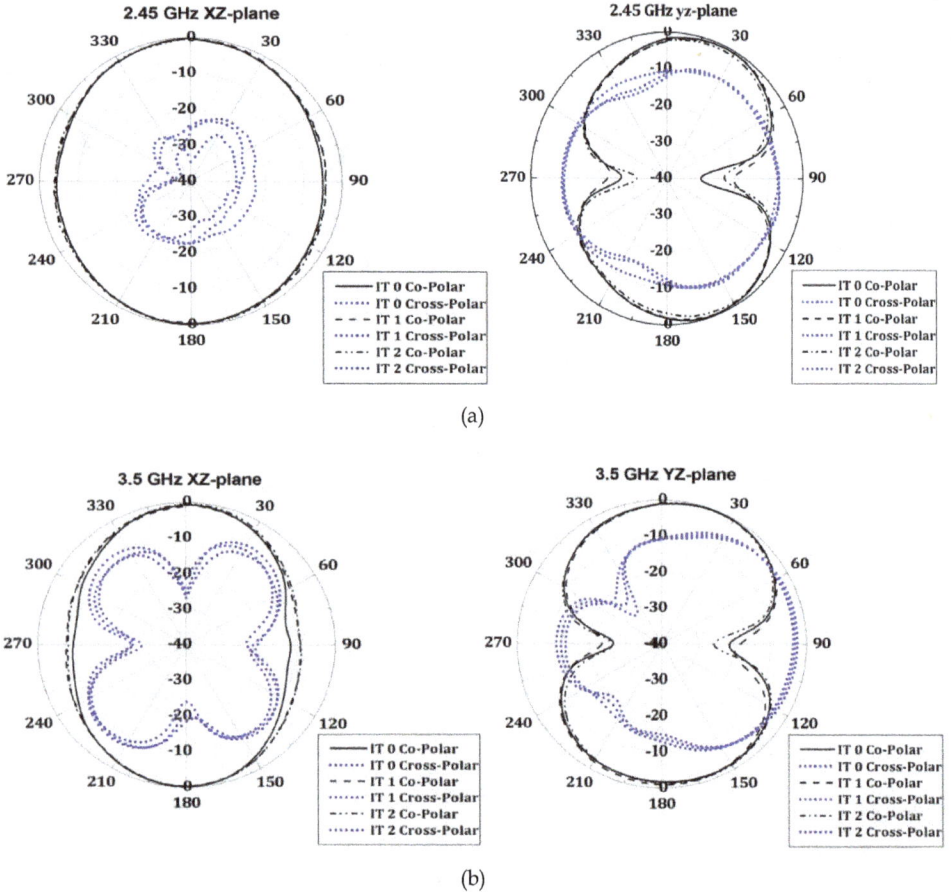

Fig. 14. Measured radiation patterns of the proposed CPW-fed slot antennas with 0, 1st and 2nd iteration fractal stubs (a) 2450 MHz and (b) 3500 MHz

This section studies CPW-fed slot antennas with fractal stubs. The return loss bandwidth of the antenna is affected by the fractal stub. It has been found that the antenna bandwidth decreases when the iteration of fractal stub increases, which it will be opposite to the conventional fractal structures. In this study, fractal models with the 0, 1st and 2nd iterations have been employed, resulting in the return loss bandwidths to be 121%, 115%, and 82%, respectively. Moreover, the radiation patterns of the presented antenna are still bidirections and the average gains of antenna are above 2 dBi for all of fractal stub iterations. Results indicate an impedance bandwidth covering the band for WiFi, WiMAX, and IMT-2000.

4. Multiband CPW-fed slot antennas

Design of antennas operating in multiband allows the wireless devices to be used with only a single antenna for multiple wireless applications, and thus permits to reduce the size of the space required for antenna on the wireless equipment. In this section, we explore the possibility of covering some the standardized WiFi and WiMAX frequency bands while cling to the class of simply-structured and compact antennas.

4.1 Dual-band CPW-fed slot antennas using loading metallic strips and a widened tuning stub

In this section, we will show that CPW-fed slot antennas presented in the previous section (Section 3.1) can also be designed to demonstrate a dual-band behavior. The first dual-band antenna topology that, we introduce in Fig. 15(a); consists of the inner rectangular slot antenna with dimensions of $w_{in} \times L_{in}$ and the outer square slot ($L_{out} \times L_{out}$). The outer square slot is used to control the first or lower operating band. On the other hand, the inner slot of width is used to control the second or upper operating band. The second antenna as shown in Fig. 15(b) combines a tuning stub with dimensions of $W_s \times L_3$ placed in the inner slot at its bottom edge. The tuning stub is used to control coupling between a CPW feed line and the inner rectangular slot. In the third antenna as shown in Fig. 15(c), another pair of loading metallic strips is added at the bottom inner slot corners with dimensions of $1 \text{ mm} \times L_2$. Referring to Fig. 15(a), if adding a rectangular slot at tuning stub with $w_{in} = 21$ mm and $L_{in} = 11$ mm to the wideband antenna (Fig. 6(a)), an additional resonant mode at about 5.2 GHz is obtained. This resonant mode excited is primarily owing to an inner rectangular slot. This way the antenna becomes a dual-band one in which the separation between the two resonant frequencies is a function of the resonant length of the second resonant frequency, the length and width of the inner slot (L_{in} and w_{in}). To achieve the desired dual band operation of the rest antennas, we can adjust the parameters, (W, L, L_1) and (w_{in}, W_s, L_2, L_3, L_{in}), of the outer and inner slots, respectively, to control the lower and upper operating bands of the proposed antennas. The measured return losses of the proposed antennas are shown in Fig. 16. It can be observed that the multiband characteristics can be obtained. The impedance bandwidths of the lower band for all antennas are slightly different, and on the other hand, the upper band has an impedance bandwidth of 1680 MHz (4840–6520 MHz) for antenna in Fig. 15(b), which covers the WiFi band at 5.2 GHz and 5.8 GHz band for WiMAX. To sum up, the measured results and the corresponding settings of the parameters are listed

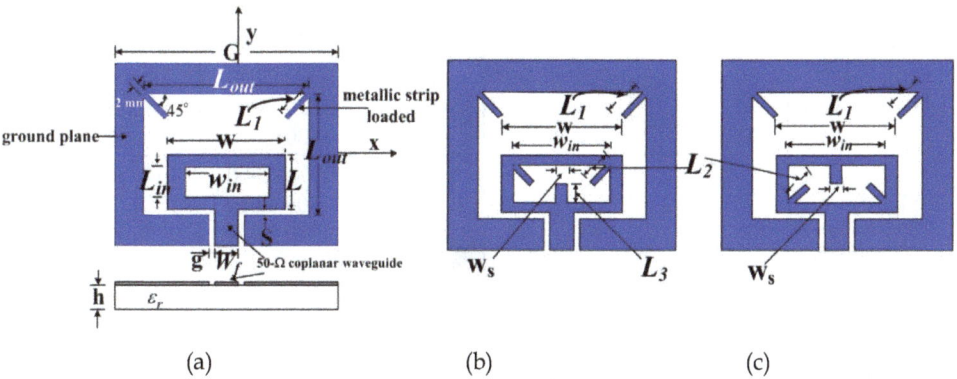

Fig. 15. Dual-band CPW-fed slot antennas with inner rectangular slot (a) without loading strip and a tuning stub, (b) with top corner loading strips and a bottom tuning stub, and (c) with bottom corner loading strips and a top tuning stub

in Table 4. Radiation patterns of the proposed antennas were measured at two resonant frequencies. Fig. 17(a) and (b) show the y-z and x-z plane co- and cross-polarized patterns at 1700 and 5200 MHz, respectively. The radiation patterns are bidirectional on the broadside due to the outer slot mode at lower frequency and the radiation patterns are irregular because of the excitation of higher order mode, the traveling wave.

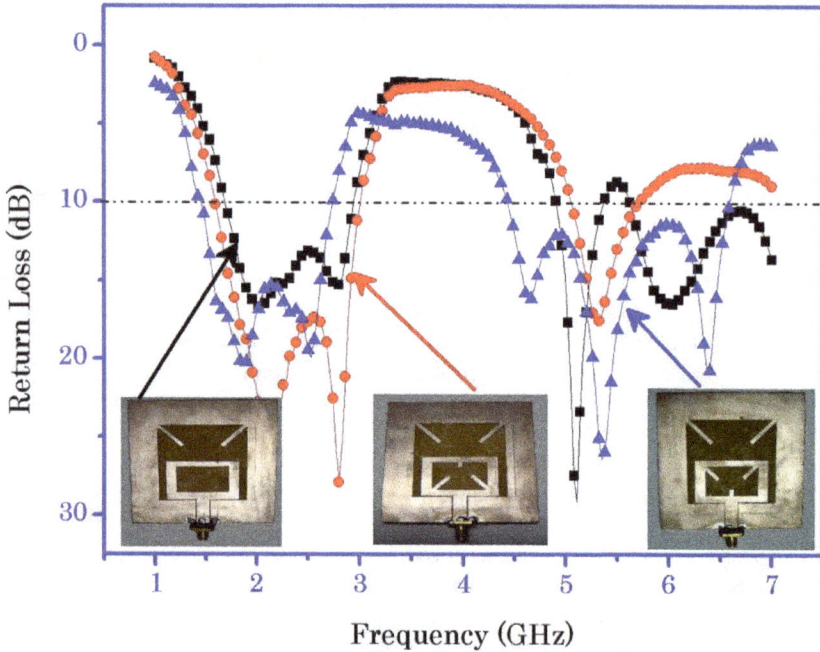

Fig. 16. Measured return losses of dual-band CPW-fed slot antennas

Antennas	Dimension (mm)					Bandwidth ($S_{11} \le$ -10 dB)	
	w_{in}	W_S	L_{in}	L_2	L_3	Lower BW(%,BW)	Upper BW(%,BW)
Fig. 15(a)	30	-	7.5	-	-	61.0, 1600–3000	7.5, 4880–5260
	30	-	6.0	-	-	58.5, 1620–2960	5.8, 5180–5490
	21	-	11.0	-	-	58.2, 1630–2970	16.1, 5040–5920
Fig. 15(b)	26	2	20	-	6.0	61.4, 1570–2960	13.2, 5200–5935
	26	2	20	-	8.0	49.4, 1600–2650	10.0, 5305–5865
	26	2	20	-	10	51.2, 1570–2650	27.9, 5060–6705
Fig. 15(c)	26	2	20	9.5	7.0	58.7,1610–2950	9.3, 4900–5380
	26	2	20	9.5	9.0	57.8, 1610–2920	9.4, 4870–5350
	26	2	20	9.5	11	37.4, 1610–2350	10.0, 4840–5350

Table 4. Performance of the proposed dual-band CPW-fed slot antennas [Figs. 15(a), 15(b), and 15(c)] for different antenna parameter values of inner slot width (w_{in}), length (L_{in}) and loading metallic strips in inner slot (W_s, L_2, and L_3) which L_{out} = 45 mm, W = 36 mm, G=72 mm, L_1=16 mm, L= 22.5 mm, h=1.6 mm, W_f=6.37 mm, and g=0.5 mm

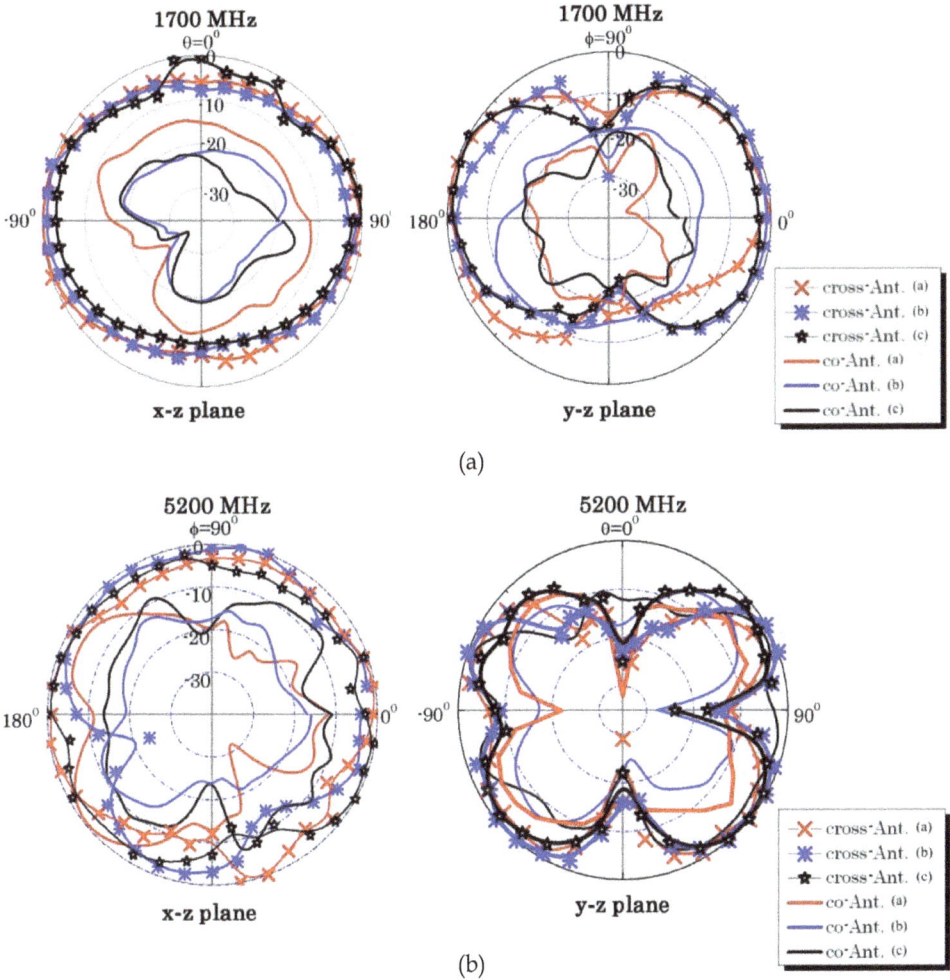

Fig. 17. Measured radiation patterns of the proposed antennas in case of optimized antennas in Table 4. (a) 1700 MHz, and (b) 5200 MHz

By inserting a slot and metallic strips at the widened stub in a single layer and fed by coplanar waveguide (CPW) transmission line, novel dual-band and broadband operations are presented. The proposed antennas are designed to have dual-band operation suitable for applications WiFi (2.4 and 5 GHz bands) and WiMAX (2.3, 2.5 and 5.8 bands) bands. The dual-band antennas are simple in design, and the two operating modes of the proposed antennas are associated with perimeter of slots and loading metallic strips, in which the lower operating band can be controlled by varying the perimeters of the outer square slot and the higher band depend on the inner slot of the widened stub. The experimental results of the proposed antennas show the impedance bandwidths of the two operating bands, determined from 10-dB return loss, larger than 61% and 27% of the center frequencies, respectively.

4.2 CPW-fed mirrored-L monopole antenna with distinct triple bands

Fig. 18 illustrates the geometry of the proposed triple-band antenna. A CPW-fed mirrored-L monopole is printed on one side (top layer) of an inexpensive FR4 dielectric substrate (dielectric constant ε_r = 4.4, thickness h = 0.8 mm). An open-loop resonator loaded with an open stub is parasitically coupled on the back-side (bottom layer) of the mirrored-L monopole. The 50-Ω CPW feed line has a width of w_f = 1.43 mm with gaps of g = 0.15 mm. Two symmetrical ground planes of size of 26 × 47 mm² are used on the top layer. The open-loop resonator has a length of about half-wavelength at 2.45 GHz but is loaded by an open-stub of 4.6 mm. The unique resonator is responsible for the generation of resonant modes at 2.5 and 3.5 GHz, whereas the mirrored-L monopole joined with the feed-line is answerable for the wideband (5.11-6.7 GHz) generation. By properly tuning the relative positions (the coupling) between the L-shaped monopole and the open-loop resonator, and the spacing to the ground plane, the antenna exhibits three distinct bandwidths that fulfilling the required bandwidths from WiFi and WiMAX standards. Throughout the study, the IE3D simulator has been used for full-wave simulations in the design and optimization phases.

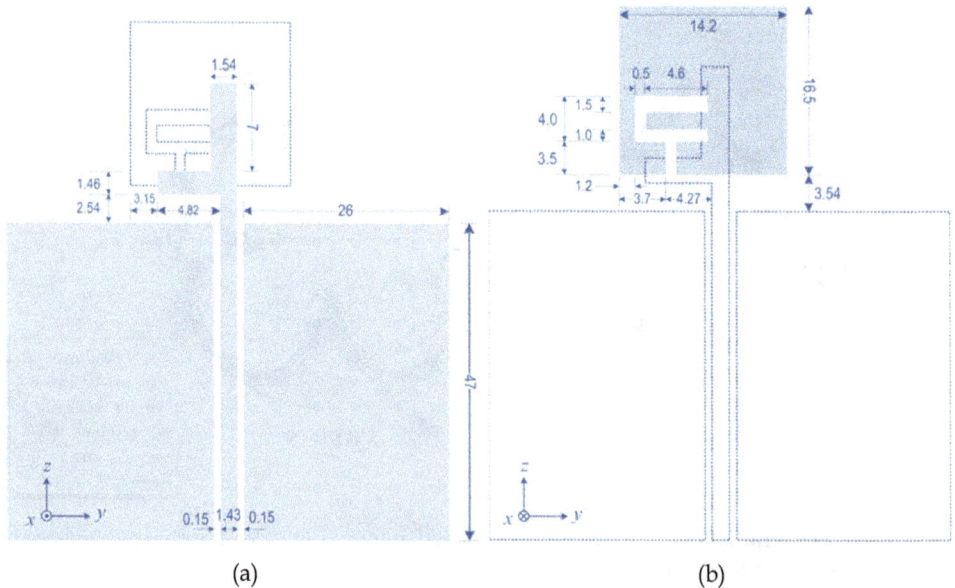

(a)   (b)

Fig. 18. Geometry of the proposed CPW-fed mirrored-L monopole antenna with dimensions in mm (a) top layer and (b) bottom layer

Based on the antenna parameters and the ground plane size depicted in Fig. 18, a prototype of this antenna was designed, fabricated and tested as shown in Fig. 19. Fig. 20 shows the measured return loss for the tri-band antenna. It is clearly seen that four resonant modes are excited at the frequencies of 2.59, 3.52, 5.56 and 6.37 GHz that results in three distinct bands. It is worthy of note that the latter two resonant modes are deliberately made in merge as a single wideband in order to cover all the unlicensed bands from 5.15 GHz to 5.85 GHz. The obtained 10-dB impedance bandwidths are 600 MHz (2.27-2.87 GHz), 750 MHz (3.4-4.15 GHz) and 1590 MHz (5.11-6.7 GHz), corresponding to the 23%, 20%, and 27%, respectively.

Obviously, the achieved bandwidths not just cover the WiFi bands of 2.4 GHz (2.4-2.484 GHz) and 5.2 GHz (5.15-5.25 GHz), but also the licensed WiMAX bands of 2.5 GHz (2.5-2.69 GHz) and 3.5 GHz (3.4 -3.69 GHz). Fig. 20 shows the measured gains compared to the simulated result for all distinct bands. For the first two bands, gains are slightly decreased with frequency increases, whereas the gains in the upper band are fallen in with the simulation. The radiation characteristics have also been investigated and the measured patterns in two cuts (x-y plane, x-z plane) at 2.59, 3.52, and 5.98 GHz are plotted in Figs. 21(a), 21(b) and 21(c), respectively. As expected, the very good omni-directional patterns are obtained for all frequency bands in the x-y plane, whilst the close to bi-directional patterns in the x-z plane are observed.

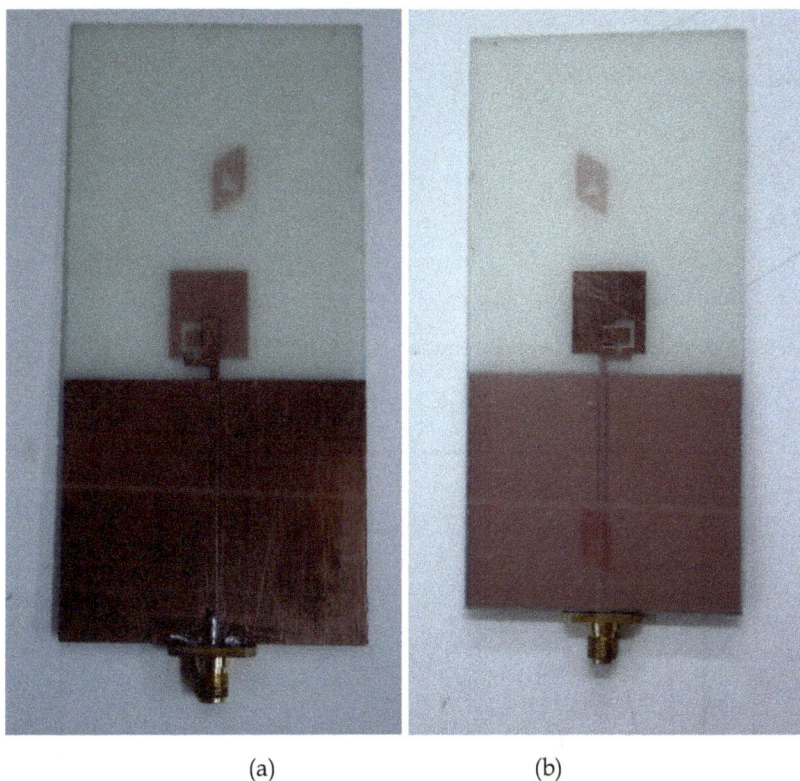

(a)   (b)

Fig. 19. Photograph of the proposed CPW-fed mirrored-L monopole antenna (a) top layer and (b) bottom layer

By coupling a stub-loaded open-loop resonator onto the back of a CPW-fed mirrored-L monopole, a novel triple-band planar antenna is achieved and presented in this section. The proposed antenna features a compact structure with reasonable gains. The measured bandwidths for the distinct triple-band are 2.27 to 2.87 GHz, 3.4 to 4.15 GHz and 5.11 to 6.7 GHz. Omni-directional radiation patterns for the three bands are observed. Simulations are confirmed by the experimental results, which ensure the proposed antenna is well suited for the WiFi and WiMAX applications.

Fig. 20. Measured return losses versus frequency

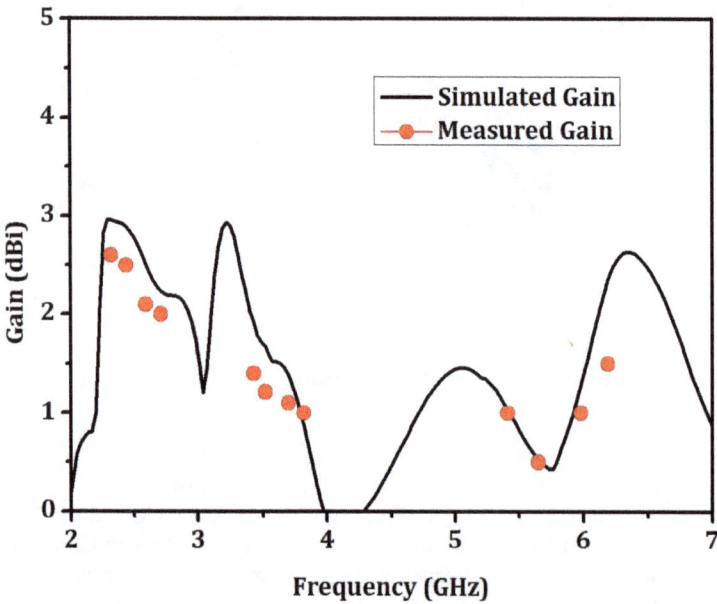

Fig. 21. Simulated and measured realized gains

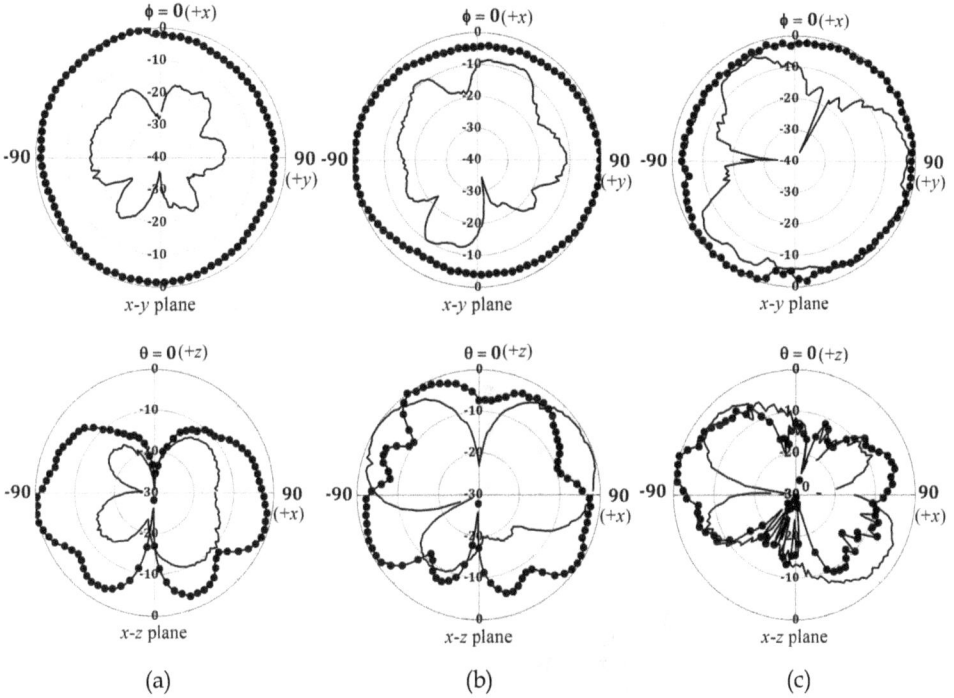

Fig. 22. Measured far-field radiation patterns in x-y plane and x-z plane (a) 2.59 GHz, (b) 3.52 GHz, and (c) 5.98 GHz

4.3 Multiband antenna with modified fractal slot fed by CPW

In this section, a fractal slot antenna fed by CPW was created by applying the Minkowski fractal concept to generate the initial generator model at both sides of inner patch of the antenna, as shown in Fig. 23. The altitude of initial generator model as shown in Fig. 24 varies with W_p. Usually, W_p is smaller than $W_s/3$ and the iteration factor is $\eta = 3W_p/W_s$; $0 < \eta < 1$. Normally, the appropriated value of iteration factor $\eta = 0.66$ was used to produce the fractal slot antenna. The configuration of the proposed antenna, as illustrated in Fig. 23, is the modified fractal slot antenna fed by CPW. The antenna composes of the modified inner metallic patch, which is fed by a 50-Ω CPW line with a strip width W_f and gap g_1, and an outer metallic patch. In the section, the antenna is fabricated on an economical FR4 dielectric substrate with a thickness of 1.6 mm (h), relative permittivity of 4.1 and loss tangent of 0.019. The entire dimensions of the antenna are 53.40mm × 75.20 mm. The 50- SMA connector is used to feed the antenna at the CPW line. The important parameters, which affect the resonant frequencies of 1.74 GHz, 3.85 GHz, and 5.05 GHz, compose of S_u, S, and S_L. The fixed parameters of the proposed antenna are following: $h = 1.6$ mm, $W_{G1} = 53.37$ mm, $W_{G2} = 38.54$ mm, $L_{G1} = 75.20$ mm, $L_{G2} = 34.07$ mm, $L_{G3} = 39.75$ mm, $W_s = 32.57$ mm, $g_1 = 0.5$ mm, $g_2 = 2.3$ mm, $W_t = 0.94$ mm, $L_t = 21.88$ mm, $W_f = 3.5$ mm, $L_f = 14.50$ mm, $W_1 = 25.92$ mm, $W_2 = 11.11$ mm, $W_3 = 16.05$ mm, $W_4 = 3.7$ mm, and $s_1 = s_2 = s_3 = 3.55$ mm.

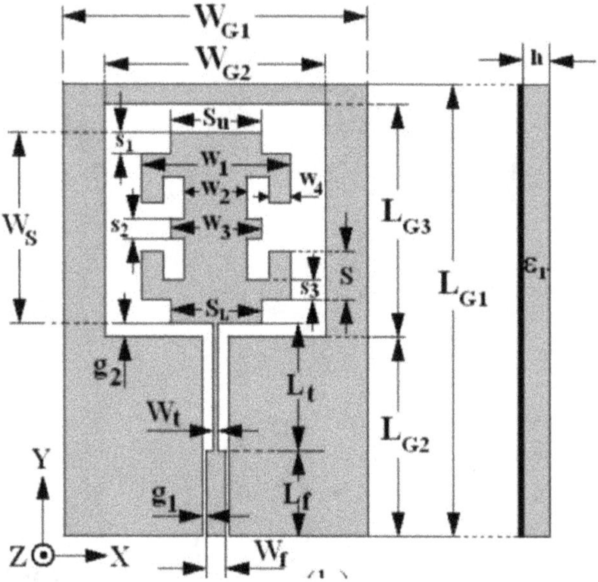

(a)

(b)

Fig. 23. (a) Configurations of the proposed fractal slot antenna and (b) photograph of the prototype

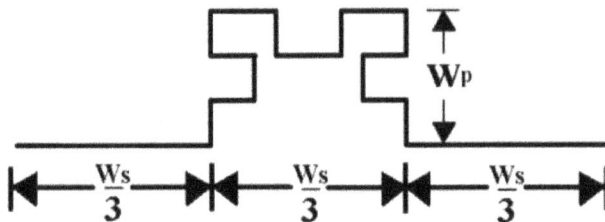

Fig. 24. The initial generator model for the proposed antenna

The suitable parameters, as following, h = 1.6 mm, W_{G1} = 53.37 mm, W_{G2} = 38.54 mm, L_{G1} = 75.20 mm, L_{G2} = 34.07 mm, L_{G3} = 39.75 mm, W_s = 32.57 mm, g_1 = 0.5 mm, g_2 = 2.3 mm, W_t = 0.94 mm, L_t = 21.88 mm, W_f = 3.5 mm, L_f = 14.50 mm, W_1 = 25.92 mm, W_2 = 11.11 mm, W_3 = 16.05 mm, W_4 = 3.7 mm, and s_1 = s_2 = s_3 = 3.55 mm, S_u = 16.050 mm, S = 4.751 mm, and S_L = 16.050 mm, are chosen to implement the prototype antenna by etching into chemicals. The prototype of the proposed antenna is shown in Fig. 23(b). The simulated and measured return losses of the antenna are illustrated in Fig. 25. It is clearly observed that the measured return loss of the antenna slightly shifts to the right because of the inaccuracy of the manufacturing process by etching into chemicals. However, the measured result of proposed antenna still covers the operating bands of 1.71-1.88 GHz and 3.2-5.5 GHz for the applications of DCS 1800, WiMAX (3.3 and 3.5 bands), and WiFi (5.5 GHz band).

This section presents a multiband slot antenna with modifying fractal geometry fed by CPW transmission line. The presented antenna has been designed by modifying an inner fractal patch of the antenna to operate at multiple resonant frequencies, which effectively supports the digital communication system (DCS1800 1.71-1.88 GHz), WiMAX (3.30-3.80 GHz), and WiFi (5.15-5.35 GHz). Manifestly, it has been found that the radiation patterns of the presented antenna are still similarly to the bidirectional radiation pattern at all operating frequencies.

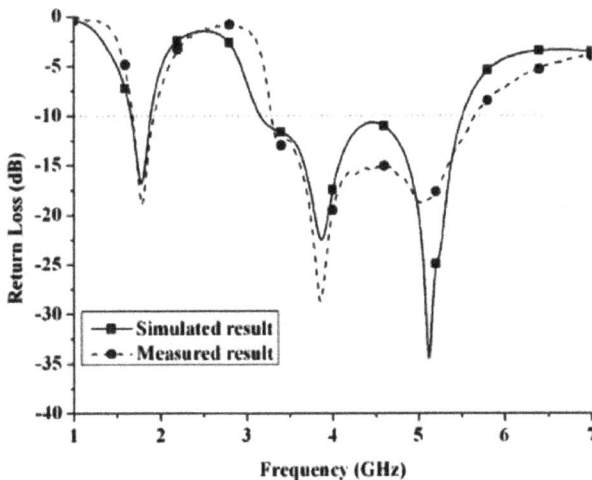

Fig. 25. Simulated and measured return losses for the proposed antenna

5. Unidirectional CPW-fed slot antennas

From the previous sections, most of the proposed antennas have bidirectional radiation patterns, with the back radiation being undesired directions but also increases the sensitivity of the antenna to its surrounding environment and prohibits the placement of such slot antennas on the platforms. A CPW-fed slot antenna naturally radiates bidirectionally, this characteristic is necessary for some applications, such as antennas for roads. However, this inherent bidirectional radiation is undesired in some wireless communication applications such as in base station antenna. There are several methods in order to reduce backside radiation and increase the gain. Two common approaches are to add an additional metal reflector and an enclosed cavity underneath the slot to redirect radiated energy from an undesired direction. In this section, promising wideband CPW-fed slot antennas with unidirectional radiation pattern developed for WiFi and WiMAX applications are presented. We propose two techniques for redirect the back radiation forward including (i) using modified the reflectors placed underneath the slot antennas (Fig. 26(a)) and (ii) the new technique by using the metasurface as a superstrate as shown in Fig 26(b).

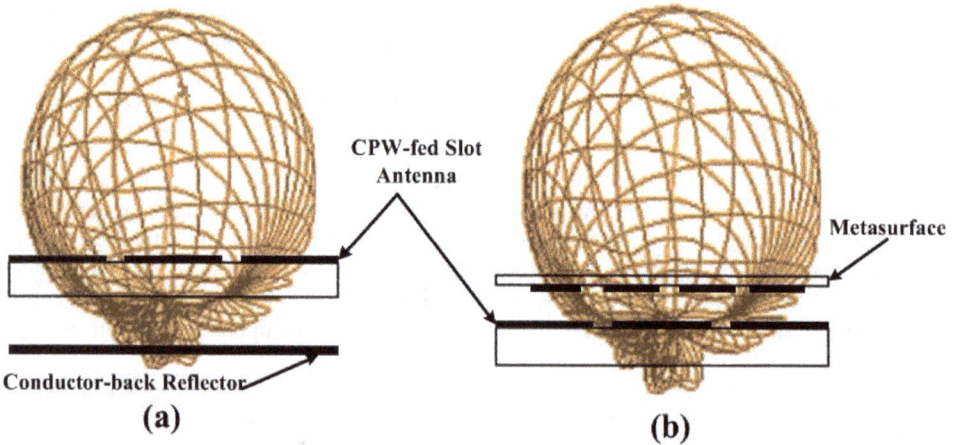

Fig. 26. Arrangement of unidirectional CPW-fed slot antennas (a) conventional structure using conductor-back reflector and (b) the proposed structure using metasurface superstrate

5.1 Wideband unidirectional CPW-fed slot antenna using loading metallic strips and a widened tuning stub

The geometry of a CPW-fed slot antennas using loading metallic strips and a widened tuning stub is depicted in Fig. 27(a). Three different geometries of the proposed conducting reflector behind CPW-fed slot antennas using loading metallic strips and a widened tuning stub are shown in Figs. 27(b), (c), and (d). It comprises of a single FR4 layer suspended over a metallic reflector, which allows to use a single substrate and to minimize wiring and soldering. The antenna is designed on a FR4 substrate 1.6 mm thick, with relative dielectric constant (ε_r) 4.4. This structure without a reflector radiates a bidirectional pattern and maximum gain is about 4.5 dBi. The first antenna, Fig. 27(b), is the antenna located above a flat reflector, with a reflector size 100×100 mm². The Λ-shaped reflector with the horizontal plate is a useful modification of the corner reflector. To reduce overall dimensions of a large corner reflector, the vertex can be cut off and replaced with the horizontal flat reflector (Wc1×Wc3). The geometry of the proposed wideband CPW-fed slot antenna using Λ-shaped reflector with the horizontal plate is shown in Fig. 27(c). The Λ-shaped reflector, having a horizontal flat section dimension of $W_{c1} \times W_{c3}$, is bent with a bent angle of β. The width of the bent section of the Λ-shaped reflector is W_{c2}. The distance between the antenna and the flat section is h_c. For the last reflector, we modified the conductor reflector shape. Instead of the Λ-shaped reflector, we took the conductor reflector to have the form of an inverted Λ-shaped reflector. The geometry of the inverted Λ-shaped reflector with the horizontal plate is shown in Fig. 27(d). The inverted Λ-shaped reflector, having a horizontal flat section dimension of $W_{d1} \times W_{d3}$, is bent with a bent angle of α. The width of the bent section of the inverted Λ-shaped reflector is W_{d2}. The distance between the antenna and the flat section is h_d. Several parameters have been reported in (Akkaraekthalin et al., 2007). In this section, three typical cases are investigated: (i) the Λ-shaped reflector with h_c = 30 mm, β =150°, W_{c1}= 200 mm, W_{c2} = 44 mm, beamwidth in H-plane around 72°, as called **72 DegAnt**; (ii) the Λ-shaped reflector with h_c = 30 mm, β =150°, W_{c1} = 72 mm, W_{c2} = 44 mm, beamwidth in H-plane around 90°, as called **90 DegAnt**; and (iii) the inverted Λ-shaped reflector with h_d = 50 mm, α = 120°, W_{d1} = 72 mm, W_{d2} = 44 mm, beamwidth in H-plane around 120°, as called **120 DegAnt.** The prototypes of the proposed antennas were constructed as shown in Fig. 28. Fig. 29 shows the measured return losses of the proposed antenna. The 10-dB bandwidth is about 69% (1.5 to 3.1 GHz) of 72DegAnt. A very wide impedance bandwidth of 73% (1.5 - 3.25 GHz) for the antenna of 90DegAnt was achieved. The last, impedance bandwidth is 49% (1.88 to 3.12 GHz) when the antenna is 120DegAnt as shown in Fig. 29. However, from the obtained results of the three antennas, it is clearly seen that the broadband bandwidth for PCS/DCS/IMT-2000 WiFi and WiMAX bands is obtained. The radiation characteristics are also investigated. Fig. 30 presents the measured far-field radiation patterns of the proposed antennas at 1800 MHz, 2400 MHz, and 2800 MHz. As expected, the reflectors allow the antennas to radiate unidirectionally, the antennas keep the similar radiation patterns at several separated selected frequencies. The radiation patterns are stable across the matched frequency band. The main beams of normalized H-plane patterns at 1.8, 2.4, and 2.8 GHz are also measured for three different reflector shapes as shown in Fig. 31. Finally, the measured antenna gains in the broadside direction is presented in Fig. 32. For the 72DegAnt, the measured antenna gain is about 7.0 dBi over the entire viable frequency band.

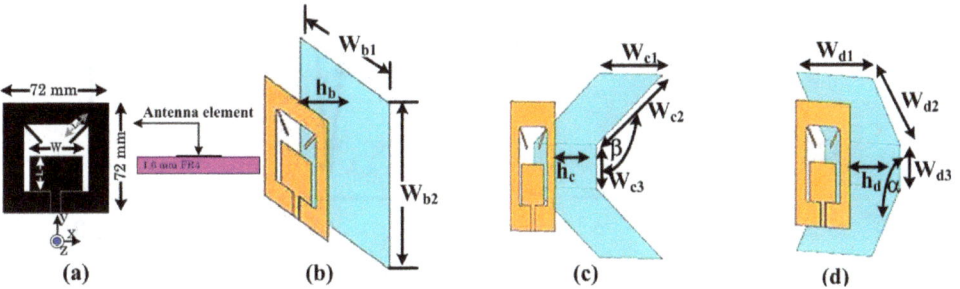

Fig. 27. CPW-FSLW (a) radiating element above, (b) flat reflector, (c) Λ -shaped reflector with a horizontal plate, and (d) inverted Λ-shaped reflector with a horizontal plate

As shown, the gain variations are smooth. The average gains of the 90DegAnt and 120DegAnt over this bandwidth are 6 dBi and 5 dBi, respectively. This is due to impedance mismatch and pattern degradation, as the back radiation level increases rapidly at these frequencies.

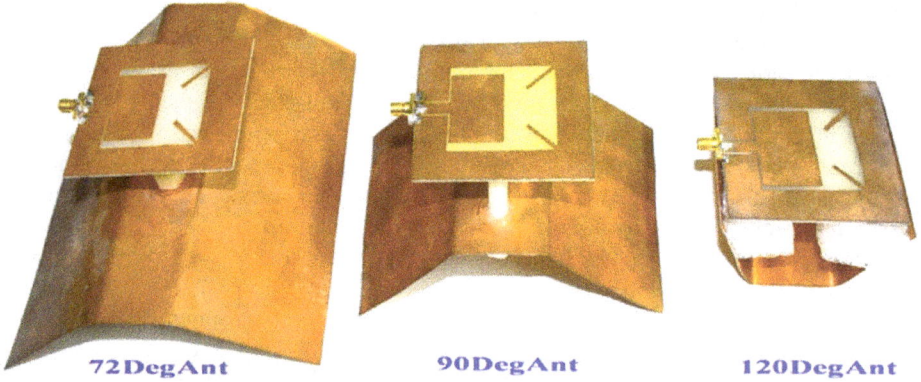

Fig. 28. Photograph of the fabricated antennas (Akkaraekthalin et al., 2007)

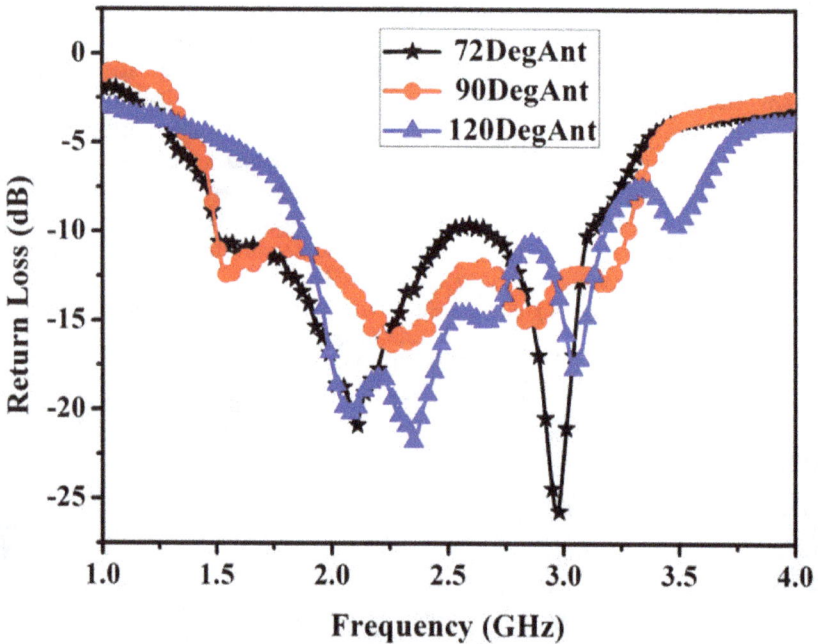

Fig. 29. Measured return losses of three different reflectors :72° (72DegAnt), 90° (90DegAnt), and 120° (120DegAnt)

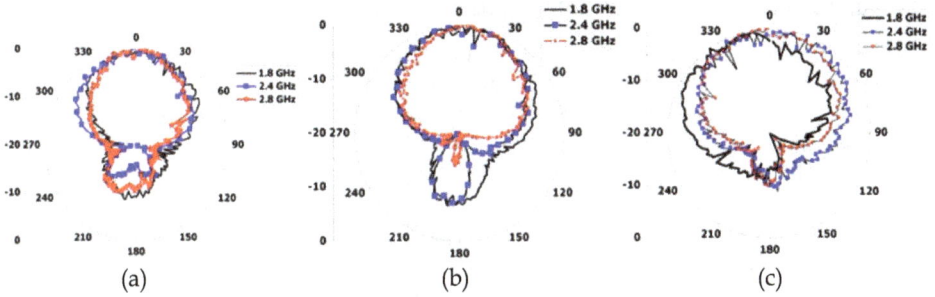

Fig. 30. Measured radiation pattern of three different reflectors, (a) 72° (72DegAnt), (b) 90° (90DegAnt), and (c) 120° (120DegAnt) (Chaimool et al., 2011)

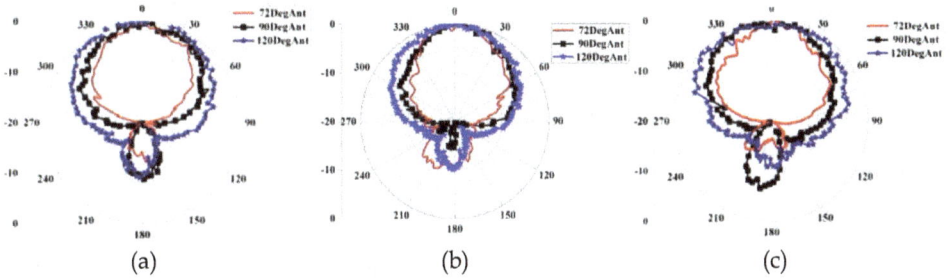

Fig. 31. Measured radiation patterns in H-plane for three different reflectors at (a) 1800 MHz, (b) 2400 MHz, and (c) 2800 MHz (Chaimool et al., 2011)

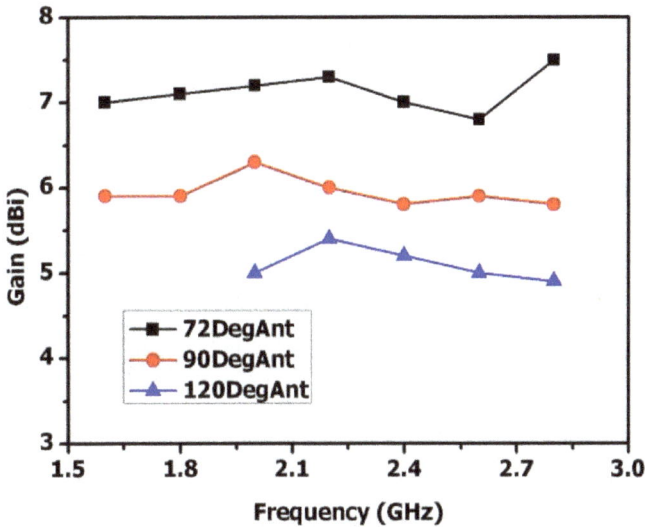

Fig. 32. Measured gains of the fabricated antennas

5.2 Unidirectional CPW-fed slot antenna using metasurface

Fig. 33 shows the configurations of the proposed antenna. It consists of a CPW-fed slot antenna beneath a metasurface with the air-gap separation h_a. The radiator is center-fed inductively coupled slot, where the slot has a length $(L-W_f)$ and width W. A 50-Ω CPW transmission line, having a signal strip of width W_f and a gap of distance g, is used to excite the slot. The slot length determines the resonant length, while the slot width can be adjusted to achieve a wider bandwidth. The antenna is printed on 1.6 mm thick (h_1) FR4 material with a dielectric constant (ε_{r1}) of 4.2. For the metasurface as shown in Fig. 33(b), it comprises of an array 4×4 square loop resonators (SLRs). It is printed on an inexpensive FR4 substrate with dielectric constant ε_{r2}= 4.2 and thickness (h_2) 0.8 mm. The physical parameters of the SLR are given as follows: P = 20 mm, a = 19 mm and b= 18 mm. To validate the proposed concept, a prototype of the CPW-fed slot antenna with metasurface was designed, fabricated and measured as shown in Fig. 34 (a). The metasurface is supported by four plastic posts above the CPW-fed slot antenna with h_a = 6.0 mm, having dimensions of 108 mm×108 mm ($0.86\lambda_0$ ×$0.86\lambda_0$). Simulations were conducted by using IE3D simulator, a full-wave moment-of-method (MoM) solver, and its characteristics were measured by a vector network analyzer. The S_{11} obtained from simulation and measurement of the CPW-fed slot antenna with metasurface with a very good agreement is shown in Fig. 34 (b). The measured impedance bandwidth (S_{11} ≤ -10 dB) is from 2350 to 2600 MHz (250 MHz or 10%). The obtained bandwidth covers the required bandwidth of the WiFi and WiMAX systems (2300-2500 MHz). Some errors in the resonant frequency occurred due to tolerance in FR4 substrate and poor manufacturing in the laboratory. Corresponding radiation patterns and realized gains of the proposed antenna were measured in the anechoic antenna chamber located at the Rajamangala University of Technology Thanyaburi (RMUTT), Thailand. The measured radiation patterns at 2400, 2450 and 2500 MHz with both co- and cross-polarization in E- and H- planes are given in Fig. 35 and 36, respectively. Very good broadside patterns are observed and the cross-polarization in the principal planes is seen to be than -20 dB for all of the operating frequency. The front-to-back ratios FBRs were also measured. From measured results, the FBRs are more than 15 and 10 dB for E- and H- planes, respectively. Moreover, the realized gains of the CPW-fed slot antenna with and without the metasurface were measured and compared as shown in Fig. 37. The gain for absence metasurface is about 1.5 dBi, whereas the presence metasurface can increase to 8.0 dBi at the center frequency.

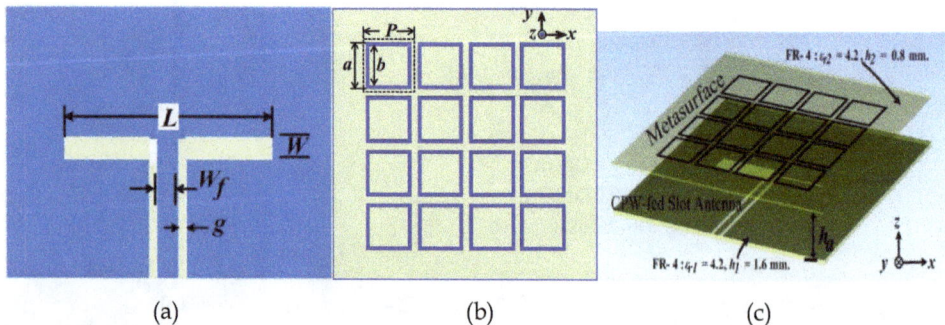

Fig. 33. Configuration of the CPW-fed slot antenna with metasurface (a) the CPW-fed slot antenna, (b) metasurface and (c) the cross sectional view

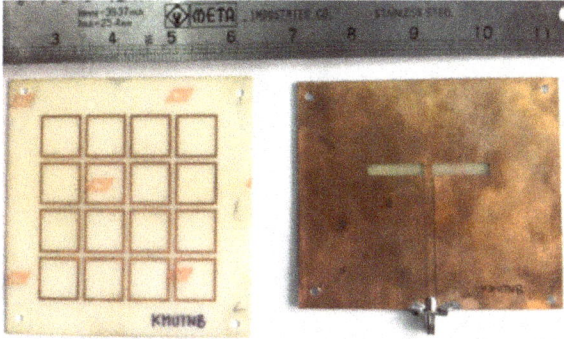

(a)

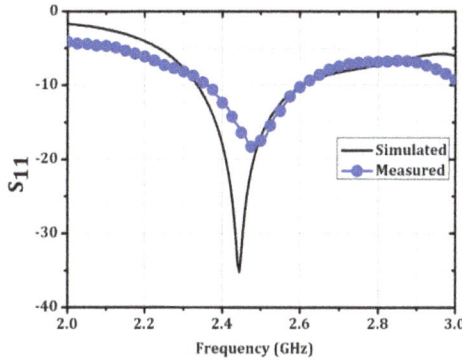

(b)

Fig. 34. (a) Photograph of the prototype antenna and (b) simulated and measured S_{11} of the CPW-fed slot antenna with the metasurface (Rakluea et al. 2011)

An improvement in the gain of 6.5 dB has been obtained. It is obtained that the realized gains of the present metasurface are all improved within the operating bandwidth.

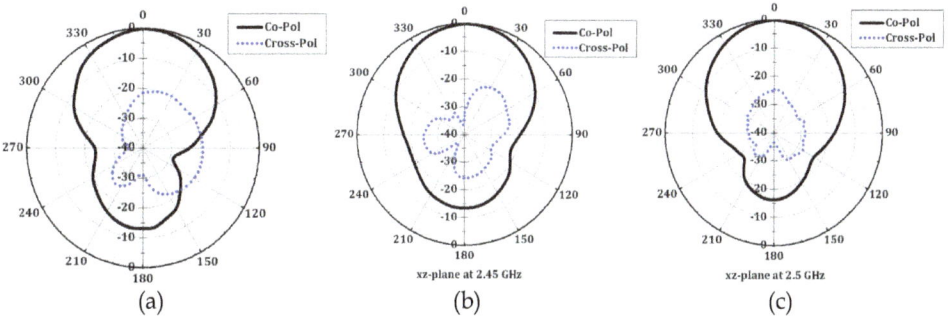

(a)

(b)

(c)

Fig. 35. Measured radiation patterns for the CPW-fed slot antenna with the metasurface in E-plane. (a) 2400 MHz, (b) 2450 MHz and (c) 2500 MHz

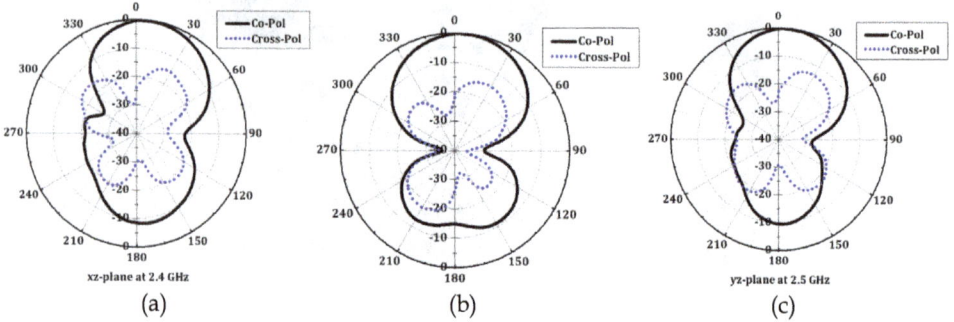

Fig. 36. Measured radiation patterns for the CPW-fed slot antenna with the metasurface in *H*-plane. (a) 2400 MHz, (b) 2450 MHz and (c) 2500 MHz

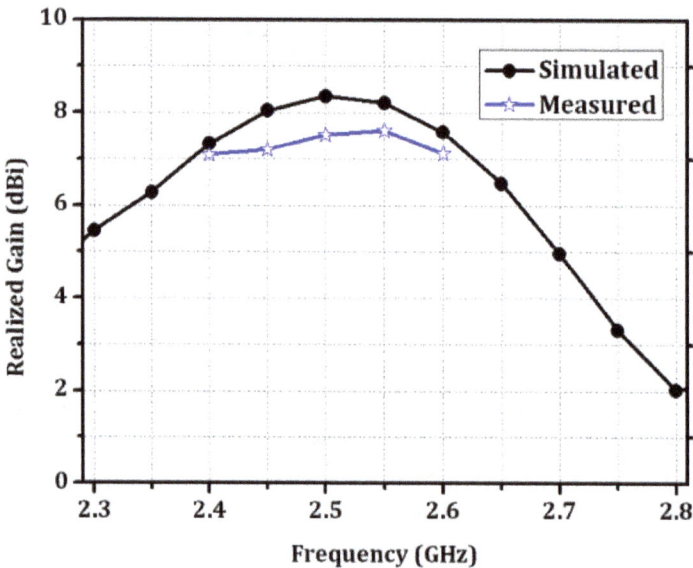

Fig. 37. Simulated and measured realized gains of the CPW-fed slot antenna with the metasurface

6. Conclusions

In this chapter, we have introduced wideband CPW-fed slot antennas, multiband CPW-fed slot and monopole antennas, and unidirectional CPW-fed slot antennas. For multiband operation, CPW-fed multi-slots and multiple monopoles are presented. In addition to, the CPW-fed slot antenna with fractal tuning stub is also obtained for multiband operations. Some WiFi or WiMAX applications such as point-to-point communications require the unidirectional antennas. Therefore, we also present the CPW-fed slot antennas with unidirectional radiation patterns by using modified reflector and metasurface. Moreover, all of antennas are fabricated on an inexpensive FR4, therefore, they are suitable for mass productions. This suggests that the proposed antennas are well suited for WiFi as well as WiMAX portable units and base stations.

7. References

Akkaraekthalin, P.; Chaimool, S.; Krairiksk, M. (September 2007) Wideband uni-directional CPW-fed slot antennas using loading metallic strips and a widened tuning stub on modified-shape reflectors, *IEICE Trans Communications,* vol. E90-B, no.9, pp.2246-2255, ISSN 0916-8516.

Chaimool, S.; Akkaraekthalin P.; Krairiksh, M.(May 2011). Wideband Constant beamwidth coplanar waveguide-fed slot antennas using metallic strip loading and a wideband tuning stub with shaped reflector, *International Journal of RF and Microwave Computer – Aided Engineering,* vol. 21, no 3, pp. 263-271, ISSN 1099-047X

Chaimool, S.; Akkaraekthalin, P.; Vivek, V. (December 2005). Dual-band CPW-fed slot antennas using loading metallic strips and a widened tuning stub, *IEICE Transactions on Electronics,* vol. E88-C, no.12, pp.2258-2265, ISSN 0916-8524.

Chaimool, S.; Chung, K. L. (2009). CPW-fed mirrored-L monopole antenna with distinct triple bands for WiFi and WiMAX applications, *Electronics Letters,* vol. 45, no. 18, pp. 928-929, ISSN 0916-8524.

Chaimool, S.; Jirasakulporn, P.; Akkaraekthalin, P. (2008) A new compact dual-band CPW-fed slot antenna with inverted-F tuning stub, *Proceedings of ISAP-2008 International Symposium on Antennas and Propagation,* Taipei, Taiwan, pp. 1190-1193, ISBN: 978-4-88552-223-9

Chaimool, S.; Kerdsumang, S.; Akkraeakthalin, P.; Vivek, V.(2004) A broadband CPW-fed square slot antenna using loading metallic strips and a widened tuning stub, *Proceedings of ISCIT 2004 International Symposium on Communications and Information Technologies,* Sapporo, Japan, vol. 2, pp. 730-733, ISBN: 0-7803-8593-4

Hongnara, T.; Mahatthanajatuphat C.; Akkaraekthalin, P. (2011). Study of CPW-fed slot antennas with fractal stubs, *Proceedings of ECTI-CON2011 8th International Conference of Electrical Engineering/Electronics, Computer, Telecommunications and Information Technology,* pp. 188-191, Khonkean, Thailand, May 17-19, 2011, ISBN: 978-1-4577-0425-3

Jirasakulporn, P. (December 2008). Multiband CPW-fed slot antenna with L-slot bowtie tuning stub, *World Academy of Science, Engineering and Technology,* vol. 48, pp.72-76, ISSN 2010-376X

Mahatthanajatuphat, C. ; Akkaraekthalin, P.; Saleekaw, S.; Krairiksh, M. (2009). A bidirectional multiband antenna with modified fractal slot fed by CPW, *Progress In Electromagnetics Research,* vol. 95, pp. 59-72, ISSN 1070-4698

Moeikham, P.; Mahatthanajatuphat, C.; Akkaraekthalin, P.(2011). A compact ultrawideband monopole antenna with tapered CPW feed and slot stubs, *Proceedings of ECTI-CON2011 8th International Conference of Electrical Engineering/Electronics, Computer, Telecommunications and Information Technology,* pp. 180-183, Khonkean, Thailand, May 17-19, 2011, ISBN: 978-1-4577-0425-3

Rakluea, C.; Chaimool, S.; Akkaraekthalin, P. (2011). Unidirectional CPW-fed slot antenna using metasurface, *Proceedings of ECTI-CON2011 8th International Conference of Electrical Engineering/Electronics, Computer, Telecommunications and Information Technology*, pp. 184-187, Khonkean, Thailand, May 17-19, 2011, ISBN: 978-1-4577-0425-3

Sari-Kha, K.; Vivek, V.; Akkaraekthalin, P. (2006) A broadband CPW-fed equilateral hexagonal slot antenna, *Proceedings of ISCIT 2006 International Symposium on Communications and Information Technologies*, Bangkok, Thailand, pp. 783-786, October 18-20, 2006, ISBN 0-7803-9741-X.

5

MicroTCA Compliant WiMAX BS Split Architecture with MIMO Capabilities Support Based on OBSAI RP3-01 Interfaces

Cristian Anghel and Remus Cacoveanu
University Politehnica of Bucharest,
Romania

1. Introduction

Modern mobile communication systems must fulfill more and more requirements received from the customers. This leads to an increase of complexity. The control part of the system becomes very important, a multi-level approach being needed. With respect to this, all BS (Base Stations) from a system are synchronized using GPS (Global Positioning System) or IEEE 1588 [1] standard, high speed synchronous interfaces are used between the BBM (Baseband Modules) and the RRU (Remote Radio Units), for example OBSAI (Open Base Station Architecture Initiative) [2, 3] or CPRI (Common Public Radio Interface) [4], and standard communication methods are provided between the control parts placed in different levels of the system.

This chapter describes the management and synchronization procedures for a WiMAX BS architecture compliant with MicroTCA standard (Micro Telecommunications Computing Architecture) [5]. The block scheme of such a BS for the case of a 3 sectors cell is presented. One can observe the main parts of the MicroTCA standard, i.e. the MCH (MicroTCA Carrier Hub) modules and the AMC (Advanced Mezzanine Card) [6] modules.

Referring now to the OBSAI RP3-01 interface, this represents an extension of the RP3 (Reference Point 3) protocol for remote radio unit use. The BS can support multiple RRUs connected in chain, ring and tree-and-branch topologies, which makes the interface very flexible. Also, in order to minimize the number of connections to RRUs, the RP1 management plan, which includes Ethernet and frame clock bursts, is mapped into RP3 messages. This solution is an alternative to the design in which the radio module collocates with the BBM. Although in such a case the interface between the radio unit and the BBM becomes less complex, the transmitter power should be increased in order to compensate the feeder loss. For the proposed WiMAX BS block scheme, some improvements can be done starting from the proprieties of OBSAI RP3-01 interface. In this proposed BS split architecture, a BBM is connected to the two RRUs in order to have multiple transmit/ receive antennas for MIMO capabilities. The connection between the two RRUs is realized using a chain topology. In order to obtain a single point failure redundancy scheme, a second BBM connected to the two RRUs is required. Only one BBM will be active at the time. There are also described the OBSAI RP3-01 Interfaces required for blocks interconnection. Note that on OBSAI RP3-01 interface of each RRU the same Transport and Application Layers serve the both Physical and Data Link Layers.

This chapter is organized as following: Section 2 describes briefly the MicroTCA standard and the most important elements of such an architecture. Section 3 proposes a simple and efficient method of synchronizing a WiMAX BS using the GPS signals. There are provided synchronization signals for the air interface, in order to avoid interferences with other BSs. Also there are obtained, based on this proposed method, synchronization signals used inside BS with the scope of aligning all the modules of the architecture, which is very important when split solution is adopted, i.e. not all the units are co-located in the same physical element. Finally, Section 4 proposes a new way of using OBSAI RP3-01 Interface in a WiMAX BS, this new implementation solution providing support for MIMO techniques and redundancy.

2. MicroTCA standard – Overview

The MicroTCA standard is created by PICMG (PCI Industrial Computer Manufacturers Group) and it defines the requirements of chassis hardware system. Such a system uses AMC (Advanced Mezzanine Card) modules interconnected by a board having a high speed interface on the backplane of the chassis. The standard defines the mechanical, electrical and management specific characteristics needed for supporting AMC standard compliant modules.

The described structure is a modular one. By the configuration and the interconnection of the AMC modules inside the chassis, a high variety of applications can be obtained. Besides this, the standard doesn't impose a certain physical configuration of the chassis and neither a mandatory signaling protocol for the backplane high speed interface. Instead, a set of communication and interconnection requirements is defined. This set of requirements should be available for any structure, providing this way a high compatibility between the equipments compliant with the standard.

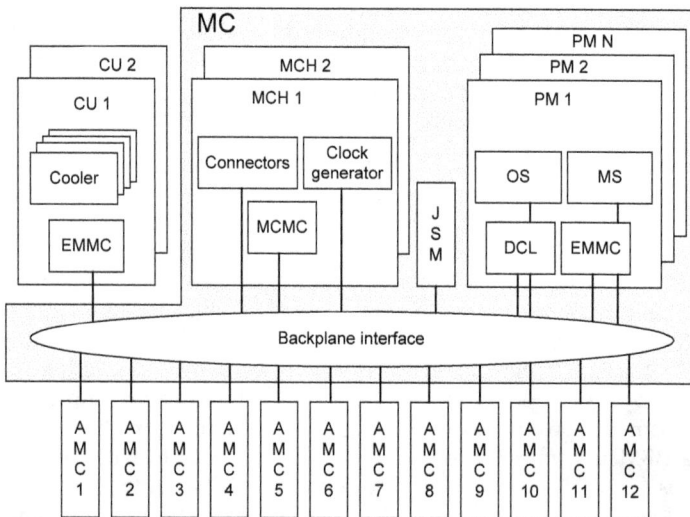

Fig. 2.1 MicroTCA – Block scheme

The proposed architecture is a modular one, as one can see from Figure 2.1. It results in a very flexible solution, allowing a high diversity in AMC modules implementation and in obtained fuctions.

A MicroTCA chassis is made of 1 up to 12 AMC modules, which will realize together the system functionality. Then there is a MCH (MicroTCA Carrier Hub) module which represents the support for the implementation of the main system management functions. On the chassis there can be found also PM (Power Modules) modules used for power supplying, CU (Cooling Unit) modules used for temperature control at system level, interconnection elements between the modules or with the external inputs / outputs (Backplane, Faceplate) plus other mechanical elements and redundant modules. A second MCH module and a second CU module can also be present in a chassis, from redundancy reasons.

2.1 AMC modules

The AMC modules are the main components of a MicroTCA chassis, containing the elements which will provide the system processing functions. There can be listed here the microcontrollers, the digital signal processors, the routers, the memory blocks, the I/O interface controllers, the base band and RF processing modules.

Initially, in AdvancedMC specifications, the AMC modules were defined as additional boards used for CB (Carrier Board) functionality extension. In MicroTCA, the AMC modules totally perform the system processing, while CB will be distributed between different architecture elements, having only a support role.

Due to the signal processing functions, the management tasks implemented on the AMC module are to be reduced as much as possible, in order to provide the maximum of the resources to the main process. This is the reason why these modules are controlled by a low level functionality entity called MMC (Module Management Controller). The set of functions this entity is performing is very simple so that it can be implemented on a low cost processor. The communication between MMC and a dedicated management entity at chassis level is made through IPMB-L (Intelligent Platform Management Bus - Local) interface, using a reduced set of requests/ confirmations specified in IPMI v2.0 (Intelligent Platform Management Interface) [7] standard.

The IPMB-L connections are isolated between each others in order to avoid the case when module issue is blocking the complete system, as in the case of bus topology.

The most important advantage introduced by this standard is the possibility of introducing/ switching in the system any module without being required to stop the power supply (Hot Swap feature) or to make any other hardware/ software modification (Plug and Play feature).

2.2 MicroTCA Carrier

MC (MicroTCA Carrier) is the novelty proposed by MicroTCA and it represents the main board, as defined by AdvancedTCA. It is responsible for the power distribution, the interconnection and the IPMI management for the 12 AMC modules. MC components are distributed in the chassis and are described in Figure 2.1.

- The power distribution infrastructure

It provides and controls the power distribution for each AMC module. The standard indicates the existence of 3 functional aspects:

- the operational supply OS providing 12V to each AMC
- the management supply MS providing 3.3V to each AMC. OS and MS are separated sub-systems in order to ensure the isolation between the processing processes and the management ones.
- the distribution control logic DCL being responsible of the protection, isolation and validation functions for each network branch

The PM units include also system surveillance functions required for the management part. The PM circuits should detect the units in the chassis, should monitor the parameters quality on each branch and should provide protection for overload. Part of these functions is made locally by a low level intelligence entity called EMMC (Enhanced Module Management Controller), but the system will be controlled by the Carrier Manager, which will compute the power budget based on the FRUI (Field Replaceable Unit Information) tables before validating the power distribution for AMC units. This process is similar with the AdvancedTCA one, and it is described in Figure 2.2.

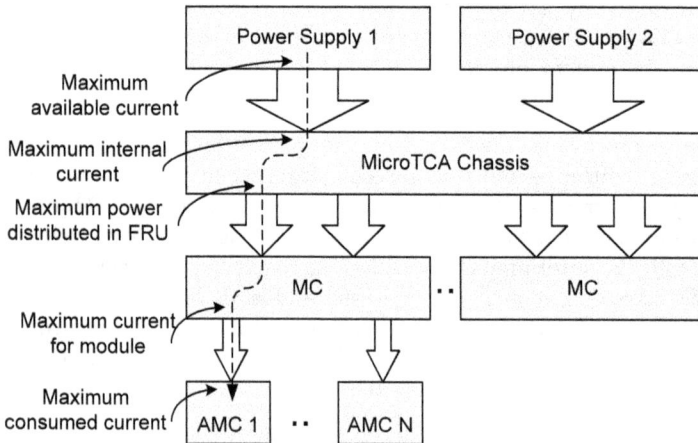

Fig. 2.2 Power distribution infrastructure

- Test infrastructure

It is represented by a JTAG switch called JSM (JTAG Switch Module). This allows the verification of the chassis, together with the modules inside, both in production period and in normal functioning period. As an alternative solution, serial asynchronous UART interfaces are implemented for the same testing scope.

2.3 MicroTCA Carrier Hub

This module combines the control and management infrastructure with the interconnection one, as depicted in Figure 2.3. It provides support for the 12 AMC units. Also, it is available for all the other modules in the chassis. Having this important role, the MCH module is a critical point of the MicroTCA architecture, and for this reason it is required to have another MCH module in the chassis for redundancy.

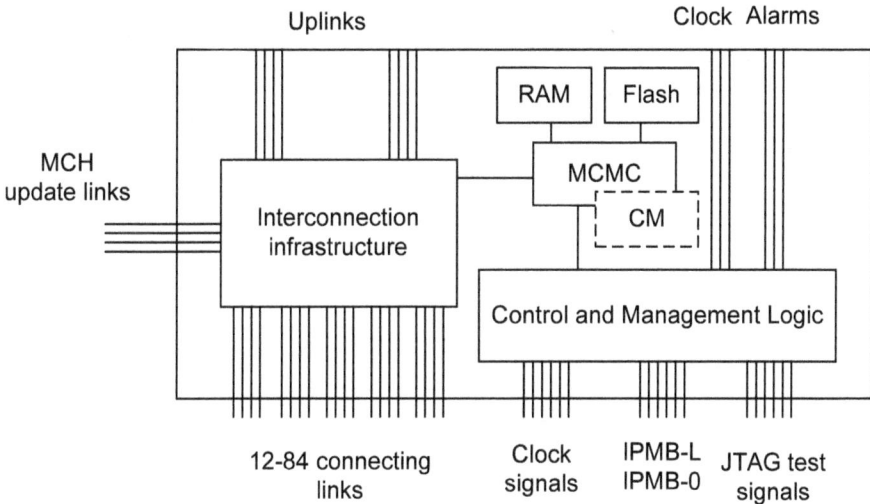

Fig. 2.3 MCH – Block scheme

The MCH module represents the physical support for a set of control and management functions called CM (Carrier Manager), which is the main authority in MicroTCA Carrier. It has to deal with the power control, the AMC network connecting management, the IPMI control and management, the E-Keying functions, the Hot Swap functions and the synchronization at system level.

The MCH structure and functionalities are based on the following modules:

-   MCMC (MicroTCA Carrier Management Controller). Using the IPMB-L interface, the AMC detection signals and the validation signals provided by the PM units, the MCMC is monitoring and controlling the AMC units through CM. On the same time, MCMC acts as MMC for the MCH.
-   Control and Management Logic block. It generates and distributes the clock signals to all the AMC units. It also provides the system level synchronization, using a reference clock received from an external module with universal synchronization capability (for example a GPS receiver) and generating a set of clock signals for the system components, with a required precision.
-   Interconnection Infrastructure. It represents the main communication path between the AMC units. It includes a switch and several high speed serial interfaces (1..10 Gb/s) which create a star network connecting the modules based on PCI Express or Ethernet protocols.

The logical link between the modules connected like this is made by the E-Keying function provided by the Carrier Manager. This function verifies that all the AMC units from a chassis are electrically compatible before giving the authorization to enter in the network.

2.4 Base station components

Based on the above description, the MicroTCA architecture totally fulfills the requirements of a mobile communication base station implementation. Properties like modularity, flexibility, high cooling capacity and low cost are supported by the standard. In addition, the MCH unit ensures an efficient control of the system elements and provides information required by the network high layers. The AMC modules have also an important role, being responsible of base band processing and RF processing that are required by a mobile communication system. A WiMAX base station example is described below based on the described MicroTCA architecture.

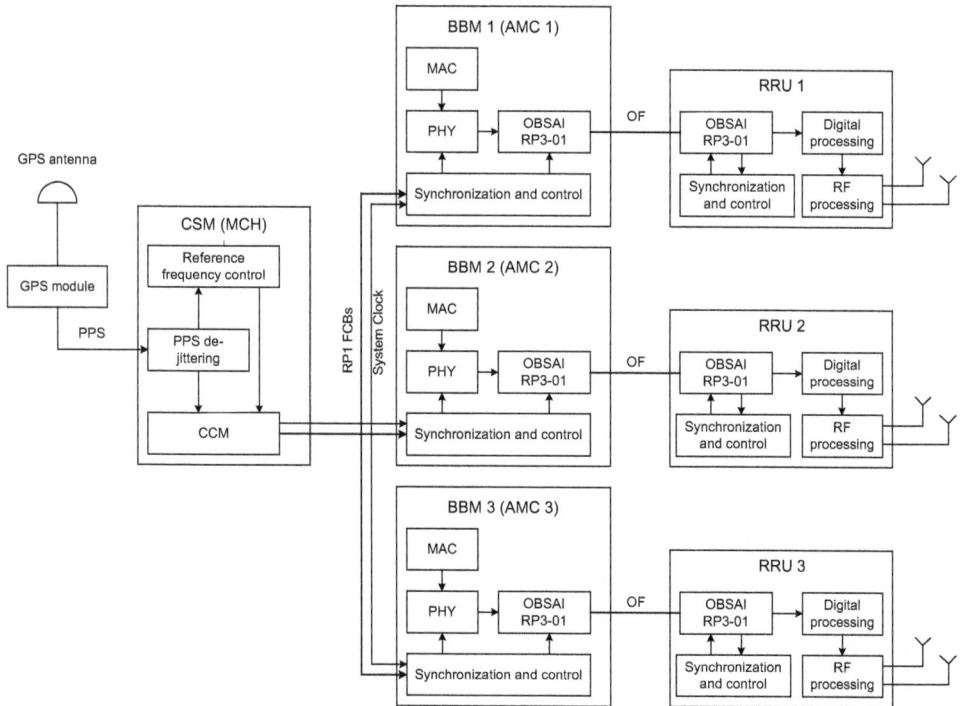

Fig. 2.4 WiMAX BS for a cell with 3 sectors

Figure 2.4 describes the WiMAX base station main components for the case of a cell with 3 sectors, each sector providing support for a certain level of diversity at transmission and reception. The main processing components are:

-   the GPS module [8] equipped with a GPS antenna: this unit is used for generating a signal called PPS (Pulse Per Second) which is synchronized with the universal reference extracted from the GPS system
-   the CSM module (Control, Synchronization and Management): it contains the algorithm used for reducing the PPS signal jitter. The new generated PPS signal controls the

system reference frequency generator. On the same time there are build in CCM (Control and Clock Module) the RP1 (Reference Point 1) synchronization signals used for OBSAI Interface and for Radio Interface. These FCBs (Frame Clock Burst) are sent time multiplexed to each BBM (Base Band Module) on a serial interface. On another serial interface all BBMs are receiving the system clock which will be used directly or to control the frequency generated by a local quart.

- the BBM module: it makes all the base band processing required for transmitting and receiving data in the system. There are included here the MAC and the PHY levels, as they are described by IEEE 802.16e standard. A block called Control and Synchronization is also part of the BBM. It is responsible of extracting the FCBs used for OBSAI interface synchronization, the FCBs used for radio interface synchronization being at this point encapsulated over the RP3 data stream inside the OBSAI RP3-01 interface and then being sent to RRUs (Remote Radio Unit) on the optic fiber.

- the RRU module: it is the external unit which, besides a possible digital processing (the decimation/ interpolation filters used for the selected channel, for example), performs all the radio domain tasks. This module may have multiple transmission/ reception paths and so, multiple antennas. The way the OBSAI interface is used for fulfilling this kind of requirements will be presented next in this chapter.

Having on one hand the WiMAX base station modules characteristics and on the other hand the MicroTCA architecture, one can identify that the CSM module has MCH specific functions, while the BBMs can be considered as being AMCs. Of course that besides these main modules, the power suppliers and the cooling units have to be added as WiMAX base station components.

3. WiMAX base station GPS based synchronization

3.1 Introduction

In communications systems using TDD (Time Division Duplex), appropriate time synchronization is critically important. In order to avoid inter-cell interference, all base stations must use the same timing reference. One solution to this problem is the Global Positioning System (GPS). The users can receive accurate time from atomic clocks and can generate themselves synchronization signals. Commonly the GPS receiver generates a Pulse per Second (PPS) signal and, optionally, a 10 MHz signal, phase synchronized with the PPS.

All the transmission over the radio channel, both on downlink and uplink, should be synchronized with the PPS signal. The RP1 synchronization burst generator, called Clock Control Manager (CCM) shall provide frame timing and time stamping for each of the air interface systems independently. The quality of the PPS signal will dictate the periodicity of these synchronization bursts. Also, algorithms for maintaining the stability of the clock reference, which can be affected by the temperature variance or by aging, can be developed based on the PPS signal. It is obvious why the PPS jitter level is a critical parameter in obtaining high synchronization performances [9].

This document will describe the digital method used for PPS de-jittering and the VCXO (Voltage Control Crystal Oscillator) oscillating frequency controlling algorithm.

3.2 Clock reference controlling scheme

The controlling scheme is a hybrid one, using both analog and digital elements. The scheme is depicted in Figure 3.1. In this application, the PPS input is sourced by a low cost GPS receiver called Resolution T, produced by Trimble.

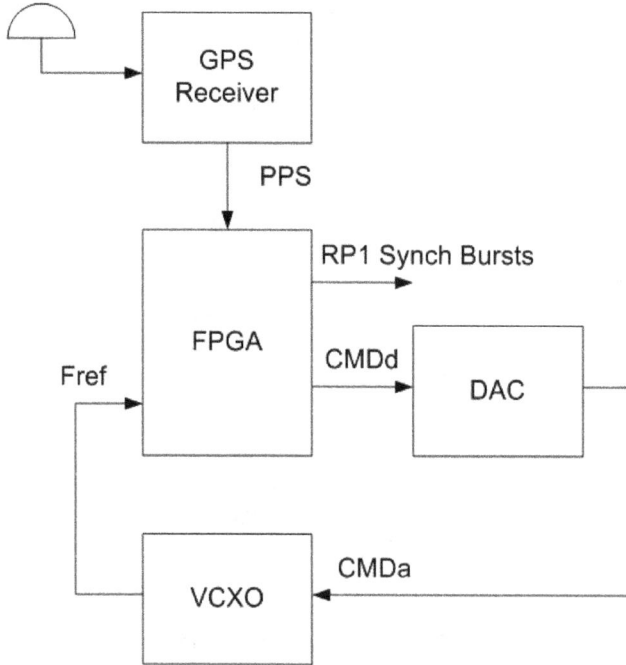

Fig. 3.1 Controlling scheme

The scheme works as follows: the Field Programmable Gate Array (FPGA), which is a XC3S500E chip representing a Spartan 3E family member produced by Xilinx, increments a counter value on every F_{ref} rising edge and resets this counter on every PPS pulse. Let's consider the nominal frequency F_{ref}^n and the counter value at a time instant $count\_val$. When a new PPS pulse is received from the GPS module, before the counter reset, his value is stored and compared with F_{ref}^n. If the two values are not equal, the digital block computes a digital command CMD_d that is converted into a voltage level by a Digital to Analog Converter (DAC). The analog command CMD_a controls the VCXO and the value of F_{ref} is changed accordingly.

As it was mentioned before, the PPS jitter can produce VCXO commands that are unnecessary or imprecise. This is the reason way the PPS signal from the GPS receiver is passed through a digital de-jittering block before it is used by the controlling algorithm and by the CCM. The bloc scheme of the digital part of the structure described in Figure 3.1 is depicted in Figure 3.2.

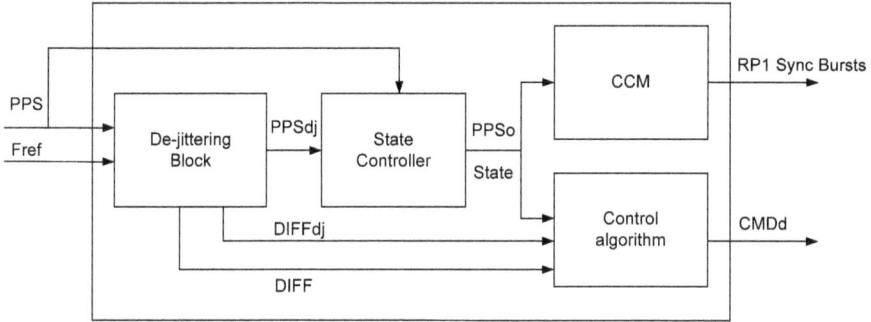

Fig. 3.2 The bloc scheme of the digital part

A. The de-jittering block

The PPS jitter characteristics are to be presented now. Figure 3.3 depicts the time instant value of the jitter. One can see from this figure that the PPS jitter is in the range ±20*ns* for the selected GPS receiver. Also is easy to observe that it does not have uniform distribution. For this reason a simple mean will not eliminate the jitter problem (see Figure 3.4).

Figure 3.5 depicts the Allan deviation. For all of these measurements, it is assumed that the function $e(t)$, representing the time error (the deviation from 1 second value), is sampled with N equally spaced samples, $e_i = e(i\tau 0)$, for $i = (1, 2, ...,N)$, and with a sampling interval $\tau 0$ of 1 second. The observation interval, τ, is given by $\tau = n\tau 0$. The Allan deviation is computed using equation 3.1:

$$ADEV(n\tau_0) = \sqrt{\frac{1}{2n^2\tau_0^2(N-2n)}\sum_{i=1}^{N-2n}\left(e_{i+2n} - 2e_{i+n} + e_i\right)^2} \tag{3.1}$$

where $n \in \left(1..\left\lceil\frac{N-1}{2}\right\rceil\right)$

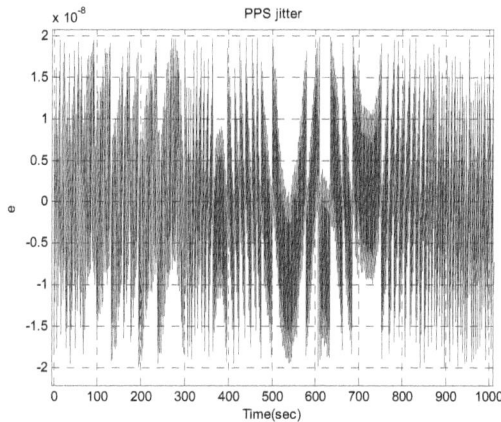

Fig. 3.3 PPS jitter

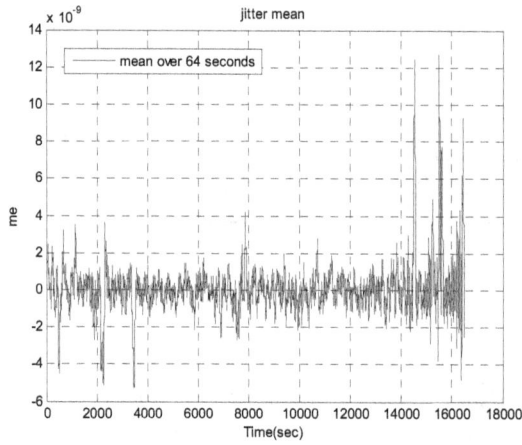

Fig. 3.4 PPS jitter mean

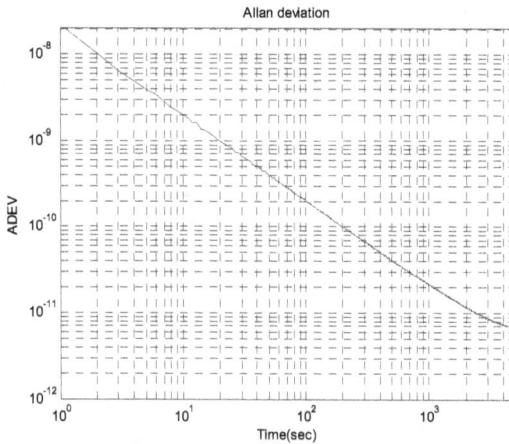

Fig. 3.5 PPS jitter Allan deviation

The slope of $ADEV(\tau)$ is τ^{-1}, which corresponds to white noise phase modulation and flicker phase modulation [1].

The de-jittering block contains a discreet-time Kalman filter. We will consider a particular algorithm of one-dimension Kalman filter intended for frequency estimation only in oscillators if GPS timing signals are used as the reference ones [10]. As it was mentioned before, on every PPS pulse we compute:

$$DIFF(n) = count\_val(n) - F_{ref}^n \qquad (3.2)$$

If the oscillator frequency is F_{ref}^n then $DIFF(n)$ will reflect only the PPS jitter. If not, the $DIFF(n)$ will contain the frequency deviation also. These values, computed every second, are used as the Kalman filter input.

Using the notations Q for process variance and R for estimate of measurement variance, the de-jittering algorithm is as follows:

$$\textit{Initialization}$$
$$\tilde{x}(1) = 0; \tag{3.3}$$
$$P(1) = 1;$$

$$\textit{for } n = 1 : N$$
$$(\textit{Time update})$$
$$\tilde{x}\_(n+1) = \tilde{x}(n)$$
$$P\_(n+1) = P(n) + Q$$
$$(\textit{Measurement update})$$
$$K(n+1) = P\_(n+1) / (P\_(n+1) + R) \tag{3.4}$$
$$\tilde{x}(n+1) = \tilde{x}\_(n+1) +$$
$$\qquad K(n+1)(DIFF(n) - \tilde{x}\_(n+1))$$
$$P(n+1) = (1 - K(n+1)) P\_(n+1)$$
$$\textit{end}$$

The $DIFFdj$ signal from Figure 3.2 is the filter output, i.e. $\tilde{x}(n)$, and it is used to compute the digital command for the VCXO. Also, the de-jittering block provides a de-jittered PPS pulse, denoted $PPSdj$ which should have a 1 second period.

B. *State Controller*

Some times, due to the lack of visibility, the GPS receiver might not transmit the PPS pulse. This situation should be detected by the State Controller by expecting the PPS pulse within a time window. This window depends on the oscillator stability. If the oscillator has a $\pm25ppm$ variation within the temperature range and a nominal frequency of 153.6 MHz, then the maximum delay of the PPS pulse can be $153.6e6 \times 25e-6 = 3840$ clock periods, i.e. the PPS pulse can be found after the previous one at $153.600.000 \pm 3840$ clock periods. If it is not so, the State Controller block confirms the absence of the PPS pulse.

The Finite State Machine (FSM) of the synchronization block is depicted in Figure 3.6. The four possible states are:

- IDLE: when the synchronization block waits for the first PPS
- TRAINING: when the synchronization block starts the Kalman filtering and waits T_{TR} seconds in order to obtain a stable output
- NORMAL: when the synchronization block works based on PPS pulses received form the GPS module

- HOLD OVER: when the PPS pulse is not received from the GPS module and the synchronization block works based on local PPS pulse.

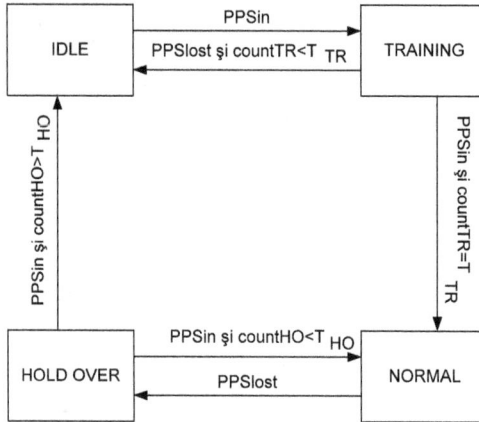

Fig. 3.6 FSM for synchronization block

After the first PPS received, the synchronization block switches from IDLE to TRAINING. In TRAINING, a counter called *countTR* is incremented on every PPS pulse. If the counter value equals the T_{TR} parameter, then a transition is made in NORMAL state. Else, if the block declares the absence of the PPS pulse and the counter value is less than T_{TR} then the new state becomes IDLE.

When the FSM is in NORMAL state and the PPS is declared to be absent, a transition to HOLD OVER state is made. In this state, from the last PPS pulse received, a counter is started in order to generate an internal PPS pulse. Also, a counter called *countHO* is incremented on every local PPS pulse, counting the number of successive absent external PPS pulses. If this counter reaches T_{HO} parameter the synchronization bloc state becomes IDLE. Else, if a new PPS pulse is detected before the counter reaches the T_{HO} value, the synchronization block returns in NORMAL state.

The values of the FSM parameters are given in Table 3.1.

Parameter	Value	Unit
T_{TR}	192	sec
T_{HO}	Depending on VCXO	sec

Table 3.1 FSM parameters

C. *Control Algorithm*

The Control Algorithm block receives the Kalman output and the state of the synchronization block. It also receives the Kalman input, as one can see from Figure 3.2. The control algorithm should compute the VCXO command. The DAC has a 16-bit input, so 2^{16} values are used to control the range of the VCXO. For a measured frequency deviation of $\pm16ppm$, result a control step of:

$$\Delta f = \frac{153.6 \, e6 \times 32 \, e - 6}{2^{16}} = 0.075 \, Hz \approx \frac{1}{13} Hz \tag{3.5}$$

The *CMDi* signal used as feedback for the controlling loop is selected as described in Figure 3.7. The *DIFF* and *DIFFdj* are expressed in clock periods per second and so the DAC value is computed as:

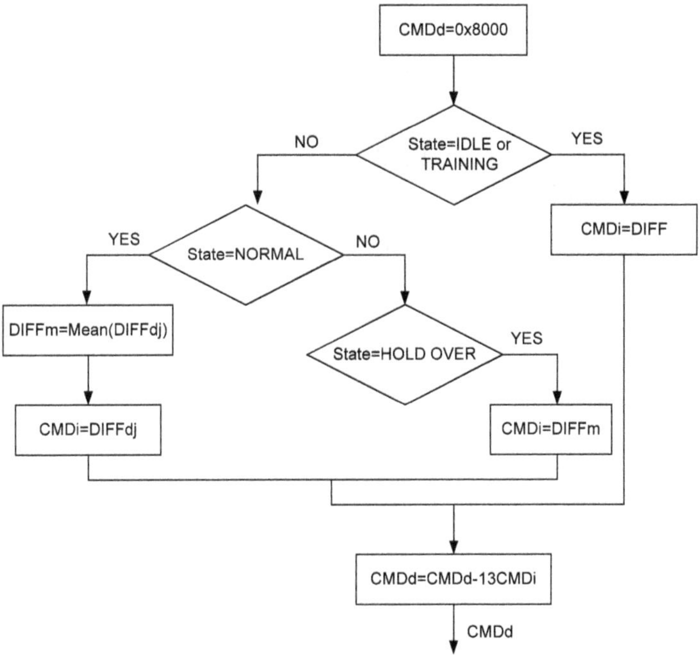

Fig. 3.7 Control Algorithm

$$CMD_d(k) = CMD_d(k-1) - 13CMD_i \tag{3.6}$$

The starting value is the central level of the DAC range, i.e. 2^{15}. In order to obtain a faster convergence, the starting value might be a DAC value saved when the synchronization block was in NORMAL state.

When the State Controller indicates IDLE or TRAINING the oscillator is controlled directly with the measured frequency deviation, in order to achieve fast convergence. In NORMAL state, the Kalman output is used for jitter reduction. The *DIFFdj* signal has a floating point format, so that frequency corrections less than 1 Hz can be produced. Also the mean of the N_m last values of *DIFFdj* signal is computed. The mean value is used when the State Controller is in HOLD OVER state and no valid *DIFFdj* values are received. Also this value is maintained for T_{HO-N} seconds when the synchronizations state returns from HOLD OVER state to NORMAL state, in order not to produce de-synchronization due to the new position of the PPS pulse.

The values of the Control Algorithm parameters are given in Table 3.2.

Parameter	Value	Unit
N_m	128	
T_{HO-N}	2	sec

Table 3.2 Control Algorithm parameters

3.3 Experimental results

The VCXO oscillating frequency is affected by the temperature variations. The controlling algorithm should provide commands fast enough not to accumulate frequency deviations. This problem is observed at the start up too. Even if the temperature is stable, the oscillator is not in a stable thermal state and the algorithm should provide frequency corrections. Figure 3.8 depicts the mean of the DAC commands, considering 0x8000 value as the reference. One can see that at start-up the oscillator has a significant frequency drift and although the temperature variation, depicted in Figure 3.9, is not important, the frequency deviation is about 10Hz per 9000 seconds.

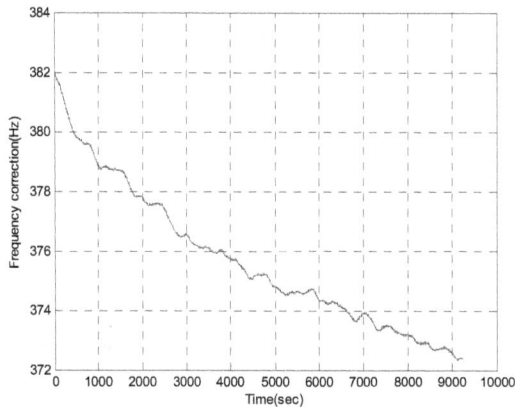

Fig. 3.8 Mean of frequency deviation

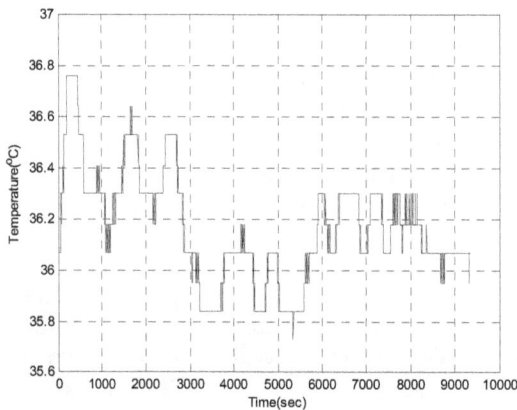

Fig. 3.9 Temperature variation

Figure 3.10 depicts the deviation measured at the input and of the Kalman filter, while Figure 3.11 depicts the deviation measured at the output of the Kalman filter One can see that the initial deviation is the jitter level ($\pm 20ns \, x \, F_{ref} \approx \pm 3$ clock periods) plus some temperature randomly added deviation. At the output of the de-jittering structure, the level of the jitter is much lower.

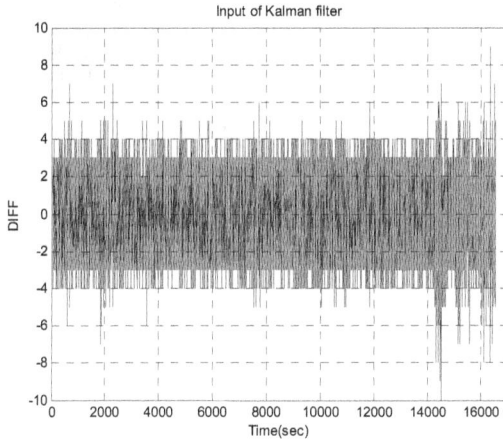

Fig. 3.10 DIFF values in clock periods

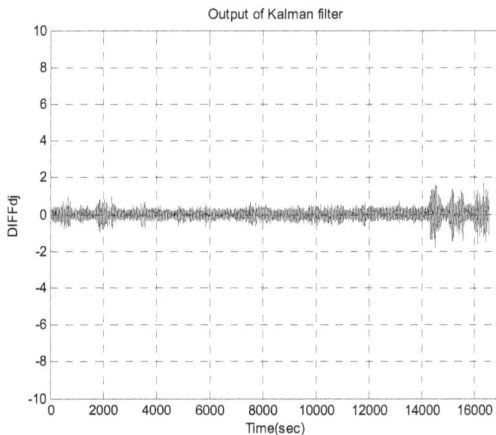

Fig. 3.11 DIFFdj values in clock periods

Figure 3.12 and Figure 3.13 depicts the deviation measured at the input and output of the Kalman filter while a frequency deviation of 1 Hz per second was added for 500 successive seconds and subtracted for the next 500 seconds. One can observe that the *DIFFdj* signal indicates the need of about 1 Hz reduction of the oscillating frequency for the first 500 seconds and then the need of about 1 Hz adding for the next 500 seconds.

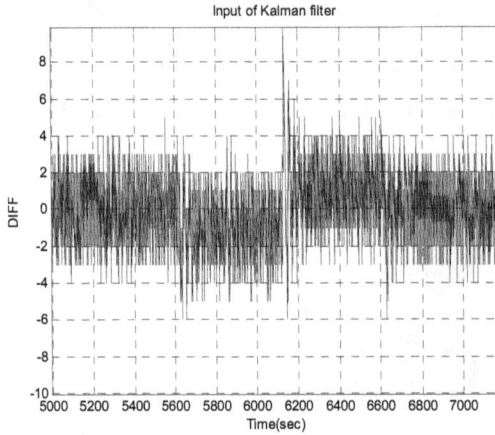

Fig. 3.12 DIFF values in clock periods

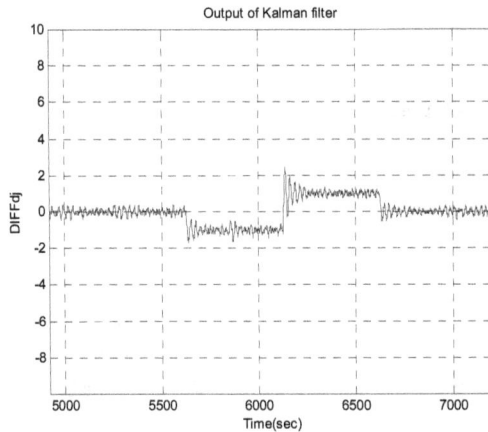

Fig. 3.13 DIFFdj values in clock periods

The results presented in this section were obtained using a hardware platform compliant with Figure 3.1. The inputs and the outputs of the algorithm were transferred to the PC using a UART interface. The pictures were obtained using Matlab.

3.4 Conclusions

This section presented a digital method of reducing the jitter level of the PPS signal generated by a GPS receiver. Also a controlling algorithm of a VCXO oscillating frequency was described. The results indicated that the frequency correction was applied only when the thermal state of the oscillator was not stable. False corrections due to the PPS jitter were almost completely eliminated.

4. WiMAX base station architecture with MIMO capabilities support based on OBSAI RP3-01 Interface

4.1 Introduction

More and more companies try to provide full solutions when it comes to wireless telecommunications systems. But unfortunately this concept, called ecosystem, is not always an easy task to realize. The costs are quite large and, due to system high complexity, the development time is also very long. That means it is very possible that parts of a system can be made by different vendors. The interconnection between different parts should be made by standard interfaces. The usage of third party intellectual property solution reduces the compatibility area. This is the reason why standards as Common Public Radio Interface (CPRI) and Open Base Station Architecture Initiative (OBSAI) were developed. The OBSAI RP3-01 interface permits the transport of data corresponding to different communications standards, such as WCDMA, GSM/EDGE, CDMA2000 and 802.16.

The OBSAI RP3-01 interface represents an extension of the Reference Point 3 protocol for remote radio unit use. The BS can support multiple RRUs connected in chain, ring and tree-and-branch topologies, which makes the interface very flexible. Also, in order to minimize the number of connections to RRUs, the RP1 management plan [3], which includes Ethernet and frame clock bursts, is mapped into RP3 messages. This solution is an alternative to the design in which the radio module collocates with the BBM. Although in such a case the interface between the radio unit and the BBM becomes less complex, the transmitter power should be increased in order to compensate the feeder loss.

This section describes the RP3-01 protocol stack and the corresponding parameters, presents the synchronization procedure. Then a BS split architecture is proposed and the implementation solutions for the most important interface layers components are provided, together with the implementation results obtained when a XC4VFX60 FPGA is targeted.

4.2 RP3-01 protocol stack

The RP3-01 interface is a high spe4ed serial interface for both up link and down link data and control transfer. The protocol stack is based on a packet concept using a layered protocol with fixed length messages.

A. Physical Layer

The transmitter Physical Layer is responsible for the 8b10b encoding, which provides a mechanism for clock recovery, and data serialization. At receiver, the mirrored functions are applied. The Physical Layer can be implemented by a dedicated device, such as an XGMII transceiver [11], or by an internal FPGA component, such as RocketIO transceiver from Xilinx Virtex 5 family [12], when a hardware implementation is considered. The supported rates are 768 Mbps, 1536 Mbps, 3072 Mbps and 6144Mbps. The 6144 Mbps rate does not concern this study.

B. Data Link Layer

The Data Link Layer contains the frame builder and the link synchronization unit. The frame builder receives from superior layers the data and control messages and generates, according to the transmitter rate, the RP3 frame. The data or control message has a fixed 19 bytes length. The message format contains 4 fields. The first one is the Address field on 13 bits, the second one is the Type field on 5 bits, the third one is the Time Stamp field on 6 bits and the last one is the Payload field on 128 bits.

The duration of RP3-01 Master Frame (MF) is fixed to 10 ms. This length corresponds to i x N_MG Master Groups (MG), where i is selected accordingly to the transfer rate, i.e. 1 for 768 Mbps, 2 for 1536 Mbps and 4 for 3072 Mbps. A MG consists of M_MG messages and K_MG idle bytes (special characters). Figure 4.1 presents an example of MF corresponding to 802.16 standard [13], when 768 Mbps rate is used. Also one can observe the values of the specified parameters.

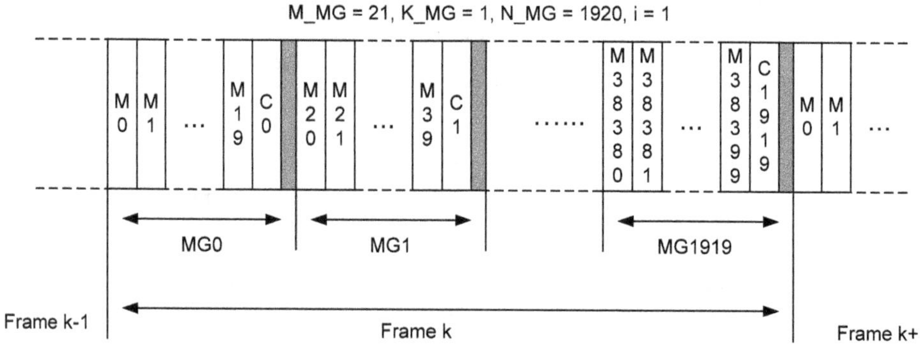

Fig. 4.1 MF for 802.16 air interface standard for the 768Mbps line rate

For the example described in Figure 4.1, the transfer rate can be computed as follows: first we compute the maximum number of bytes per frame, and then, having in mind the 8b10b line encoder and the MF duration, we calculate the transfer rate, as in (4.1). In order to generate the MF, the frame builder uses two counters: for data messages and for control messages.

$$Rate = i \cdot N\_MG(M\_MG \cdot 19 + K\_MG \cdot 1) \cdot 1e3 \tag{4.1}$$

The link synchronization unit contains transmit and receive Finite State Machine (FSM). Both the transmitter and receiver FSMs contribute to physical and logical link synchronization. The physical synchronization is based on special characters, K28.5 and K28.7, which mark the end of message groups and MF, while the logical synchronization is based on fixed MF structure.

C. Transport Layer

The Transport Layer is responsible for the end-to-end delivery of the messages, which could be simply routing of messages. The routing is based on the first 13 bits of the message, which represent the address field. The first 8 bits of the address represent the Node address and the other 5 bits represent the Sub-node address. These fields are used in a hierarchical addressing scheme where the first field identifies the bus nod and the second one selects the corresponding module from the device. The message routing is made based on a Routing Table which indicates the correspondence between the used addresses and the node ports. The table content is defined by the initial configuration procedure of the interface.

Another block of the Transport Layer is the Message Multiplexer/ Demultiplexer. It performs time interleaving/ deinterleaving of messages from N RP3 input links into one RP3 output link. Several multiplexing/ demultiplexing tables are defined as functions of the number of input links and their corresponding rates and the rate of the output link.

D. Application Layer

The Application Layer builds the data messages. It maps data and control information into the message payload and attaches the message header. The payload is represented by concatenation of signal samples in the baseband. For the case of the 802.16 air interface, the format of a data message payload field is shown in Figure 4.2.

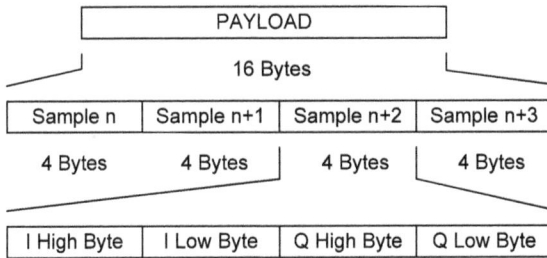

PAYLOAD
16 Bytes

Sample n	Sample n+1	Sample n+2	Sample n+3
4 Bytes	4 Bytes	4 Bytes	4 Bytes

I High Byte	I Low Byte	Q High Byte	Q Low Byte

Fig. 4.2 802.16 payload mapping

All transfers over the bus are performed over paths. A path represents a connection between a source node and a destination node. The connection is made by a set of bus links defined by the routing tables. Physically, a path consists of a set of message slots per MF. Paths are defined before bus initialization and they remain fixed during operation, i.e. the transfer between two nodes is made on the same bus links using the same message slots. For each path, a message transmission rule is applied. There are two types of rules: mandatory low level rules, using modulo computation over message slot counters and optionally high level rules, using the Bit Map (BM) concept. In the second case each MG is formed as groups of messages from baseband channels. Each group may contain Ethernet frames as well.

4.3 RP3-01 synchronization procedure

In frame based communications systems, appropriate frequency/ time synchronization is critically important. In order to avoid inter-cell interference, all base stations must use the same frequency/ time reference. The transmission over the radio channel, both on downlink and uplink, should be synchronized with the same reference. The RP1 synchronization burst generator, located at Clock Control Manager (CCM) level, shall provide frame timing and time stamping for each of the air interface standard independently, based on that reference. The PR1 Frame Clock Bursts (FCB) is mapped into RP3 messages.

The synchronization algorithm uses several inputs and is realized based on information collected both at BBM and RRU. The first one is the propagation delay (PD) between BBM and RRU. This time interval is measured on BBM side using special message transmission called Round Trip Time (RTT) measurement message. The second input is the receiving time of the RP1 FCB. This interval is measured from the beginning of the last MF until the last bit of the FCB using a counter called *C1*. After measuring this time interval, the Frame

Clock Synchronization (FCS) message is generated. FCS contains information from FCB and the $C1$ value. The last input is the detection time of a FCS inside a MF. This time interval is measured by RRU using a counter called $C2$. Having all this information, RRU computes the buffering time (Brru). Figure 4.3 describes such an example. The time intervals are not at real scale. They are expressed in Time Units (TU). One can observe that at RRU side, the end of recovered FCB corresponds to the beginning of RF(k+4).

Using the formulas from [2], we obtain the Bruu as in (4.2), where k equals to 2 from some computing conditions.

$$Brru = k \cdot RFd + (C1 - PD) - FCBd = 11\,\mathbf{TU} \tag{4.2}$$

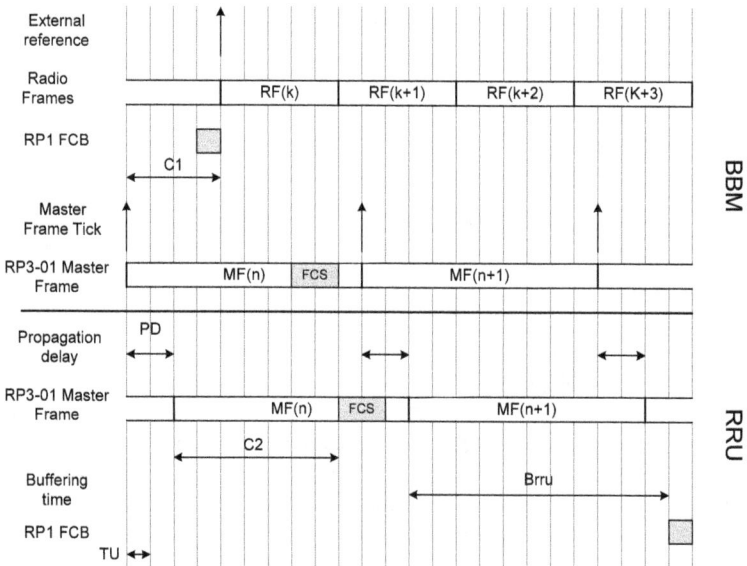

Fig. 4.3 Timing principle in RP1 frame clock burst transfer

4.4 Proposed implementation scheme

The proposed BS split architecture is presented in Figure 4.4. A BBM is connected to the two RRUs in order to have multiple transmit/ receive antennas for MIMO capabilities. The connection between the two RRUs is realized using a chain topology. In order to obtain a single point failure redundancy scheme, a second BBM connected to the two RRUs is required. Only one BBM will be active at the time. Figure 4.4 depicts also the OBSAI RP3-01 Interfaces required for blocks interconnection. Note that on OBSAI RP3-01 interface of each RRU the same Transport and Application Layers serves the both Physical and Data Link Layers.

The RP3-01 connections between each BBM and RRU or between RRUs are bidirectional. On downlink (DL) direction (from BBM to RRU), the data stream from a BBM can contain multiple data streams interleaved/ multiplexed for the two Transceivers (e. g. in order to provide Space Time Coding or MIMO) or can contain data streams only for one Transceiver.

By selecting the right node/ sub-node address, the Application Layer from BBM OBSAI RP3-01 Interface selects the desired RRU. The Transport Layer from RRU1 OBSAI RP3-01 Interface directs the data streams to its own Application Layer when RRU1's address is used, otherwise forwards the data streams to the second Data Link Layer from the OBSAI RP3-01 Interface. In uplink (UL) direction (from RRU to BBM) the procedure is similar. Both receivers can be used (e. g. receive diversity or collaborative MIMO) or only one receiver can be active. In addition to these data and control streams that should be treated by the OBSAI RP3-01 Interface as RP3 streams, an Ethernet stream will be also transmitted between BBM and RRUs in order to connect the corresponding Control & Management (CM) modules to RRU. This stream should be treated by the OBSAI RP3-01 Interface as RP1 stream.

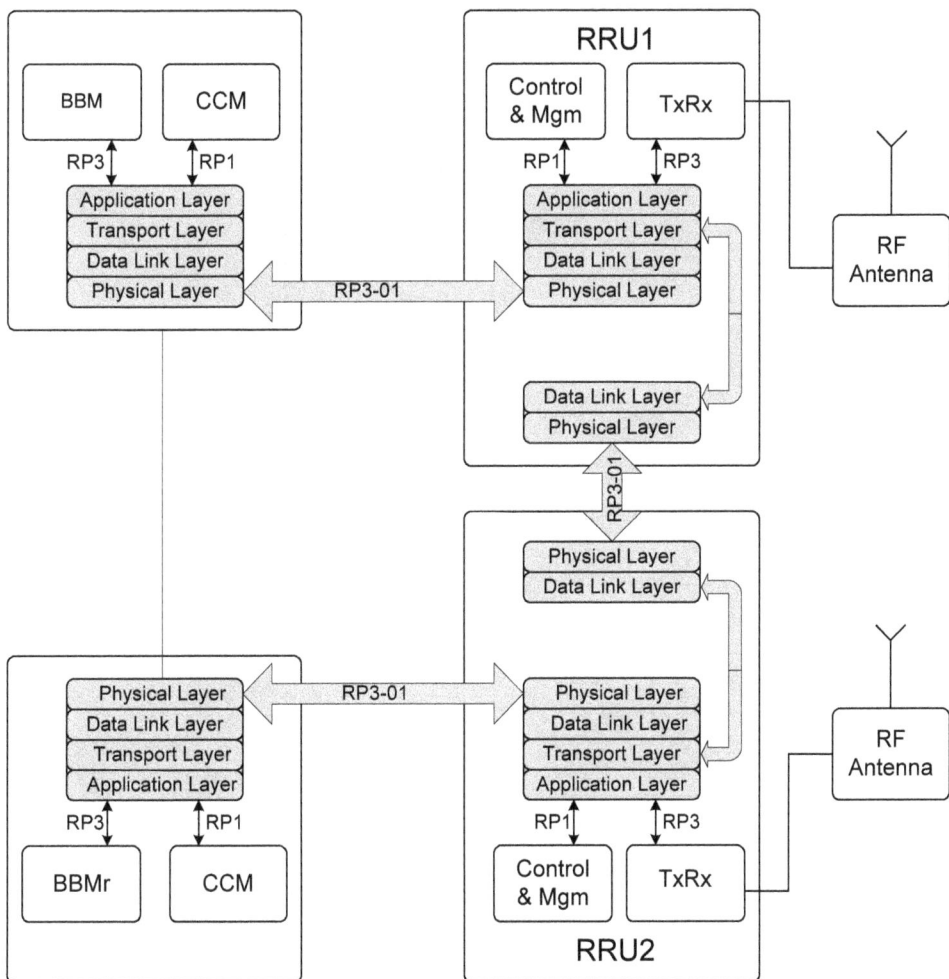

Fig. 4.4 Proposed BS split architecture with interface protocol stack

Figure 4.5 presents the generation of RP3-01 stream at BBM side, while Figure 4.6 corresponds for RRU1 side. Each RP3-01 node has its own address known from the initial configuration procedure. The DL node addresses is represented with normal fonts and with bold italic fonts the UL node addresses. For the proposed system architecture, in DL direction the RP3-01 link is used to connect a source node with two destination nodes, so two paths exist over DL connection. Even if only one RRU transmitter is used at the time, the two paths will exist and the Transport Layer from BBM OBSAI RP3-01 Interface will place its messages over the message slots corresponding to selected RRU (consider twice the rate on link between BBM and RRU1 comparing with the one between RRU1 and RRU2). In the downlink direction, a point-to-point message transfer is applied, while in uplink direction, the same message is multicast to the two BBMs. Only the active BBM uses the received information.

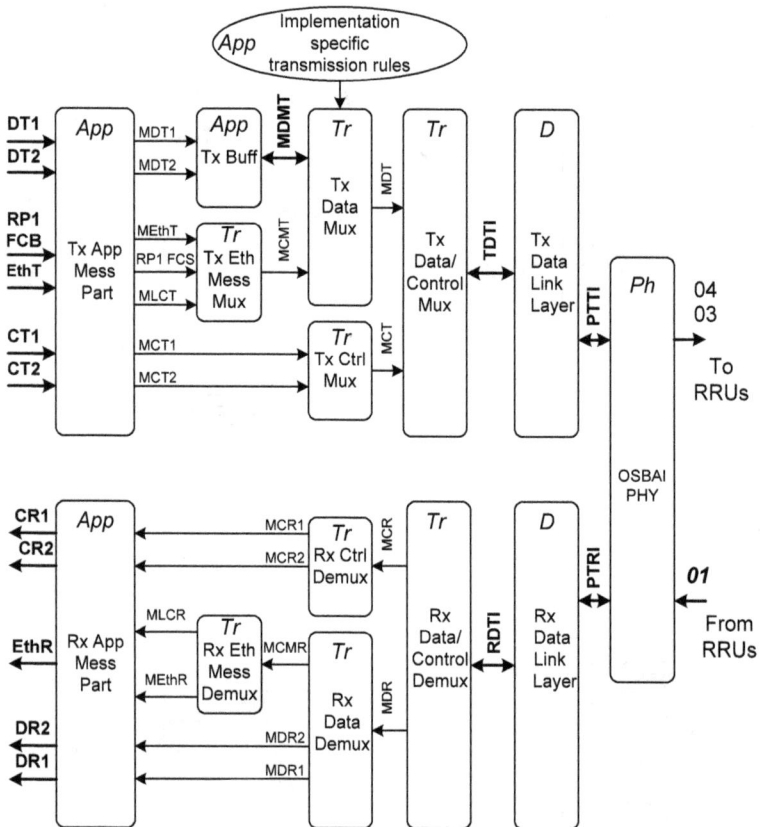

Fig. 4.5 Block scheme for BBM RP3-01 interface

Each path will be considered having a data path and a control path. Bus manager will provide separate message transmission rules for the paths utilizing data and control message slots. We will explain the notations used in Figure 4.5 and Figure 4.6 and we will describe the steps made for RP3-01 interface generation.

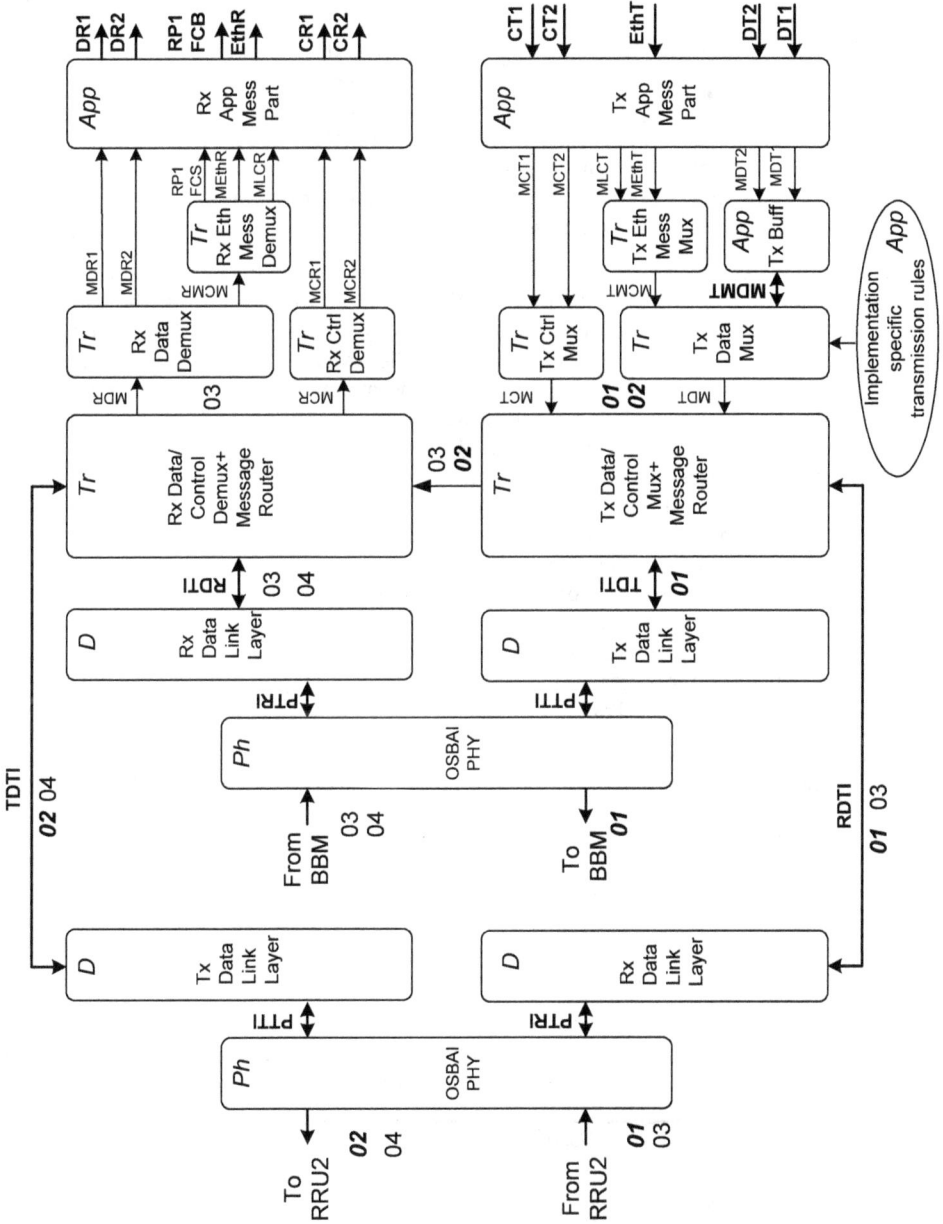

Fig. 4.6 Block scheme for RRU1 RP3-01 interface

In DL direction, on BBM side, the Application Layer receives two data streams for the two RRUs, DT1 and DT2 and two control streams, CT1 and CT2. Also, an Ethernet stream for management called EthT and the RP1 FCBs are received. The Application Layer generates the corresponding messages steams, i.e. MDT1, MDT2, MCT1, MCT2, MEthT and RP1 FCS. Also specific RP3-01 link control messages are generated. This stream is called MLCT. Beside the message generator function, Application Layer is responsible for buffering the data paths. A buffer is required for each 802.16 signal (antenna-carrier) in order to compensate the jitter caused by message transmission. Finally, the Application Layer has to provide to Transport Layer the implementation specific message transmission rules. These rules could include the lower layer message transmission rules and/ or extra rules for mapping the RP1 traffic to RP3 data message slots.

The Transport Layer has four blocks with interleaving/ multiplexing function. First, the management messages, including RP1 FCS, MLCT and MEthT are put on the same flow called MCMT, based on a priority list. The two control messages streams MCT1 and MCT2 are multiplexed in the MCT flow. The data flow, called MDT is obtained by interleaving/ multiplexing the data messages from buffers and the management messages from MCMT stream. Finally, the Transport Layer multiplexes the MDT and the MCT streams based on the TDTI interface with the Transport Layer. Using this interface, the frame builder from Transport Layer requires data or control messages and increments the corresponding counters for each successful transfer. The generated RP3-01 frame, including also the special characters, is transferred to Physical Layer on PTTI interface. The DL continues on RRU side with the receiving chain form Figure 4.6. The Transport Layer works out the RP3-01 flow, first on data stream MDR and control stream MCR, and then the data path is split into data streams MDR1 and MDR2, respectively management messages RP1 FCS, MEthR and MLCR, while the control path is split into MCR1 stream and MCR2 stream. The Application Layer receives all these messages and extracts the corresponding payload. One interesting observation is that at RRU, the Transport Layer has also the message router function, as one can see from Figure 4.6.

In UL direction the procedure follows the same steps as the one described for DL.

For implementation we considered a XC4VFX60 device from Xilinx Virtex 5 family. The functional tests were made using ModelSIM 6.2g and the synthesis results were obtained using Xilinx ISE 9.2i. From the proposed architecture depicted in Figure 4.4 one can see that the RRU Interface contains two Data Link and two Physical Layers. The implementation results obtained for the Data Link Layer are critical for the global resources, while the Physical Layer implementation cost reflects in the number of used RocketIO Transceivers.

For this reason we start our implementation with the Data Link Layer. Figure 4.7 depicts the main blocks of Data Link Layer on the transmit chain, respective on the receive chain of the interface. The area and speed reports are presented in Table 4.1.

Component	No. of slices (from 25280)	Speed (MHz)
FSM Sync Tx	8	407
Framer Tx	197	194
FSM Sync Rx	8	407
Framer Rx	274	185

Table 4.1 Area and speed reports

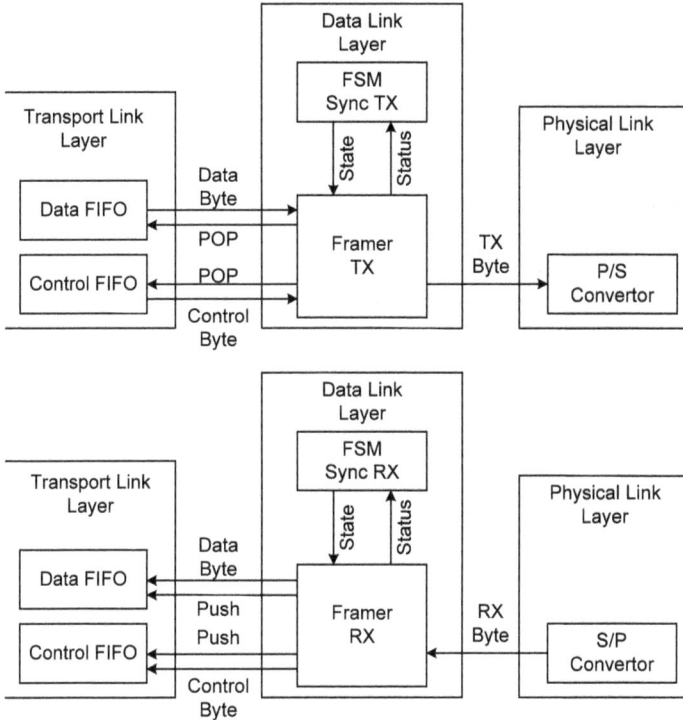

Fig. 4.7 Interface Tx and Rx chains

4.5 Conclusions

This section presented an overview of an OBSAI RP3-01 Interface implementation. It was described the main functions of the interface layers and there were presented interface block schemes for both BBM and RRU sides based on the protocol stack. Some examples were made for 802.16 air interface standard without reducing the generality of presentation. Base station split architecture was proposed, with support for redundancy and multiple transmit and receive antennas.

5. References

[1] John C. Eidson, „Measurement, control and communication using IEEE 1588", Springer 2006
[2] "OBSAI Reference Point 3 Specification", version 4.1, July 2008
[3] "OBSAI Reference Point 1 Specification", version 1.0, October 2003
[4] "CPRI Specification v4.0", June 2008
[5] "PICMG MicroTCA Base Specification R1.0", July 2006
[6] "PICMG Advanced Mezzanine Card Base Specification R 2.0", November 2006
[7] "Intelligent Platform Management Interface Specification – v2.0", February 2006
[8] "Trimble Resolution T system designer reference manual", www.trimble.com

[9] L. Gasparini, O. Zadedyurina, G. Fontana, D. Macii, A. Boni, Y. Ofek, "A Digital Circuit for Jitter Reduction of GPS-disciplined 1-pps Synchronization Signals", AMUEM 2007 – International Workshop on Advanced Methods for Uncertainty Estimation in Measurement Sardagna, Trento, Italy, 16-18 July 2007

[10] Yu. S. Shmaliy, A. V. Marienko and A.V. Savchuk, "GPS-Based Optimal Kalman Estimation of Time Error, Frequency Offset and Aging", 31st Annual Precise Time and Time Interval Meeting

[11] "PMC Sierra PM8358 QuadPHY 10GX" datasheet, March 2005

[12] "Virtex 5 FPGA RocketIO GTP Transceiver. User Guide", http://www.xilinx.com/support/documentation/user_guides/ug196.pdf

[13] IEEE Standard 802.16e, 2005

Reduction of Nonlinear Distortion in Multi-Antenna WiMAX Systems

Peter Drotár, Juraj Gazda, Dušan Kocur
and Pavol Galajda
Technical University of Kosice
Slovakia

1. Introduction

Multiple-input multiple-output (MIMO) techniques in combination with orthogonal frequency-division multiplexing (OFDM) have already found its deployment in several standards for the broadband communications including WiMAX or 3GPP proposal termed as Long Term Evolution (LTE). The MIMO-OFDM allows to substantially increase the spectral efficiency, link reliability and coverage of the signal transmission. With recent advent of the hardware processing enhacements, the processing requirements of MIMO-OFDM might be accomodated in the portable units and thus, it is widely expected that this technology will dominate over the next years in the wireless communications.

Despite of its undoubted benefits, MIMO-OFDM transmission systems are also characterized by the large envelope fluctuation of the transmitted signal Drotar et al. (2010a). This requires the application of the high Input Back-off (IBO) at the nonlinear High Power Amplifier (HPA) stage that subsequently results in an inefficient use of HPA and limitation of the battery life in the user mobile stations.

In is important to note that nonlinear amplification manifests itself in the form of Bit-Error-Rate (BER) degradation at the receiver side and simultaneously, in the form of the out-of-band radiation Deumal et al. (2008). An intuitive solution to supress the out-of-band radiation and thus, occupy the area within the spectral mask of the transmission is is to deactivate subcarriers at the borders of the used MIMO-OFDM spectrum. However, this approach impairs the spectral efficiency of the transmission and may not be convenient for the high data rate applications. Therefore it is feasible to look for the additional technique that aim to reduce the out-of-band emissions and to maintain the specific spectral mask of the transmission Khan (2009).

The possible solution is to design MIMO-OFDM systems such that the signal is less sensitive to the nonlinearity impairments. Lower fluctuation of the signal envelope can be achieved by modifying the transmitted signal prior to the transmission. However, this approach requires additional hardware and signal processing at the transmitter, which is not feasible in some applications. For these applications, the receiver based compensation is of more interest.

In the following sections, we will review the details of the most favourite methods reducing envelope fluctuation, which are intended to be used in Single-Input Single-Output (SISO) OFDM and MIMO-OFDM systems. Moreover, we will introduce two novel techniques that aim to supress the effects of the nonlinearities in MIMO-OFDM. The former will significantly reduce the envelope fluctuation by using the null subcarriers occuring in the transmission and the latter will improve the BER performance of MIMO-OFDM by means of the iterative detection.

Specially, the salient advantage of employing the nonlinear detector scheme in WiMAX is that, since it is implemented at the base station, it does not increase the computational complexity of the mobile terminal, thus neither increasing the cost nor reducing the battery life. On the other hand, using the null subcarriers for the envelope fluctuation reduction does not reduce the data rate, nor the spectral efficiency of the transmission and therefore its application is also vital in WiMAX.

2. MIMO-OFDM system model

Given we have N_t transmit antennas and N_c OFDM subcarriers, at each time instant t a block of symbols is encoded to generate space-frequency codeword. The space-frequency block-code (SFBC) codeword is then given by

$$
\mathbf{X} = \begin{pmatrix} x_1^1 & x_2^1 & \cdots & x_{N_t}^1 \\ x_1^2 & x_2^2 & \cdots & x_{N_t}^2 \\ \vdots & \vdots & \ddots & \vdots \\ x_1^{N_c} & x_2^{N_c} & \cdots & x_{N_t}^{N_c} \end{pmatrix}, \tag{1}
$$

where n-th column is the data sequence for n-th transmit antenna.

Space-frequency codeword is generated by grouping subcarriers and applying space time block code only across the sub-carriers in the same group Giannakis et al. (2007); Jafarkhani (2005); Liu et al. (2002). If SFBC is designed carefully, such a grouping will not degrade the diversity gain of the proposed coding scheme. Moreover, the subcarrier grouping reduces the complexity and allows the design of code matrices per subsystem since space-frequency coding constructs $\mathbf{X}_g$ separately as in (2) instead of constructing the entire $\mathbf{X}$ as in (1).

$$
\mathbf{X} = \begin{pmatrix} \mathbf{X}_0 \\ \mathbf{X}_1 \\ \vdots \\ \mathbf{X}_{N_g-1} \end{pmatrix}, \tag{2}
$$

where N_g is number of sub-blocks equal to $N_g = N_c/N_s$ and N_s is number of the time slots required to transmit one codeword.

3. Problem formulation

The discussion in this chapter assumes the single antenna system. However, the extension to MIMO-OFDM is straightforward and will be used with advantage later in the sections.

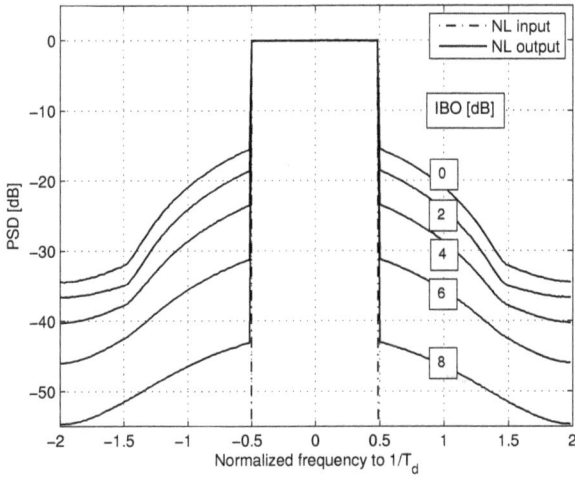

Fig. 1. Power spectral density at the output of the transmitter at various IBO.

If there is a non-constant envelope signal (e.g OFDM signal) at the input of HPA, the nonlinear amplification might result in the significant nonlinear distortion that consequently affects the system performance. The resulting effect of the nonlinear distortion can be divided into the two types: the out-of-band distortion and the in-band distortion. The in-band distortion produces inter-carrier interference increasing BER, or equivalently reducing the system capacity or operational range. The out-of-band distortion appears as the spectral regrowth, hence causing the interference in the adjacent channels.

The spectral regrowth can be easily explained by the intermodulation product introduced by the nonlinearity. Intermodulation products may potentially lay outside the transmission bandwidth, what means that some portion of energy is generated into the neighbouring channel. However, these channels are usually occupied by the adjacent user and so the operation point of HPA has to be chosen very carefully to meet the spectral mask constrains. Employing higher IBO values leads to the suppression of the out-of-band radiation, but at the cost of reduced HPA efficiency. Figure 1 shows the PSD curves for the OFDM signal employing $N_c = 256$ subcarriers and soft limiter model of HPA at various IBO levels. As can be seen from the figure, there is a significant out of band radiation at low IBO levels, but it decreases towards larger IBO. As the result, by applying larger IBO, HPA operates in the linear region of its characteristic. The spectral regrowth and out-of-band distortion is treated in more detail in e.g. Baytekin & Meyer (2005); Zhou & Raich (2004).

Next, the BER performance degradation caused by the nonlinear amplification is considered. In the following we assume that the distortion caused by the HPA can be modelled as an additive Gaussian noise (AWGN) whose variance depends on the input signal and the nonlinear HPA characteristics. Note that, even though this is the most common assumption in the literature Dardari et al. (2000); Ochiai & Imai (2001); Tellado (2000), there are some cases, e.g low number of subcarriers or low clipping levels, when this assumption is inaccurate and does not hold.

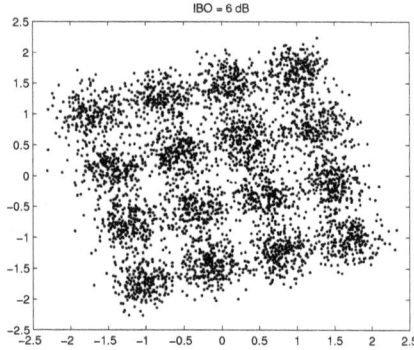

Fig. 2. The effects of nonlinear distortion on 16-QAM OFDM constellation for different IBO values

Assuming that the nonlinear distortion is additive and Gaussian, the OFDM signal at the output of the nonlinearity can be written as

$$\bar{x} = G_{HPA}x + d, \tag{3}$$

where the term $\bar{x}$ is the distortion free input signal vector. G_{HPA} is the complex scaling term that is responsible for the attenuation and rotation of the constellation. The term d is responsible for clouding of the constellation is the function of the modulated symbol vector x and the nonlinear transfer function $g(\cdot)$. Moreover, if the symbol size is large and so the number of nonzero distortion terms, the distortion term will be approximately Gaussian random variable, as was already pointed out in Tellado (2000).

The constellation of an exemplary distorted 16-QAM symbol alphabet for selected IBO values is shown in Figure 2. Figure 2 also confirms that in-band nonlinear distortion behaves as an additive Gaussian noise.

4. Review of selected PAPR reduction methods

In this section, we provide the brief overview of the most well-known PAPR reduction methods. Formerly, they were designed for conventional SISO systems, but the extension to MIMO systems is in the most cases straightforward.

4.1 Clipping of the transmitted signal

The simplest technique for reduction of the envelope fluctuation is clipping. In clipping all the samples exceeding a given threshold are forced to this maximum value. This is similar to the passing signal through soft limiter nonlinearity. The major disadvantage of clipping technique is that it introduces distortion and increase both BER and out-of-band radiation. In order to improve BER performance, the receiver needs to estimate clipping that has occurred and conversely, compensate received OFDM signal accordingly.

Authors in Kwon et al. (2009) propose the new low complexity SFBC transmitter for clipped OFDM signals, which preserves orthogonality of transmitted signals. Furthermore, clipping reconstruction method for SFBC/STBC-OFDM system based on iterative amplitude reconstruction (IAR) in Kwon & Im (2006) is presented. Another approach for improving the performance of clipped MIMO-OFDM systems with (quasy)-OSTBC is to use the statistics of the clipping distortions to develop maximum likelihood (ML) decoding Li & Xia (2008). For the case of spatially multiplexed systems, the soft correction method of Bittner et al. (2008) is applicable.

4.2 The selected mapping scheme

In the selected mapping technique, the transmitter generates several candidate data blocks from the original data block. Subsequently, the one with the lowest envelope fluctuation is transmitted. The candidate data blocks are generated as follows. First, U different phase sequences of length N are generated, $\mathbf{b}^{(u)} = (b_0^{(u)}, b_1^{(u)}, \ldots, b_{N-1}^{(u)})$, $u = 1, \ldots, U$, where normally $\mathbf{b}^1$ is set to be all-one vector of the length N in order to include also unmodified block into the set of candidate data blocks. Then candidate data blocks are generated by element-wise multiplication of the frequency-domain OFDM symbol by each $\mathbf{b}^{(u)}$, IFFT is applied and the resulting block with the lowest envelope fluctuations is selected for the transmission. The information about the selected phase sequence has to be transmitted to the receiver in the form of the side information. At the receiver, the reverse operation is performed to recover the original data block.

The straight-forward implementation of conventional SISO SLM is similar to PTS. In this scheme the SISO SLM technique is applied separatelly to each of the N_t transmitting antennas in the MIMO-OFDM system. For each of the parallel OFDM frames, the best phase modification out of the U possible ones is individually selected.

Concurrent SLM approach ensures higher reliability of side information. This is achieved through the spatial diversity by transmitting the same side information on different antennas Lee et al. (2003).

In Fischer & Hoch (2006), authors introduce *directed* SLM (*d*SLM) scheme that uses advantage of multiple antennas. The PAPR decreasing abilities of this method improve with increasing number of antennas, however this comes at the expense of higher number of side information (SI) bits.

In order to improve bandwidth efficiency number of transmitted SI bits has to be decreased. Therefore *small-overhead* SLM was proposed in Hassan et al. (2009). This scheme not only improve bandwidth efficiency but achieves also substantially better BER performance compared to *d*SLM or SISO SLM applied on multiple antennas.

The computational complexity of SLM method is relatively high therefore there is strong need for low-complexity solutions. The promising approach that require only one FFT operation was introduced in Wang & Li (2009). It exploits the time-domain signal properties of MIMO-OFDM systems to achieve a low-complexity architecture for candidate signal generation.

4.3 The partial transmit sequence technique

In the partial transmit sequence technique Han & Lee (2006); Muller & Huber (1997), the original data block of length N is partitioned into V disjoint subblocks, $\mathbf{s}_v = (s_{v,0}, s_{v,1}, \ldots, s_{v,N-1})$, $v = 1, \ldots, V$ such that $\sum_{v=1}^{V} \mathbf{s}_v = \mathbf{s}$. The subcarriers in each subblock are weighted by a phase factor, $b_v = e^{j\Phi_v}$, $v = 1, 2, \ldots, V$, for vth subblock. Such phase factors are selected in the way that the envelope fluctuation of the combined signal is minimized. The time domain signal after applying PTS can be expressed as

$$s = \sum_{v=1}^{V} b_v \cdot s_v, \tag{4}$$

where $\{b_1, b_2, ..., b_V\}$ is the selected set of phase factors.

The *straight-forward* implementation of the PTS technique for MIMO-OFDM is the independent application of PTS to each transmit antenna. It is just simple application of single antenna PTS. *Simplified approach* provides advantage over straight-forward implementation by decreasing required side information. The input data symbols are converted into the several parallel streams and the conventional PTS technique for single antenna OFDM is applied for each antenna with the sets of the phase factors being equal for all transmit antennas. Since the side information is the same for all transmit antennas, the amount of the side information per transmit antenna is reduced.

Another, *directed PTS*, approach is based on directed Selected Mapping technique Fischer & Hoch (2006). The idea of this technique is to increase number of possible alternative signal representations. In order to keep the complexity similar to the straight-forward or the simplified approach, not all possible candidates are evaluated for each transmitt antenna. The algorithm concentrates on the antenna exhibiting the highest PAPR and aims to reduce it Siegel & Fischer (2008).

In contrast to afore mentioned approaches, *Spatial shifting* provides additional way to exploit presence of multiple antenna by cyclically shifting the partial sequences between antennas Schenk et al. (2005). In other words, instead of using weighting factors for generating the different signal representations cyclic shifting of the partial sequences between the antennas is used. The advantage of this technique is its possible implementation as transparent version, where no side information needs to be transmitted.

Recently, *Siegel* and *Fischer* proposed *Spatially permuted PTS* that is more general permutation compared to cyclic shifting, described above Siegel & Fischer (2008).

Similarly as for SLM method, complexity remains significant issue also for PTS methods. The approach with reduced complexity, named *Polyphase interleaving and inversion*, for SFBC MIMO-OFDM can be find in Latinović & Bar-Ness (2006).

Finally, interesting comparison of PTS and SLM can be found in Siegel & Fischer (2008).The comparison is based on equal computational complexity of both schemes and presented analysis indicate better performance (in terms of PAPR reduction) of PTS method.

4.4 Active constellation extension

In the active constellation extension strategy, some of the outer constellation points of each OFDM block are extended toward the outside of the original constellation such that the envelope variations of OFDM signal are reduced. By doing this, some constellation points are set to be further from the decision boundaries than the nominal constellation points that slightly reduce BER.

The advantages of ACE are that it is transparent to the receiver, there is no loss of the data rate and no need for the side information. On the other hand, it increases the total transmitted power, that has to be considered in system design.

Recent work in this area includes extensions of the concept of ACE using a modified smart gradient-project (SGP) algorithm for MIMO-OFDM systems Krongold et al. (2005) and extension of the efficient ACE-SGP method to STBC, SFBC and V-BLAST OFDM systems Tsiligkaridis & Jones (2010).

4.5 Tone reservation

The basic idea of the tone reservation is to add data-block dependent time domain signal u_n to the original OFDM signal s_n with aim to reduce its peaks. In case of Tone Reservation (TR), the transmitter does not use the small subset of subcarriers that are reserved for the correcting tones. These reserved subcarriers are then stripped off at the receiver. TR similarly to the ACE technique has the slight drawback of the increase in the power of the transmission.

The extension of TR for MIMO-OFDM is straightforward, but it does not take advantage of MIMO potential. TR technique tailored for eigenbeamformed multiple antenna systems has been proposed in Zhang & Goeckel (2007), where authors introduced so-called mode reservation as analogy for TR of SISO-OFDM. Nevertheless this technique requires a perfect CSI at the transmitter.

5. Tone reservation for OFDM SFBC using null subcarriers

Now, let us assume SFBC-OFDM system with the code rate $r = 3/4$ corresponding to the selected code $\mathbf{C}_{334}$ Jafarkhani (2005) , equipped with $N_t = 3$ transmitting and N_r receiving antennas. Furthermore we assume that system employs N_c sub-carriers and M-QAM based-band modulation. The data symbol vector $\mathbf{s} = [s_0, s_1, \ldots, s_{r \cdot N_c - 1}]$ is encoded with the space-frequency encoder producing three vectors $\mathbf{x}_1, \mathbf{x}_2, \mathbf{x}_3$ as

$$\mathbf{x}_1 = [s_0, -s_1^*, s_2^*, 0, \ldots, s_{rN_c-3}, -s_{rN_c-2}^*, s_{rN_c-1}^*, 0] \tag{5}$$

$$\mathbf{x}_2 = [s_1, s_0^*, 0, s_2, \ldots, s_{rN_c-2}, s_{rN_c-3}^*, 0, s_{rN_c-1}] \tag{6}$$

$$\mathbf{x}_3 = [s_2, 0, -s_0^*, -s_1^*, \ldots, s_{rN_c-1}, 0, -s_{rN_c-3}^*, s_{rN_c-2}^*] \tag{7}$$

The vectors $\mathbf{x}_1, \mathbf{x}_2, \mathbf{x}_3$ corresponds to the columns of (1). After the mapping according to the orthogonal design on several streams associated with the transmit antennas, a simple serial to parallel converter is used for each transmit antenna, followed by IFFT processing, cyclic prefix insertion and amplification. A simplified block diagram is shown in Figure 3.

From the above discussion it is clear that due to the SFBC coding scheme, there will be uniformly distributed zero tones at the input of IFFTs. Let us define the positions of the correcting signals $\mathcal{Q}_{R,n}$ for $n = 1, \ldots, N_t$ by these zero subcarriers. The proposed method consists of adding the correcting tones at the subcarrier indices occupied by the zero symbols according to $\mathcal{Q}_{R,n}$, instead of reserving the set of the subcarriers from the data bearing tones. By doing so, we can avoid an important drawback of the tone reservation technique-bandwidth expansion.

It is clear that adding correcting signal to the SFBC encoded signals $\mathbf{x}_1, \mathbf{x}_2, \mathbf{x}_3$ may result in loss of the orthogonality, thereby eventually increasing the probability of erroneous detection. The correcting signal represents additive distortion for the decision variables in the receiver. Conversely, in order not to increase BER, the amplitude of the correcting tones must be

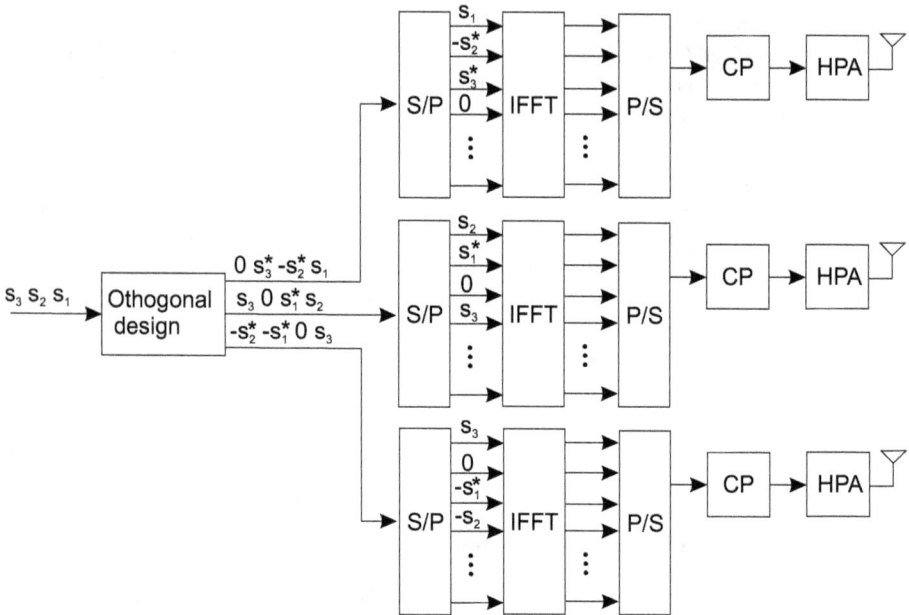

Fig. 3. Transmitter of MIMO SFBC-OFDM employing C_{334} code

controlled. Maximal amplitude that does not result in increase in BER depends on both, the baseband modulation scheme and the in-band nonlinear distortion introduced by HPA.

Figure 4 shows maximum allowed amplitude vs. IBO for various modulations. All curves fulfill the following condition: $BER_{TR} \leq BER_{conv}$ i.e. BER of TR based SFBC-OFDM system is lower or equal to that of the conventional system. This figure can be used by system designer as upper bound for the amplitude of the reserved tones in the different system setups. As it can be appreciated, these results are in compliance with our previous assumptions. We can go for higher amplitudes of peak-reduction tones and achieve large out-of-band radiation reduction without BER penalty when QPSK and 16 QAM or coded 64 QAM are adopted for the transmission. The presumptions of the amplitude constraints when uncoded 64 QAM is used are of more relevance, especially for lower IBO. In other words, when applying the uncoded higher modulation schemes (e.g. 64 QAM), the amplitude of the correcting tones is constrained to the very low power, leading to poorer performance of the proposed method performing at the low IBO. However, it should be noted that for low IBO achieved BER of the original system is very poor, characterized by the occurrence of the error floor, thus this performance is not of our interest. Because of this, designer must go for the higher IBO.

Figure 5 shows the PSD of original and TR-reduced OFDM signals when a soft limiter operating at IBOs of 4dB or 5dB is present at the output of the transmitter. In order to prevent the BER performance degradation resulting from the broken space orthogonality among transmitted signals, the maximum amplitude γ is constrained to be $\gamma = 0.2$. That corresponds to the power of reserved tones being more than 14 dB lower than the average signal power. It allows for obtaining the reduction in terms of the out-band-radiation while keeping the BER performance of the system at the same or even better level than BER of the

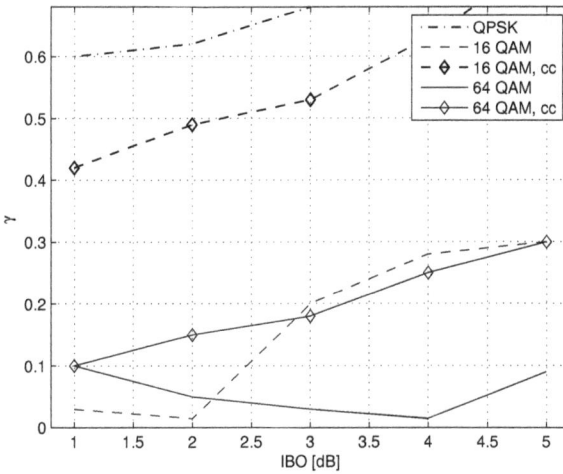

Fig. 4. Maximal normalized amplitude of reserved tones for various IBO satisfying $BER_{TR} \leq BER_{conv}$

conventional system without the application of TR. Moreover, such a value is suitable for most of the system setup implementations. It can be seen in Figure 5 that the spectrum at the center of the adjacent channel is reduced by 2.7 dB and 4.3 dB when the nonlinearity is operating at IBO = 4dB and 5dB respectively. Based on the analytical results introduced in Deumal et al. (2008) it can be stated that the amount of the out-of-band radiation is independent on the mapping scheme. Therefore by applying the proposed technique here, the same out-of-band radiation suppression can be observed for all modulation formats which make the application of the proposed technique robust in general.

6. Iterative nonlinear detection

This novel method aims to improve the system performance of SFBC OFDM based transmission system affected by the nonlinear amplification by means of the iterative decoding. It will be showed that the BER performance could be significantly improved even after the first iteration of the decoding process and thus, does not require the large computation processing. Moreover, also the second and the third iteration might be beneficial, especially in the strong nonlinear propagation environment.

Now, we would like to express the input signal of the receiver in the frequency domain. Let $\mathbf{Y}$ be the $N_c \times N_r$ matrix containing received signal after CP removal and OFDM demodulation. Similarly to the transmitter case, we can divide $\mathbf{Y}$ into N_g sub-blocks yielding $\mathbf{Y} = \left[\mathbf{Y}_0, \mathbf{Y}_1, \ldots, \mathbf{Y}_{N_g-1} \right]$. Then, the SFBC-OFDM system follows input-output relationship

$$\mathbf{Y}_g = \overline{\mathbf{X}}_g \mathbf{H}_g + \mathbf{W}_g, \tag{8}$$

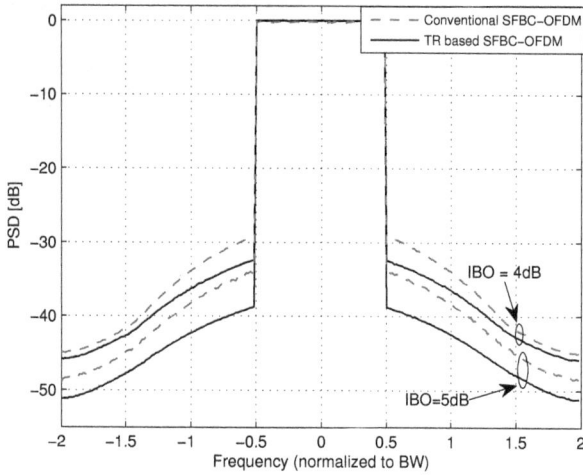

Fig. 5. PSD of a conventional and a TR-based SFBC-OFDM system obtained when a soft limiter is present. IBO={4, 5} dB.

for $g = 0, 1, \ldots, N_g - 1$. The $\mathbf{W}_g$ is $N_s \times N_r$ matrix containing noise samples with variance σ_n^2 and $\mathbf{H}_g$ is $N_t \times N_r$ matrix of path gains h_n between $n - th$ transmit and receive antenna at subcarrier frequency $g \cdot N_s$.

From (3) and (8), the signal in the frequency domain at the output of OFDM demodulator can be rewritten as

$$\mathbf{Y}_g = (\mathbf{X}_g + \mathbf{D}_g)\mathbf{H}_g + \mathbf{W}_g, \tag{9}$$

where noise term $\mathbf{D}_g$ is the frequency domain representation of nonlinear distortion. Hence, the maximum likelihood sequence detector has to find codeword $\check{\mathbf{X}}_g$ that minimises frobenius norm as

$$\tilde{\mathbf{X}}_g = \arg\min_{\forall \check{\mathbf{X}}_g} \left\| \mathbf{Y}_g - \left(\check{\mathbf{X}}_g \mathbf{H}_g + \mathbf{D}_g \mathbf{H}_g \right) \right\|_F, \tag{10}$$

where $\check{\mathbf{X}}_g$ is any possible transmitted codeword Drotár et al. (2010b). Using a full search to find the optimal codeword is computationally very demanding. However, if we assume that receiver knows NLD it can be compensated in decision variables. Since $\mathbf{D}_g$ is deterministic it does not play any role in ML detector. Orthogonal SFBC coding structure that we have considered make it possible to implement a simpler per-symbol ML decoding Giannakis et al. (2007); Tarokh et al. (1999). It can be shown Drotár et al. (2010b) that transmitted symbols to be decoded separately with small computational complexity as follows

$$\tilde{s}_{g,k} = \arg\min_{\forall \check{s}} \left\| \tilde{y}_{g,k} - d_{g,k} - \kappa \sum_{n=1}^{N_t} |h_n|^2 \check{s}_{g,k} \right\|. \tag{11}$$

Here, $\tilde{y}_{g,k}$ is $k - th$ entry of $\tilde{\mathbf{Y}}_g$ and $d_{g,k}$ is $k - th$ entry of $\mathbf{d}_g$ computed as

$$\mathbf{d}_g = \mathbf{D}_g' \mathbf{H}_g^H. \tag{12}$$

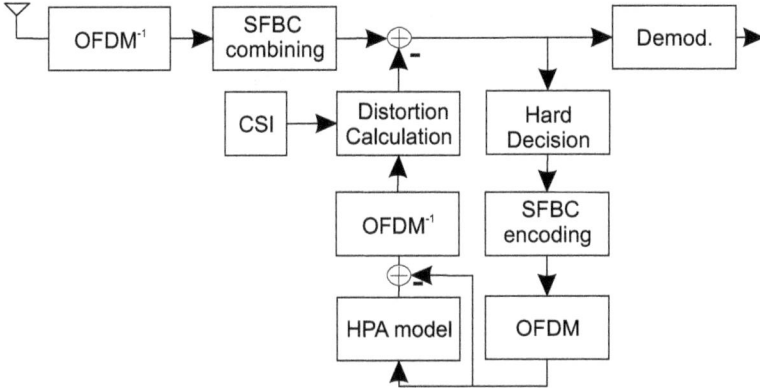

Fig. 6. Proposed SFBC-OFDM receiver structure for iterative detection of nonlinearly distorted signals

Term $\mathbf{D}'_g$ is obtained from $\mathbf{D}_g$ by conjugating second half of $\mathbf{D}_g^{(\mathrm{H})}$ entries. In practice the receiver does not know $\mathbf{D}_g^{(\mathrm{H})}$. However, if receiver knows the transmit nonlinear function, it can be estimated from the received symbol vector $\mathbf{Y}_g$.

Let us assume, that complex characteristics of HPA $g(\cdot)$ and channel frequency responses are known. Then, taking into account these assumptions, the nonlinear iterative detection procedure will consist of the following steps:

1. Compute the estimation $\tilde{s}_{g,k}^{(i)}$ of the transmitted symbol $s_{g,k}$ by the hard decisions applied to signals at the output of SFBC decoder according :

$$\tilde{s}_{g,k}^{(i)} = \left\langle \tilde{y}_{g,k} - \tilde{d}_{g,k}^{(i-1)} \right\rangle \tag{13}$$

The symbols $< \cdot >$ and i denote the hard decision operation and the iteration number, respectively. The estimated distortion terms $\tilde{d}_{g,k}^{(i)}$ are assumed to be zero for $i = 1$.

2. Compute the estimation $\tilde{\mathbf{D}}_g$ of the nonlinear distortion terms $\mathbf{D}_g$

$$\tilde{\mathbf{D}}_g = FFT\left(\tilde{\tilde{\mathbf{X}}}_g - \tilde{\mathbf{X}}_g\right)$$

where $\tilde{\mathbf{X}}_g$ is obtained by taking the IFFT of block $\tilde{\mathbf{s}}_g^{(i)} = \left[\tilde{s}_{g,0}^{(i)}, \ldots, \tilde{s}_{g,K-1}^{(i)}\right]$ after SFBC encoding and $\tilde{\tilde{\mathbf{X}}}_g = g\left(\tilde{\mathbf{X}}_g\right)$.

3. Go to step 1 and compute $\tilde{s}_{g,k}^{(i+1)}$.

The block scheme of the proposed iterative receiver is depicted in Fig. 6. The iterative process is stopped if $BER(i+1) = BER(i)$ or if the BER is acceptable from an application point of view.

Figure 7 shows the performance of the proposed method for different iterations with {16, 64}-QAM and Rapp model of HPA operating at IBO = 5 dB. We assume convolutionaly coded system. Most of the performance improvement is achieved with first and second iteration for 16-QAM and 64-QAM, respectively. When more iterations are applied, no further performance improvement is observed. Incremental gains diminish after the first for 16-QAM and second iteration for 64-QAM, respectively. This can be explained by the reasoning that some OFDM blocks are too badly distorted for the iterative process to converge and more iterations will not help.

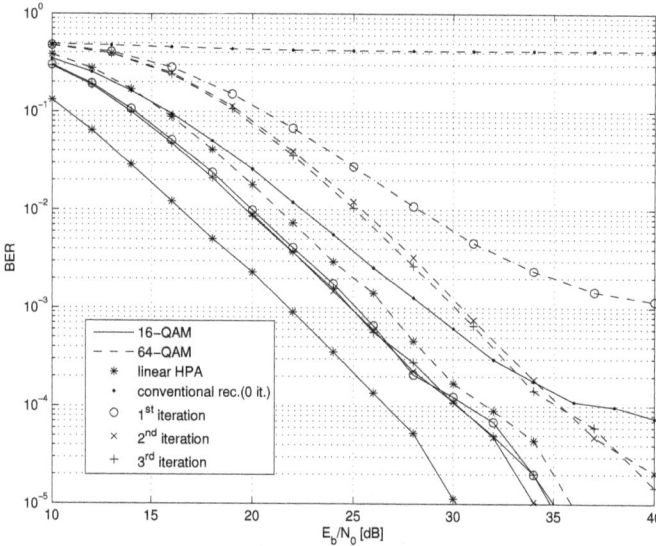

Fig. 7. BER performance of a coded SFBC-OFDM system with a Rapp nonlinearity operating at IBO=5 dB for {16, 64}-QAM and for {1, 2, 3 } of iterations. HPA characteristics is perfectly known at the receiver.

7. Extension of iterative nonlinear detection

7.1 Spatial multiplexing

In the previous section, we have assumed MIMO SFBC-OFDM systems. However, if our aim is to increase capacity of system better solution is to use Spatial Multiplexing (SM) MIMO-OFDM systems. Unfortunately, as long as the fundamental operation of SM MIMO-OFDM remains identical to conventional OFDM, the SM MIMO-OFDM transmitted signal suffers from nonlinear distortion.

It was shown that we can estimate distortion term by using received signal and characteristic of HPA. The estimated distortion term can be afterwards cancelled from the received distorted signal. When the estimation is quite accurate cancellation results in reduction of in-band nonlinear distortion. The very similar approach can be taken also for SM MIMO-OFDM systems.

The procedure of iterative detection is illustrated in Figure 8 and can be described as follows:

1. First, received signal is processed in OFDM demodulator followed by equalisation technique such as zero forcing or minimum mean square error.

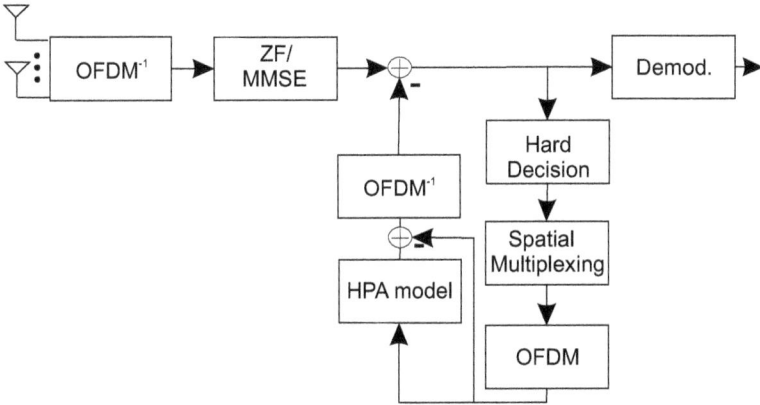

Fig. 8. Proposed receiver structure for iterative detection of nonlinearly distorted signals in SM MIMO-OFDM.

2. The estimation of transmitted symbol is computed by means of hard decision applied to symbol at the output of the detector.

3. Further, transmitter processing is modelled in order to obtain estimate of transmitted symbol that allows to compute distortion term, when HPA characteristics is known at the receiver.

4. Finally, distortion term in frequency domain is subtracted from the signal at the output of detector.

5. Whole procedure can be repeated to obtain additional improvement.

To evaluate the performance of the proposed detection, let us consider the coded SM MIMO-OFDM system with $N_c = 128$ subcarriers and 2 transmit and 2 receive antennas performing with Rapp nonlinearity. Figure 9 shows the simulation results for Rapp nonlinearity operating at IBO=4 dB using 16-QAM. The results are reported for $1, 2, 3$ iterations of proposed cancellation technique. The results of conventional receiver are also shown as a reference. It can be seen that proposed technique provides a serious performance improvement even with the first iteration.

7.2 Application to improve BER of tone reservation for SFBC OFDM using null subcarriers

As was indicated in section 5 addition of correcting signal to the SFBC encoded signals may result in loss of orthogonality, thereby eventually degrade BER performance of the system. The probability of erroneous detection is increased because correcting signal represents additive distortion - tone reservation distortion (TRD). In this section, we attempt to cancel this distortion at the receiver side of SFBC-OFDM transmission system.

Let us recall from section 5, the SFBC coded signal vectors $\mathbf{x}_n$, for $n = 1, \ldots, N_t$ to be transmitted from N_t antennas in parallel at N_c subcarriers. These signals carry zero symbols at subcarriers positions defined by $\mathcal{Q}_{R,n}$. The correcting signal in frequency domain $\mathbf{u}_n$ is added to the data signal. The position of nonzero correcting symbols in $\mathbf{u}_n$ is given by $\mathcal{Q}_{R,n}$. Therefore, the signal to be transmitted from n-th antenna can be described as

$$\mathbf{x}_n + \mathbf{u}_n. \tag{14}$$

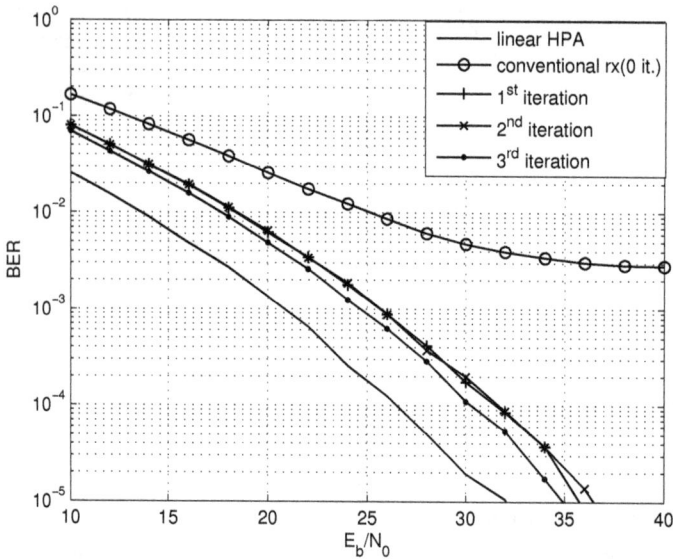

Fig. 9. BER performance of a coded SM MIMO-OFDM system with a Rapp nonlinearity operating at IBO=4 dB, 16-QAM and for {1, 2, 3 } iterations. HPA characteristic is perfectly known at the receiver.

Let us assume only one receive antenna. Then, the received signal in the frequency domain is

$$\mathbf{Y} = \sum_{n=1}^{N_t} (\mathbf{x}_n + \mathbf{u}_n + \mathbf{d}_n) \odot \mathbf{h}_n + \mathbf{w}_n. \tag{15}$$

Here $\mathbf{d}_n$ represents the in-band nonlinear distortion, $\mathbf{h}_n$ is the channel frequency response between n-th transmit and receive antenna, $\mathbf{w}$ is vector of AWGN noise samples and $\odot$ stands for element-wise multiplication. The best way how to limit the influence of TRD, represented by $\mathbf{u}_n$, on decision variable is to cancel it from received signal. However, in order to subtract TRD from received signal correcting signal has to be known. The feasible approach is to obtain the estimate of correcting signal by means of iterative estimation and then cancel it from received signal. The background and details of process of iterative estimation and cancellation were treated in detail in the section 6 for the matter of nonlinear distortion. Now, we will apply the same concept in the straight-forward manner for TRD.

Similarly to Figure 4, in Figure 10 we show the maximal available amplitudes of correcting signal, that can be used in conjunction with TRD cancellation technique. As it can be seen from Figure 10 the combination of TRD cancellation and convolutional coding for 64-QAM leads to higher affordable amplitudes in comparison with only coding application. Moreover, the combination of these approaches makes it possible to use TR technique with no spectral broadening also for 256-QAM modulation.

Finally, we present performance results for uncoded SFBC-OFDM employing three transmit antennas and $\mathbf{C}_{334}$ code. Rapp model of the HPA operating at IBO=5 dB is assumed. In this

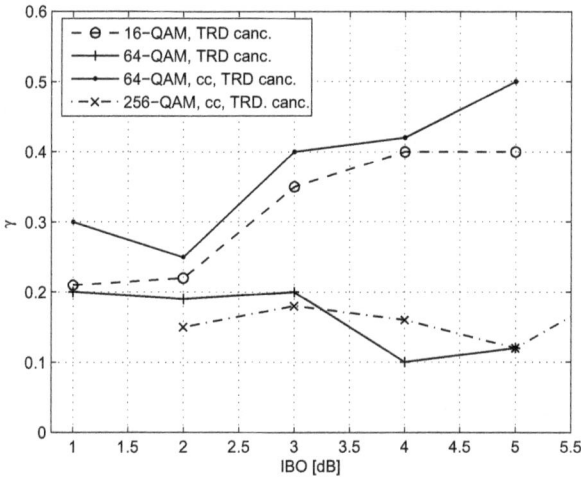

Fig. 10. Maximal normalized amplitude of reserved tones for various IBO satisfying $BER_{TR} \leq BER_{conv}$, TRD cancellation technique applied at the receiver

case, the both techniques for reduction of nonlinear distortion introduced in this thesis i.e. tone reservation with no spectral broadening and the iterative receiver technique are applied. BER curves for assumed scenario are depicted in Figure 11. As reported results indicate the best BER performance is achieved when the iterative receiver for estimation and cancellation

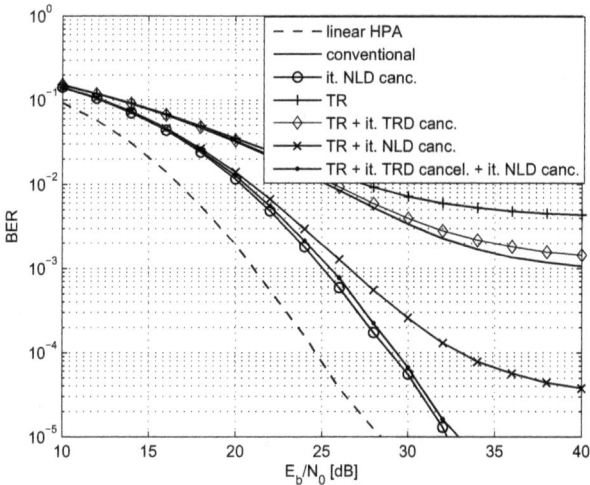

Fig. 11. BER vs. E_b/N_0 for uncoded SFBC-OFDM employing three transmit antennas and C_{334} code. Rapp model of HPA operating at IBO=5. HPA characteristics is perfectly known at the receiver.

of NLD (it. NLD canc.) is used. This is illustrated by a curve with circle marker. However, applying only the receiver technique does not bring any reduction in out-of-band radiation at the transmitter side. Therefore, TR with no spectral broadening was applied at the transmitter. Amplitude of correcting tones was constraint to $\gamma = 0.2$, but this results in increased BER for the Rapp nonlinearity operating at IBO=5 dB. Increase in BER is noticeable for TR with no spectral broadening when compared to the conventional system and also for application of TR together with iterative NLD cancellation compared to iterative NLD cancellation without TR. Fortunately, this can be solved by application of the receiver cancellation of TRD. Then, the dotted marker BER curve represents results for the application of both the transmitter and the receiver based methods. As can be seen from the figure significant BER performance reduction is obtained, moreover out-of-band radiation reduction is also achieved.

8. Conclusion

This chapter deals with the nonlinear impairments occuring in OFDM MIMO transmission. We present the brief overview of several PAPR reduction methods. The major contribution of this chapter is the introduction of two strategies, capable of mitigating the nonlinear impairments occuring in MIMO OFDM based transmission system. The fundamental idea of the former one is to use the null subcarriers for the reduction of the out-of-band radiation. The latter method, employed in the detector, improves significantly the BER performance of the MIMO-OFDM system degradaded by HPA nonlinearities. Finally, we present their joint impact on overall performance of MIMO-OFDM sytem operating over nonlinear channel. We show that the application of these methods is specially vital in the broadcast cellular standards, such as WiMAX, and therefore we believe that this contribution might be of interest to the readers and researchers working in this area.

9. Acknowledgments

Work was supported by VEGA Advanced Signal Processing Techniques for Reconfigurable Wireless Sensor Networks, VEGA 1/0045/10, 2010 Ű 2011.

10. References

Baytekin, B. & Meyer, R. G. (2005). Analysis and simulation of spectral regrowth in radio frequency power amplifiers, *IEEE J. Solid-State Circuits* 40: 370–381.

Bittner, S., Zillmann, P. & Fettweis, G. (2008). Equalisation of MIMO-OFDM signals affected by phase noise and clipping and filtering, *Proc. IEEE Int. Conf on Communications*, Beijing, China, pp. 609–614.

Dardari, D., Tralli, V. & Vaccari., A. (2000). A theoretical characterization of nonlinear distortion effects in OFDM systems, *IEEE Trans. Commun.* 48: 1755–1764.

Deumal, M., Behravan, A., Eriksson, T. & Pijoan, J. (2008). Evaluation of performance capabilities of PAPR reducing methods, *Wireless Personal Communications* 47(1): 137–147.

Drotar, P., Gazda, J. & et. al. (2010a). Receiver based compensation of nonlinear distortion in MIMO-OFDM, *Proc IEEE Int. Microwave Workshop Series on RF Front-ends for Software Defined and Cognitive Radio Solutions*, Aveiro, Portugal, pp. 53–57.

Drotár, P., Gazda, J. & et. al. (2010b). Receiver technique for iterative estimation and cancellation of nonlinear distortion in MIMO SFBC-OFDM systems, *IEEE Trans. Consum. Electron.* 56: 10–16.

Fischer, R. F. & Hoch, M. (2006). Directed selected mapping for peak-to-average power ratio reduction in MIMO OFDM, *IEE Electronics Lett.* 42: 1289–1290.

Giannakis, G. B., Liu, Z., Ma, X. & Zhou, S. (2007). *Space-time coding for broadband wireless communications*, John Wiley & Sons, Hoboken, USA.

Han, S. H. & Lee, J. H. (2006). PAPR reduction of OFDM signals using a reduced complexity PTS technique, *IEEE Signal Process. Lett.* 11: 887–890.

Hassan, E., El-Khamy, S., Dessouky, M., El-Dolil, S. & El-Samie, F. A. (2009). Peak-to-average power ratio reduction in spaceÜtime block coded multi-input multi-output orthogonal frequency division multiplexing systems using a small overhead selective mapping scheme, *IET Communications* 3: 1667–1674.

Jafarkhani, H. (2005). *Space - Time Coding: Theory and Practice*, Cambridge University Press, New York, USA.

Khan, F. (2009). *LTE for 4G mobile broadband*, Cambridge University Press, Cambridge.

Krongold, B. S., Woo, G. R. & et. al. (2005). Fast active constellation extension for MIMO-OFDM PAR reduction, *Proc. IEEE Int. Conference on Communications*, pp. 1476 – 1479.

Kwon, U. I., Kim, K. & Im, G.-H. (2009). Amplitude clipping and iterative reconstruction of MIMO-OFDM signals with optimum equalization, *IEEE Trans. Wireless Commun.* 8(1): 268–277.

Kwon, U. K. & Im, G. H. (2006). Iterative amplitude reconstruction of clipped OFDM signals with optimum equalization, *IEE Electronics Lett.* 42: 1189–1190.

Latinović, Z. & Bar-Ness, Y. (2006). SFBC MIMO-OFDM peak-to-average power ratio reduction by polyphase interleaving and inversion, *IEEE Commun. letters* 10: 266–268.

Lee, Y.-L., You, Y.-H. & et.al. (2003). Peak-to-average power ratio in MIMO-OFDM systems using selective mapping, *IEEE Commun. letters* 7: 575–577.

Li, Z. & Xia, X.-G. (2008). Single-symbol ML decoding for orthogonal and quasi-orthogonal stbc in clipped MIMO-OFDM systems using a clipping noise model with gaussian approximation, *IEEE Trans. Commun.* 56: 1127–1137.

Liu, Z., Xin, Y. & Giannakis, G. B. (2002). Space-time-frequency coded OFDM over frequency-selective fading channels, *IEEE Trans. on Signal Processing* 50: 2465–2476.

Muller, S. H. & Huber, J. B. (1997). OFDM with reduced peak-to-average power ratio by optimum combination of partial transmit sequences, *IEE Electronics Lett.* 33: 36–39.

Ochiai, H. & Imai, H. (2001). On the distribution of the peak-to-average power ratio in OFDM signals, *IEEE Trans. Commun.* 49: 282–289.

Schenk, T., Smulders, P. & Fledderus, E. (2005). Peak-to-average power reduction in space division multiplexing based OFDM systems through spatial shifting, *Electronic letters* 41: 860–861.

Siegel, C. & Fischer, R. F. H. (2008). Partial transmit sequences for peak-to-average power ratio reduction in multiantenna OFDM, *EURASIP Journal on Wireless Communications and Networking* 2008: 1–11.

Tarokh, V., Jafarkhani, H. & Calderbank, A. R. (1999). Space - time block codes from orthogonal designs, *IEEE Trans. Inf. Theory* 45: 1456–1467.

Tellado, J. (2000). *Peak-to-average Power Reduction for Multicarrier modulation*, PhD thesis, Stanford Univeristy, Stanford.

Tsiligkaridis, T. & Jones, D. L. (2010). PAPR reduction performance by active constellation extension for diversity MIMO-OFDM systems, *Journal of Electrical and Computer Engineering* 2010: 5–9.

Wang, S.-H. & Li, C.-P. (2009). A low-complexity PAPR reduction scheme for SFBC MIMO-OFDM systems, *IEEE Signal Proc. letters* 16: 941–945.

Zhang, H. & Goeckel, D. L. (2007). Peak power reduction in closed-loop MIMO-OFDM systems via mode reservation, *IEEE Commun. Lett.* 11: 583–586.

Zhou, G. T. & Raich, R. (2004). Spectral analysis of polynomial nonlinearity with applications to rf power amplifiers, *EURASIP Journal on Applied Signal Processing* 12: 1831–1840.

Part 2

Physical Layer Models and Performance

7

On Efficiency of ARQ and HARQ Entities Interaction in WiMAX Networks

Zdenek Becvar and Pavel Mach
Czech Technical University in Prague, Faculty of Electrical Engineering,
Czech Republic

1. Introduction

During data exchange in wireless networks, an error in transmission can occur. Corrupted data cannot be further processed without a correction. A technique based on either Automatic Repeat reQuest (ARQ) or Forward Error Correction (FEC) is conventionally used to repair erroneous data in wireless networks. The ARQ is backward mechanism that uses a feedback channel for the confirmation of error-free data delivery or to request a retransmission of corrupted data. This method can increase a network throughput if radio channel conditions are getting worse (Sambale et al., 2008). On the other hand, the ARQ method increases the delay of packets due to the retransmission of former unsuccessfully received packets. The FEC can increase user's data throughput over the channel with poor quality despite the fact that additional redundant bits are coded together with users' data at the transmitter side. The method combining both above mentioned methods is called Hybrid ARQ (HARQ). All three error correction mechanisms are implemented on physical and/or Medium Access Control (MAC) layer.

The performance of ARQ defined in standard IEEE 802.16e (IEEE802.16e, 2006) depends on the setting of several parameters such as size of user data carried in a frame, size of ARQ block, size of PDU (Protocol Data Unit), limit of retransmission timeout timer, or type of packet acknowledgement (Lee & Choi, 2008). Evaluation of the type of packet acknowledgment for different channel condition is presented in (Kang & Jang, 2008). In other paper, the authors evaluate the ARQ performance for different ARQ parameters (Tykhomyrov et al., 2007). This work is later on enhanced by analysis of the impact of PDU size on IEEE 802.16e networks performance while ARQ mechanism is used (Martikainen et al., 2008). Further, a comparison of ARQ and HARQ performance in IEEE 802.16 networks is presented in (Sayenko et al., 2008). This paper also compares the amount of overhead generated by ARQ and HARQ. The optimal PDU size and MAC overhead due to the packets retransmission is analyzed in (Hoymann, 2005). The authors in (Sengupta et al., 2005) propose to adjust the MAC PDU size depending on the channel state to achieve the best ARQ performance. The paper is extended for analysis of a combination of error correction techniques such as ARQ, FEC or MAC PDU aggregation on the VoIP speech quality (Sengupta et al., 2008). Authors proof the improvement of VoIP speech quality by using these techniques. In (Chen & De Marca, 2008), the authors investigate an optimization of ARQ parameter setting from the link throughput point of view.

In conventional WiMAX network, the ARQ and HARQ work independently on each other (IEEE802.16e, 2006). In standalone ARQ process, a number of blocks received with errors increases as the link quality between the transmitter and the receiver decreases. Thus, if the Block Error Rate (BLER) is more significant, the amount of retransmitted blocks is higher as well. It can be assumed that if the channel quality is high, most of the blocks are transferred without errors and the number of unsuccessfully received blocks is kept to minimum. In such case, the transmission of positive acknowledgement (ACK) of correctly delivered blocks appears more often than negative acknowledgement (NACK) of corrupted blocks. This assumption is considered in (Becvar & Bestak, 2011), where authors propose to send only NACKs to significantly reduce signaling overhead introduced by ARQ mechanism.

On the other hand, the HARQ is able to detect and correct the most of the radio channel errors. However, due to the limitation of a number of retransmissions, some data may not be delivered without errors if only HARQ is utilized. Consequently, these data have to be retransmitted by ARQ process. The conventional ARQ has to acknowledge all data independently on the result of HARQ procedure. In order to significantly reduce signaling overhead, an interaction of both ARQ and HARQ methods should be utilized, see, e.g., (Maheshwari et al., 2008)).

The contributions of this chapter are as follows. Firstly, the results of improved ARQ scheme according to (Becvar & Bestak, 2011) cooperating with HARQ is compared to the results achieved by the conventional ARQ scheme with enabled and disabled cooperation between both entities. Secondly, while only one hop communication is assumed when data are sent only between a mobile station (MS) and a base station (BS) in (Becvar & Bestak, 2011), this chapter analyzes the impact of relay stations (RS), defined in IEEE 802.16j (IEEE802.16j, 2009), on the performance of individual methods. The extended simulations are performed considering various setting of parameters. The amount of generated overhead is the metric for the performance assessment.

The rest of this chapter is organized as follows. In the next section, the principle of ARQ and HARQ used in WiMAX are described. In addition, the optimization of conventional ARQ scheme according to (Becvar & Bestak, 2011) is presented in this section. In the section 3, the overhead of HARQ algorithm in WiMAX networks is evaluated. The section 4 provides an overview on simulation model and contemplates the parameters applied in simulator. The section 5 presents the simulation results. Last section gives our conclusions.

2. ARQ and HAQR in WiMAX

This section provides overview on conventional ARQ and HARQ used in WiMAX networks. Further, the innovative ARQ proposed in (Becvar & Bestak, 2011) is also described to enable easy understanding of results presented in next sections.

2.1 Conventional ARQ

The principle of conventional ARQ method according to the IEEE 802.16e standard and the structure of user's information carried in the frame are depicted in Fig. 1.

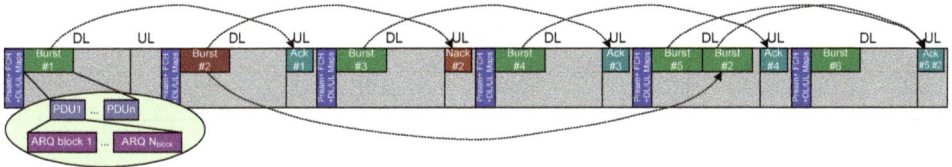

Fig. 1. Principle of conventional ARQ

In WiMAX, each data burst generated either by the MS or the BS is segmented into PDUs. These PDUs are further mapped into MAC frame. A PDU usually consists several blocks N_{block}, which number is given by following equation:

$$N^{i,k}_{blocks} = \frac{S^{i,k}_{data}}{S^{i,k}_{ARQ-blocks}}$$ (1)

where $S^{i,k}_{data}$ is a total size of data of *i-th* user in *k-th* frame, and similarly $S^{i,k}_{ARQ-blocks}$ represents a block size defined by parameter denoted in the standard as ARQ_Block_Size (IEEE802.16e, 2006). This parameter is carried in TLV (Type/Length/Value) section of registration messages (REG-REQ/RSP) exchanged between the BS and MS (see (IEEE802.16e, 2006)). The parameter ARQ_Block_Size can take values from the following range: 16, 32, 64, 128, 256, 512 and 1024 bytes. During a transmission, a sequence of consecutive blocks is sent in the PDU. After that the receiver evaluates whether the data are received correctly or not and sends an appropriate feedback message to the transmitter. Note that all transmitted blocks (N_{block}) have to be confirmed by ACK or NACK even if all blocks are received without errors. The IEEE 802.16e standard defines four types of acknowledgments: Selective ACK entry, Cumulative ACK entry, Cumulative with Selective ACK entry and Cumulative with Block Sequence ACK entry.

The first type of acknowledgment uses selective maps to provide feedback to the transmitter. In the selective map, each bit corresponds to one ARQ block. A bit set to "1" indicates error-free reception of the corresponding ARQ block. The second type, Cumulative ACK entry, is based on the utilization of sequence maps. A sequence map defines a group of consecutive blocks where each group includes a sequence of only erroneous blocks or sequence of only error free blocks. The sequence maps can contain two or three sequences with a length of 64 or 16 blocks respectively. The third type of ACK combines the previous two types. Finally, the last type combines the second type with ability to acknowledge ARQ blocks in the form of block sequences.

The ACK or NACK is sent through above mentioned feedback message. The feedback is transmitted in the next frame after the data transmission. The feedback message contains 8 bit field indicating Message ID and the rest of the message is dedicated to field consisting ARQ_Feedback_Payload. The ARQ payload can be carried either via standalone ARQ feedback message or by piggybacking the ARQ payload to the user's data block. The payload is always carried in a single PDU. The ARQ_Feedback_Payload includes one or more ARQ_Feedback_IE (see Table 1) where IE stands for an Information Element.

Syntax	Size	Notes
CID	16 bits	Connection ID
Last	1 bit	Identify the last IE in ARQ_Feedback
ACK Type	2 bits	0x0...Selective ACK 0x1...Cumulative ACK 0x2...Cumulative with Selective 0x3...Cumulative with Block Sequence
BSN	11 bits	Block Sequence Number (0...2047)
Number of ACK Map	2 bits	Number of Maps (M) = 1,2,3 or 4
Maps	M x 16 bits	Selective (16 blocks) or Cumulative maps (2 x 64 blocks / 3 x 16 blocks) Cumulative maps: 1 bit sequence format (2 or 3 blocks), 2/3bits Sequence ACK (ACK/NACK of sequence), (2x6) / (3x4) bits Sequence length

Table 1. Structure of ARQ_Feedback_IE (IEEE802.16e, 2006)

The size of an IE of each ARQ feedback message can be calculated according to equation:

$$Size_{ARQ\_FB\_IE} \ [bits] = 32 + (M \times 16) \tag{2}$$

where M represents the number of maps carried in one ARQ_Feedback_IE (see Table 1). Consequently, the overall size of whole feedback message is given by following formula:

$$Size_{ARQ\_FB} \ [bits] = 8 + \sum_{i=1}^{N_{IE}} Size_{ARQ\_FB\_IE_i} \tag{3}$$

where N_{IE} corresponds to the amount of information elements carried in one ARQ Feedback message and the first eight bits represents the ARQ feedback message overhead (i.e., Message ID field). The overhead transmitted in all considered frames (N_{frame}) is equal to the sum of partial overheads over the N_{frame}:

$$OH_{ConvARQ} \ [bits] = \sum_{i=1}^{N_{frame}} Size_{ARQ\_FB_i} \tag{4}$$

As indicated in Fig. 1, the retransmission of erroneous blocks cannot be accomplished before the third frame after the original transmission since the transmitter receives NACK in the next frame after transmission (2nd frame). Hence a request for additional resources can be created earliest at the uppcomming frame (3rd frame). Therefore, the dedicated resources are not available before the 4th frame. The retransmission of data (burst #2 in Fig. 1) can be scheduled either together with normally ordered data (burst #5 in Fig. 1) or the new data (burst #5 in Fig. 1) can be delayed by one frame. It causes a delay of retransmitted packets with duration that corresponds to at least 3 times of frame duration (e.g., if the frame duration is 10 ms, the packet delay is at least 30 ms).

The new and retransmitted data are sent within the same frame only if the requested capacity (new data plus retransmitted data) is available. The WiMAX technology implements Stop-and-Wait mechanism that requests a confirmation of the previous block before transmitting subsequent blocks. The number of blocks that can be unconfirmed before a transmission of the consequent blocks is defined in the standard by the parameter ARQ_Window_Size.

2.2 Innovative ARQ

The innovative ARQ takes into consideration that the number of blocks received with errors increases as the link quality between transmitter and receiver decreases. This scheme adaptively selects one of three different ways of data delivery confirmation: i) conventional ARQ, ii) transmission of only NACK (ARQ Scheme I), iii) retransmission of only corrupted blocks (ARQ Scheme II).

The first type of data acknowledgement, the conventional ARQ, was already explained before.

The second type (ARQ Scheme I) assumes ARQ feedback message and ARQ_Feedback_IEs with the same structure as the conventional IEEE 802.16 ARQ feedback message. However in this proposal, the ARQ feedback is sent only if a received PDU contains at least one erroneous block. If all blocks in the PDU are error free, no feedback is sent. The PDU is assumed to be correctly transferred if the transmitter receives no feedback in the following W frames after the transmission. If the feedback with NACK is not delivered, the data conveying the delay sensitive services (e.g., VoIP) are assumed to be lost since the delay caused by repeated ARQ retransmission is significant. In case of services not sensitive to delay, data belonging to lost NACK can be retransmitted using upper layer protocols, e.g., TCP (Transmission Control Protocol). As the probability of lost packet or packet with errors together with the NACK feedback is very low, the increase of overhead due to upper layer protocols is negligible.

The third way of data acknowledgement (ARQ Scheme II) is based on the same assumptions as the previous one. The ACK feedback is likewise transmitted only if there is at least one block with errors. A block is assumed to be error-free if no feedback is received in one of the following W frames after the transmission of appropriate data frame.

The overhead generated by the innovative ARQ (denoted as ARQ PIII) by a user in one frame can be calculated according to the following equation:

$$Size_{ARQ\_FB\_III} = 8 + 18 + min\left\{\sum_{N_{IE}} 16 + 16 \times M_{N_{IE}}, 10 + B \times 11\right\} + res \tag{5}$$

where N_{IE} is the number of IEs carried in one ARQ feedback message, M_{NIE} corresponds to the number of ACK maps in ARQ_Feedback_IE, B stands for the number of BSNs included in one message and *res* is the number of bits used for an alignment of the feedback message length to integer number of bytes. The overhead generated by new ARQ scheme is given by the following equitation:

$$OH_{SchemeIII} = \sum^{N_{frame}} Size_{ARQ\_FB\_III_{N_{frame}}} \qquad (6)$$

2.3 HARQ

The utilization of ARQ with support of FEC is known as HARQ. The HARQ method uses not only retransmitted packets to reconstruct the original error free packets, but it also utilizes the packets received with errors. The original packet can be reconstructed by a combination of several versions of packet with errors. The HARQ described in (IEEE802.16e, 2006) uses two different types of reconstruction: Chase Combining (CC) and Incremental Redundancy (IR).

The first version of HARQ is denoted as Type I HARQ Chase Combining. In this case, blocks of data together with a CRC code are encoded using a FEC coder before transmission. If the channel quality is low and errors of data are identified, the data block is not discarded however it is kept in the memory. In the next phase, the receiver requests for retransmission of this data block. The retransmitted block of data is then combined with the previous blocks received with errors. Combining more versions of the data blocks improves the probability of correct decoding even if all of them are received with errors.

Optionally, the IEEE802.16 standard also supports type II HARQ, which is known as Incremental Redundancy. In case of IR HARQ, the FEC coder codes one packet into several subpackets. Each of subpacket is coded with different code ratio. The subpackets are distinguished by 2-bits SubPacket IDentifier (SPID). If the packet is transmitted for the first time, the subpacket with SPID=00 is sent. The successful receive of the packet at the destination station is indicated by ACK. Otherwise, the transmitter sends a NACK and the transmitter has to send another packet carrying one of four subpackets. Both received packet (the first transmission and retransmissions) are again combined by receiver to increase the probability of correct decoding.

The overhead introduced by HARQ in WiMAX depends on the HARQ Type as follows. Firstly, the acknowledgment of HARQ bursts by modification of AI_SN (HARQ Identifier Sequence Number) of appropriate ACID (HARQ Channel ID) is assumed (for more information, see (IEEE802.16e, 2006)). The AI_SN is included in HARQ DL or UL. The size of HARQ map can be described by the subsequent formula:

$$HARQOH_{DL} = \begin{cases} 64 + SubB + res \dots Re\,gionID\_ON \\ 40 + SubB + res \dots Re\,gionID\_OFF \end{cases}$$

$$\qquad (7)$$

$$HARQOH_{UL} = \begin{cases} 48 + SubB + res \dots Re\,gionID\_ON \\ 24 + SubB + res \dots Re\,gionID\_OFF \end{cases}$$

where *SubB* is a size of management overhead according to a sub-burst. The amount of management overhead also depends on the utilization of *Region ID* (see (IEEE802.16e,

2006)). In the simulations performed in this chapter, the *Region ID* is not considered. The actual amount of bits of *SubB* depends on the HARQ Type. Based on the (IEEE802.16e, 2006), the size of message according to the sub-bursts is following:

$$SubB_{CC} = 8 + N_{sub} \times (RCID + 20 + DIUC)$$
$$SubB_{IR-CTC} = 8 + N_{sub} \times (RCID + 20)$$
$$SubB_{IR-CC} = 8 + N_{sub} \times (R + 22 + DIUC)$$

(8)

where N_{sub} is a number of sub-bursts; *RCID* represents a size of Reduced CID; and *DIUC* represents the size of optional field, denoted as DIUC, containing 8 bits if included.

For the case when a low number of bursts are transmitted within a frame, the utilization of so called Compact HARQ DL/UL maps enables to reduce an overhead (see (IEEE802.16e, 2006)). The overhead generated by compact version of maps is not dependent on the HARQ type. The amount of overhead can be expressed by the next equations:

$$HARQOHcomp_{DL} = 12 + RCID + HCI + CCI$$
$$HARQOHcomp_{UL} = 12 + RCID + HCI$$

(9)

where *HCI* is a size of HARQ control IE (8 bits if HARQ is enabled and 4 bits if HARQ is temporary disabled); *CCI* is a size of CQICH control IE (16 bits if CQICH information are included and 4 bits if the information are not included).

The simple evaluation of equations for full and compact HARQ maps enables to determine which kind of maps generates minimum management overhead over the number of HARQ sub-bursts (see Fig. 2). As the results show, the compact version of maps is profitable for all numbers of sub-bursts in UL as well as for up to 12 sub-bursts in DL over all length of Reduced CID.

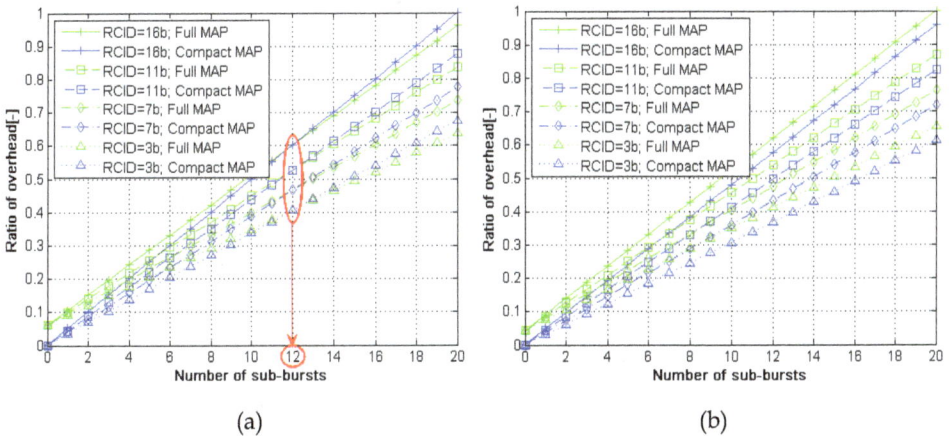

(a)   (b)

Fig. 2. Comparison of the overhead generated by compact and full HARQ maps for DL *(a)* and UL *(b)*

The relation between BLER (for ARQ confirmation) and PER (for HARQ confirmation) is defined according to (Provvedi et al., 2004) by following equation:

$$PER = 1 - (1 - BLER)^{N_{blocks}} \qquad (10)$$

2.4 Cooperation of ARQ with HARQ to reduce signaling overhead

The mutual interaction consists in exchanging of information on successful packets transmission between ARQ and HARQ entities (see Fig. 3). Therefore, the data confirmed by HARQ need not to be confirmed again by ARQ process.

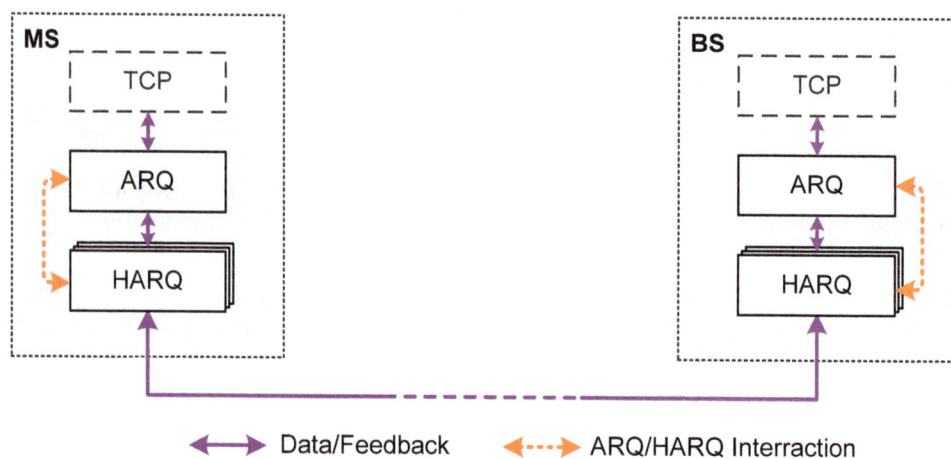

Fig. 3. Principle of ARQ and HARQ cooperation

At the side of transmitter, both ARQ and HARQ are implemented and applied on data. Similarly, the receiving side evaluates both ARQ and HARQ as well. However, ARQ process on the receiving side need not to transmit all requests related to the corrupted data if these data are already requested to be retransmitted by HARQ. The same way is applied for acknowledgement of data. In other words, data confirmed by HARQ are not further confirmed by ARQ. The information on ACK/NACK is delivered to HARQ processes at the side of original transmitter and HARQ just provides information of ACK/NACK data to the ARQ process at transmitter. Therefore, a part of overhead due to duplicated confirmation of data delivery is saved.

3. System model and simulation parameters

The simulator, developed in MATLAB, focuses on the evaluation of overhead generated by ARQ and HARQ procedure in the uplink direction by one user (see Fig. 4). In simulations, we assume direct communication between MS and BS as well as multihop communication using RSs.

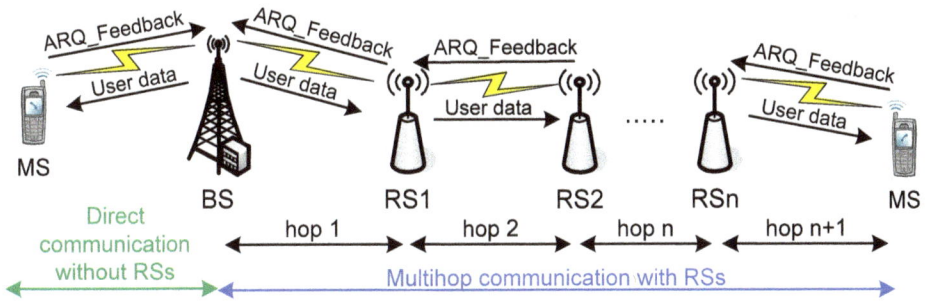

Fig. 4. Link level simulation scenario

Each packet is transmitted either directly to the BS or over particular number of hops. The probability of block error between two stations is the same over all hops. Therefore, the overall BLER of all hops (between the MS and the BS) is calculated according to the following formula:

$$BLER_{MS-BS} = \left(1 - BLER_{hop}\right)^{N_{hops}} \qquad (11)$$

where $BLER_{hop}$ represents a BLER over each hop and N_{hops} is the number of overall hops between the MS and the BS. Note that $N_{hops}=n+1$, where n is the number of RS in the communication chain.

If RSs are considered, the absolute level of transmitted overhead rises n times comparing to the direct communication without RSs. This is due to the fact that feedback information is transmitted individually over each hop.

The setting of simulation parameters is depicted in Table 3. The evaluation is performed for BLER up to 10% per one hop. For higher BLER level, the channel is nearly unusable due to high error rate. Note that BLER of overall path from the MS to the BS is significantly increasing with rising number of hops (see (11)). The BLER of whole path from the MS to the BS is 27% if three hops are taken into account and if BLER of a hop is 10%.

For more precise evaluation, the overhead of upper layer is also considered. The TCP protocol is assumed for an error correction by upper layer.

The user's data are transmitted in a number of frames transmitted from the BS to the MS. The overhead size is evaluated per all transmitted frames. A frame consists of one or several PDUs and a PDU itself contains one or several ARQ blocks. The frames are subsequently sent by the BS to the MS. A vector indicating positions of blocks with/without errors is created for each frame based on the given value of BLER. The MS responds to the BS by sending ARQ feedback message that includes selected ARQ scheme, ACK Type, and a vector of errors in the transmission. According to the feedback message, the BS retransmits erroneous blocks as soon as possible, but not before the third frame after the original transmission. The size of user's data in a DL frame is kept the same within each simulation drop (1024 bytes or 4096 bytes).

This process is repeated until all frames are sent to the MS and the MS confirms error-free reception of all blocks. The same vectors indicating positions of blocks with/without errors are considered in all ARQ schemes.

Parameter	Value
Number of frames	2000
BLER per hop [%]	0 – 10
Number of hops	1, 3
ARQ_Block_Size [bytes]	16 – 1024
PDU size [blocks]	1 – 16
ARQ ACK Types	Selective, Cumulative
Max. HARQ retransmissions	2, 4
HARQ Type	CC, IR-CTC
HARQ packet/burst size	1 PDU
RCID [bits]	7
Size of data in each DL frame [bytes]	1024

Table 2. Simulation parameters for ARQ and HARQ

The maximum number of HARQ retransmissions is set to 2 and 4. Both types of HARQ, Chase Combining (CC) and Incremental Redundancy (IR) are considered in evaluations. The Convolutional Turbo Code (CTC) is considered in evaluation if IR HARQ is performed.

4. Results

The results are separated into several groups according to the number of hops (left-hand and right-hand figures corresponds to one and three hops respectively), HARQ Type (CC HARQ in Fig. 5 - Fig. 10 and IR-CTC in Fig. 11 - Fig. 16), and maximum number of HARQ retransmissions for higher clarity. The figures are grouped into set of six figures with the same HARQ Type, with the same maximum number of retransmissions, and further, with varying number of hops, ARQ_Block_Size, and PDU Size. The results are presented in form of figures showing the overhead generated due to ACK/NACK by HARQ and ARQ for 2000 continuously transmitted frames. The expressed overhead is normalized to the overhead generated by conventional IEEE802.16e ARQ (in figures noted as *Conv. ARQ*) for error free channel, using Selective ACK (in figures marked as *SACK*) together with HARQ while no interaction between both is considered. The cumulative ACK (*CACK*) is also taken into account in figures. All figures also depict results for both techniques while interaction is not enabled (without interaction - in figures denoted *w/o int.*) and while the interaction is enabled (with interaction - in figures noted as *w int.*). The overhead for the same cases is presented also if HARQ and innovative ARQ scheme proposed in (Becvar & Bestak, 2011) (in figures *ARQ PIII*) are simultaneously utilized.

As can be observed from Fig. 5 - Fig. 16, the scenario where ARQ and HARQ interact outperforms all other scenarios. Additional minor improvement is achieved by using innovative ARQ instead of conventional ARQ. However, this improvement is noticeable only as long as ARQ_Block_Size is low (e.g., 16 bytes), PDU Size is higher (e.g., 16 blocks) and mutual interaction of ARQ and HARQ is considered.

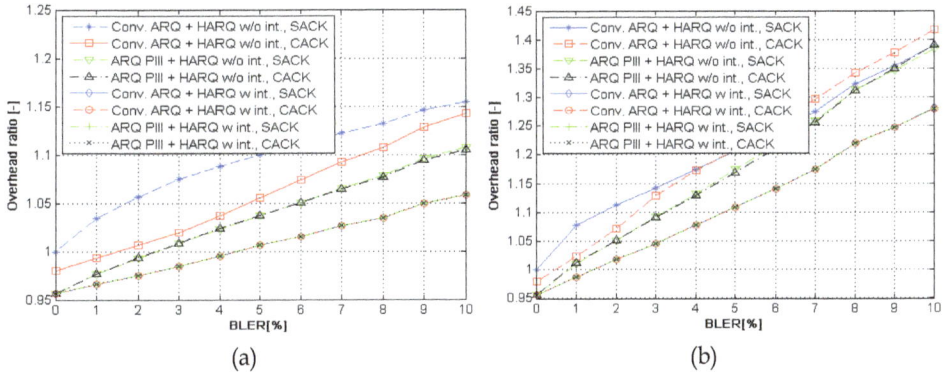

Fig. 5. ARQ & HARQ Overhead vs.BLER for ARQ_Block_Size = 16 B, PDU Size = 1, Size of user data = 1024 B/frame, 4 HARQ retrans., HARQ Type: CC, 1 hop *(a)* / 3 hops *(b)*

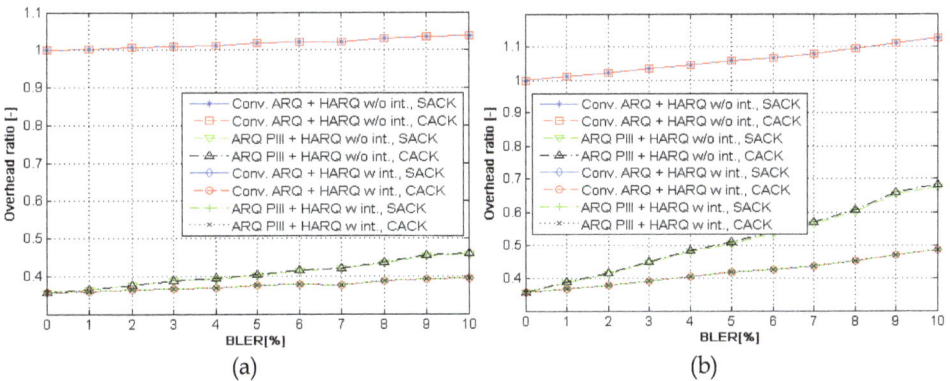

Fig. 6. ARQ & HARQ Overhead vs. BLER for ARQ_Block_Size = 1024 B, PDU Size = 1, Size of user data = 1024 B/frame, 4 HARQ retrans., HARQ Type: CC, 1 hop *(a)* / 3 hops *(b)*

While no interaction between HARQ and ARQ entities is enabled, the difference between conventional innovative ARQ is more significant. The reduction of overhead is more appreciable for lower number of hops or higher ARQ_Block_Size. The improvement achieved by innovative ARQ in comparison to scenario using conventional ARQ without interaction is due the fact that the ARQ PIII generates lower overhead while the packets are delivered without errors.

The first group of figures shows the results of CC HARQ for maximum four HARQ retransmissions.

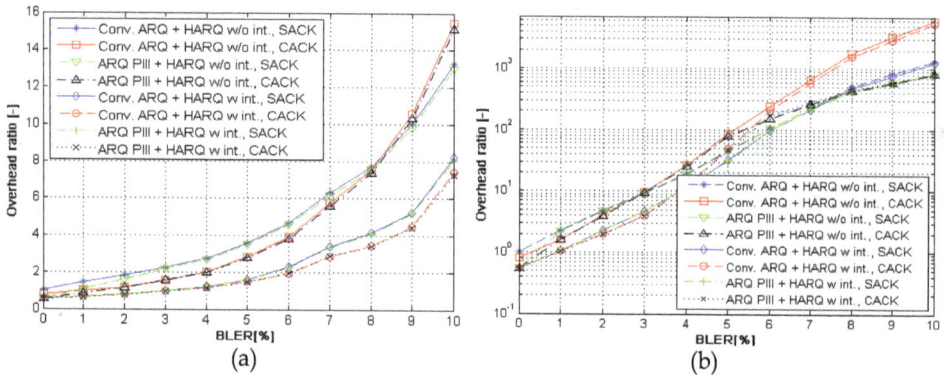

Fig. 7. ARQ & HARQ Overhead vs. BLER for ARQ_Block_Size = 16 B, PDU Size = 16, Size of user data = 1024 B/frame, 4 HARQ retrans., HARQ Type: CC, 1 hop (a) / 3 hops (b)

The next group of figures depicts the results of CC HARQ for maximum two HARQ retransmissions.

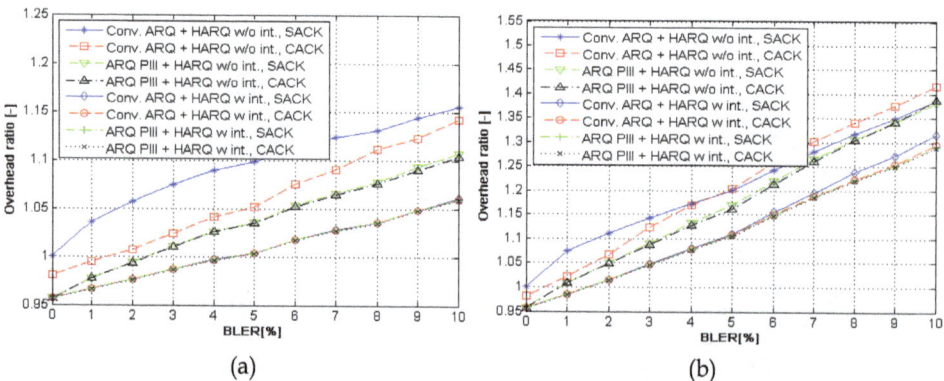

Fig. 8. ARQ & HARQ Overhead vs. BLER for ARQ_Block_Size = 16 B, PDU Size = 1, Size of user data = 1024 B/frame, 2 HARQ retrans., HARQ Type: CC, 1 hop (a) / 3 hops (b)

The following group of figures represents the results of IR_CTC HARQ for maximum four HARQ retransmissions.

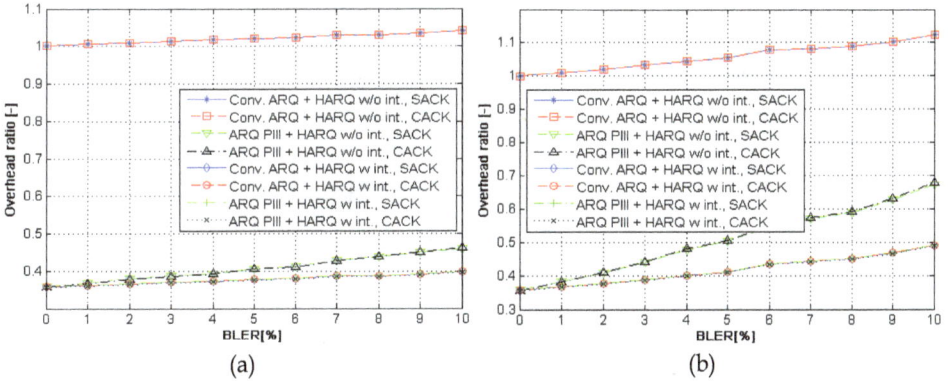

Fig. 9. ARQ & HARQ Overhead vs. BLER for ARQ_Block_Size = 1024 B, PDU Size = 1, Size of user data = 1024 B/frame, 2 HARQ retrans., HARQ Type: CC, 1 hop *(a)* / 3 hops *(b)*

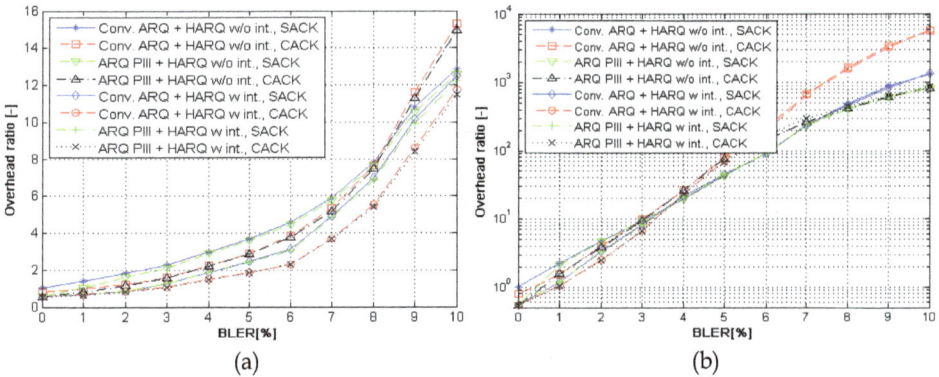

Fig. 10. ARQ & HARQ Overhead vs. BLER for ARQ_Block_Size = 16 B, PDU Size = 16, Size of user data = 1024 B/frame, 2 HARQ retrans., HARQ Type: CC, 1 hop *(a)* / 3 hops *(b)*

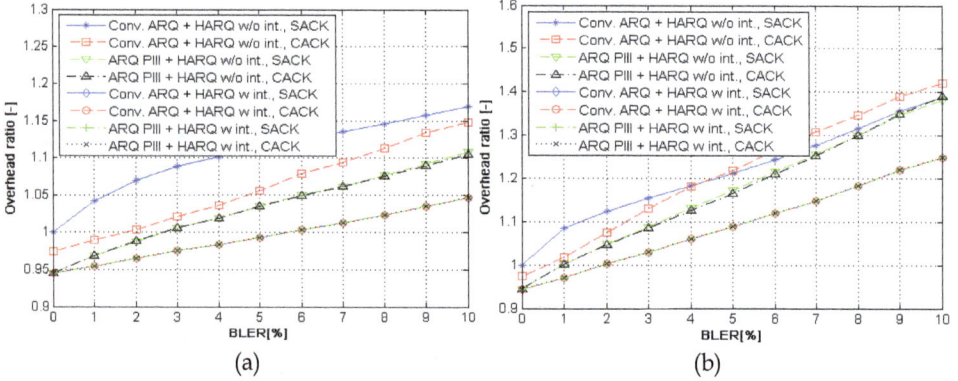

Fig. 11. ARQ & HARQ Overhead vs. BLER for ARQ_Block_Size = 16 B, PDU Size = 1, Size of user data = 1024 B/frame, 4 HARQ retrans., HARQ Type: IR, 1 hop *(a)* / 3 hops *(b)*

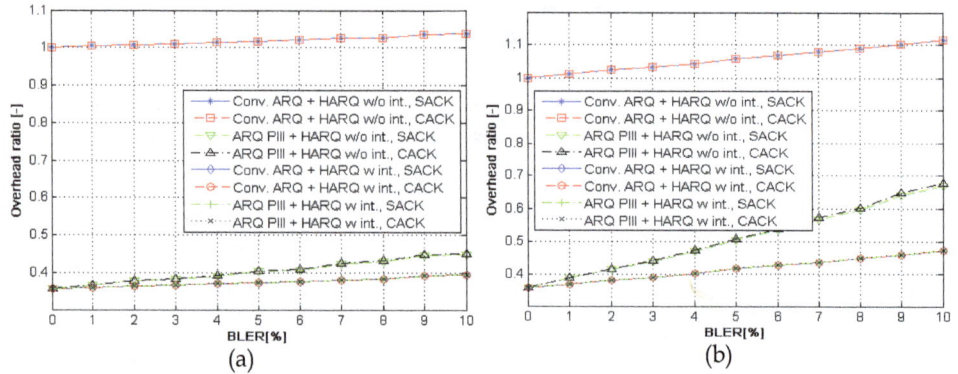

Fig. 12. ARQ & HARQ Overhead vs. BLER for ARQ_Block_Size = 1024 B, PDU Size = 1, Size of user data = 1024 B/frame, 4 HARQ retrans., HARQ Type: IR, 1 hop *(a)* / 3 hops *(b)*

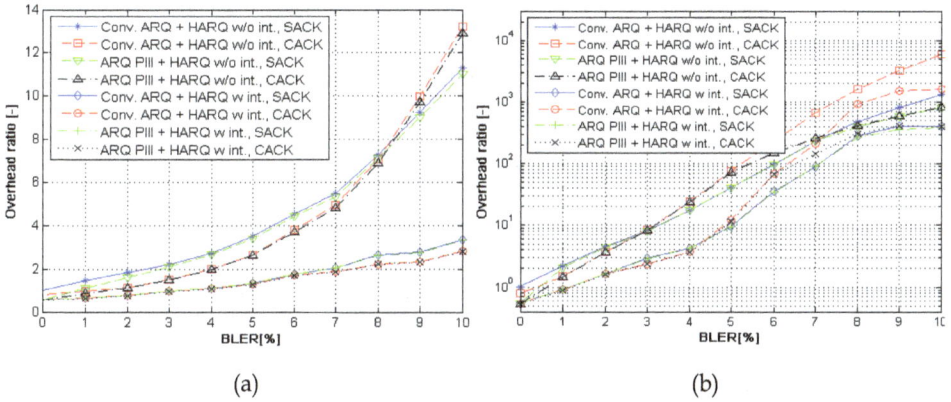

Fig. 13. ARQ & HARQ Overhead vs. BLER for ARQ_Block_Size = 16 B, PDU Size = 16, Size of user data = 1024 B/frame, 4 HARQ retrans., HARQ Type: IR, 1 hop *(a)* / 3 hops *(b)*

The last group of figures represents the results of IR_CTC HARQ for maximum two HARQ retransmissions.

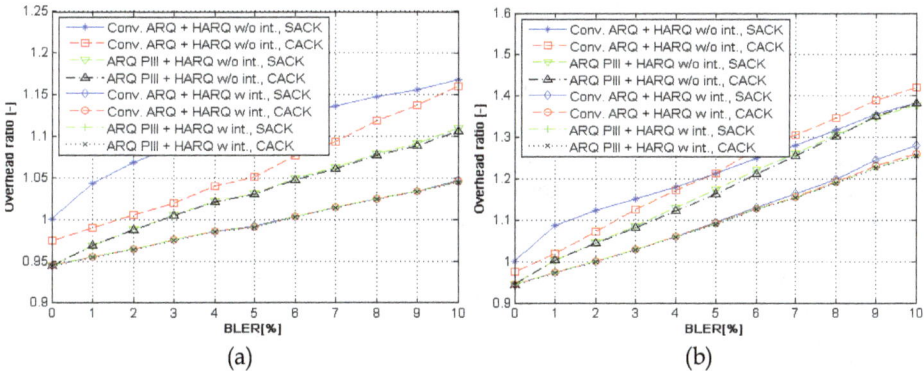

Fig. 14. ARQ & HARQ Overhead vs. BLER for ARQ_Block_Size = 16 B, PDU Size = 1, Size of user data = 1024 B/frame, 2 HARQ retrans., HARQ Type: IR, 1 hop *(a)* / 3 hops *(b)*

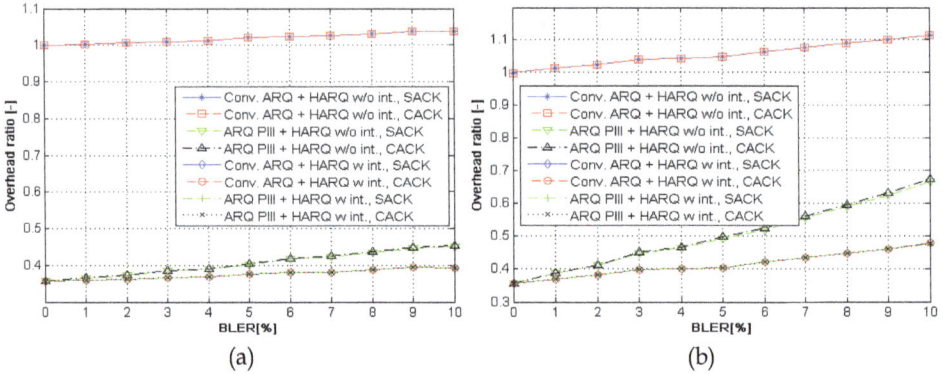

Fig. 15. ARQ & HARQ Overhead vs. BLER for ARQ_Block_Size = 1024 B, PDU Size = 1, Size of user data = 1024 B/frame, 2 HARQ retrans., HARQ Type: IR, 1 hop *(a)* / 3 hops *(b)*

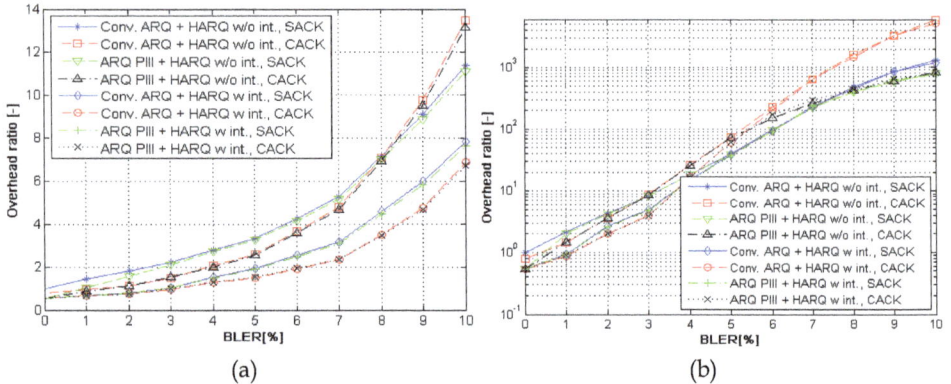

Fig. 16. ARQ & HARQ Overhead vs. BLER for ARQ_Block_Size = 16 B, PDU Size = 16, Size of user data = 1024 B/frame, 2 HARQ retrans., HARQ Type: IR, 1 hop *(a)* / 3 hops *(b)*

The impact of individual parameters observed from the previous figures on the efficiency of overhead reduction can be summarized into the following concluding remars:

- **Number of hops**: the efficiency of innovative ARQ scheme is decreasing with higher number of hops if no interaction is considered; however the reduction of the overhead is not influenced by a number of hops if interaction is enabled.
- **ARQ_Block_Size**: the more significant reduction of overhead is achieved by utilization of innovative ARQ with HARQ even without interaction for higher ARQ_Block_Size; additional reduction of overhead by increase of this parameter is enabled by the interaction for both conventional ARQ as well as for innovative ARQ; the level of this additional reduction is getting higher with BLER since the higher BLER increases the amount of packets not corrected by HARQ.
- **PDU Size**: the overall overhead of ARQ and HARQ rises considerably with PDU size as more blocks have to be corrected by ARQ since the probability that certain part of PDU

is delivered with errors is increasing as well; however influence of the level of overhead reduction by this parameter is only minor.

- **Maximum number of retransmissions**: the impact of a number of retransmissions on the overhead is negligible in most of scenarios; nevertheless the noticeable overhead reduction if four retransmissions occur is achieved only for low ARQ_Block_Size together with high PDU Size when interaction is enabled; the reason is that number of uncorrected errors by two retransmissions do not differ to much from four retransmissions. Hence the ARQ overhead in similar for both cases.

- **Type of HARQ (CC vs. IR)**: the IR slightly outperforms CC, however the difference in overall overhead between both HARQ types is also negligible with exception of scenarios with low ARQ_Block_Size, high PDU_Size and enabled interaction. The reason for this conclusion is the same as explained in the previous bullet.

5. Conclusions

The chapter investigates the efficiency of ARQ and HARQ mechanism used in WiMAX. The conventional ARQ, innovative ARQ, and HARQ are described. In addition, their cooperation is contemplated for stand alone operation of both ARQ and HARQ as well as for cooperation between both.

The results demonstrate that if the HARQ and ARQ are enabled and no mutual interaction between both entities is considered, the difference between conventional ARQ and innovative ARQ is significant. The exact level of overhead reduction depends heavily on the setting of the ARQ and HARQ parameters. The local interaction between ARQ and HARQ enables additional reduction of the overhead. If interaction is considered, the significant improvement by using innovative ARQ instead of conventional ARQ is achieved only while ARQ_Block_Size is low and PDU Size is high.

6. Acknowledgment

This work has been performed in the framework of the FP7 project ROCKET IST-215282 STP, which is funded by the European Community. The Authors would like to acknowledge the contributions of their colleagues from ROCKET Consortium (http://www.ict-rocket.eu).

7. References

Becvar, Z. & Bestak, R. (2011). Overhead of ARQ mechanism in IEEE 802.16 networks. *Telecommunication Systems*, Vol.46, No.4, (March 2010), pp. 353-367, ISSN 1018-4864

Hoymann, C. (2005). Analysis and performance evaluation of the OFDM-based metropolitan area network IEEE 802.16. *Computer Networks*, Vol.49, No.3, pp. 341-363, ISSN 1389-1286

IEEE 802.16e. (2006). Air Interface for Fixed and Mobile Broadband Wireless Access Systems: Amendment for Physical and Medium Access Control Layers for Combined Fixed and Mobile Operation in Licensed Bands. Standard IEEE

IEEE 802.16j. (2009). Air Interface for Broadband Wireless Access Systems, Amendment 1: Multihop Relay Specification. Standard IEEE

Kang, M. S. & Jang, J. (2006). Performance evaluation of IEEE 802.16d ARQ algorithms with NS-2 simulator, *Proceeding of Asia-Pacific Conference on Communications.* Busan, Republic of Korea, 2006

Lee, B.G. & Choi,S. (2008). *Broadband Wireless Access and Local Networks: Mobile WiMAX and WiFi.* USA: Artech House

Maheshwari, S., Boariu, A. & Bacciccola, A. (2008). ARQ/HARQ inter-working to reduce the ARQ feedback overhead. Contribution to IEEE 802.16m No. C802.16m-08/1142

Martikainen, H.; Sayenko, A.; Alanen, O. & Tykhomyrov, V. (2008). Optimal MAC PDU size in IEEE 802.16, *Proceeding of 4th International Telecommunication Networking Workshop on QoS in Multiservice IP Networks,* pp. 66-71

Nuaymi, L. (2007). *WiMAX: Technology for Broadband Wireless Access.* West Sussex: Wiley&son Ltd.

Provvedi, L.; Rattray, C.; Hofmann, J. & Parolari, S. (2004). Provision of MBMS over the GERAN: technical solutions and performance. *Proceeding of Fifth IEEE International Conference on 3G Mobile Communication Technologies,* pp. 494- 498, 2004.

Sambale, K.; Becvar, Z. & Ulvan, A. (2008). Identification of the MAC/PHY key reconfiguration parameters, ICT ROCKET project milestone 5M2, ICT-215282 STP

Sayenko, A.; Martikainen, H. & Puchko, A. (2008). Performance comparison of HARQ and ARQ mechanisms in IEEE 802.16 networks, *Proceeding of International symposium on Modeling, analysis and simulation of wireless and mobile systems.* Vancouver, Canada

Sengupta, S.; Chatterjee, M. & Ganguly, S. (2008). Improving Quality of VoIP Streams over WiMax. *IEEE Transactions on Computers,* Vol.57, No.2, 145-156

Sengupta, S.; Chatterjee, M.; Ganguly, S. & Izmailov, R. (2005). Exploiting MAC Flexibility in WiMAX for Media Streaming, *Proceedings of World of Wireless Mobile and Multimedia Networks,* pp. 338-343, Taormina - Giardini Naxos, Italy, 13-16 June 2005

Tykhomyrov, V.; Sayenko, A.; Martikainen, H.; Alanen, O. & Hämäläinen, T. (2007). Performance Evaluation of the IEEE 802.16 ARQ Mechanism, *Proceeding ofNext Generation Teletraffic and Wired/Wireless Advanced Networking,* pp. 148-161

8

Performance Evaluation of WiMAX System Using Different Coding Techniques

M. Shokair, A. Ebian, and K. H. Awadalla

El-Menoufia University,
Egypt

1. Introduction

In this chapter, we introduce a new class of coding technique that belongs to product code family. This technique is based on convolutional code. The use of convolutional code in the product code setting makes it possible to use the vast knowledge base for convolutional codes as well as their flexibility.

Product codes studied thus far have been constructed using linear block codes, such as Hamming [1], Bose–Chaudhuri–Hocquenghem (BCH) [2] and [3], Reed Solomon codes [4] and single parity check (SPC) [5]. These types of the product codes are traditionally constructed by linear block codes that have structure with a time varying property [6].

The product code proposed in this chapter is constructed by using time-invariant convolutional code. Its component codes' trellis structure does not vary in time as in product codes constructed with Hamming, BCH, and Reed Solomon block codes. Moreover, the number of states in the trellis structure of a block code may grow exponentially with the difference of codeword and data block lengths, whereas the number of states in a convolutional code can be set as desired.

The time invariant trellis structure of convolutional codes makes them more convenient for implementation. In addition, numerous practical techniques such as trellis coded modulation and puncturing can be simply utilized with convolutional codes as opposed to linear block codes.

Multi-input multi-output (MIMO) techniques are quite important to enhance the capacity of wireless communication systems. Space-time trellis codes provide both diversity and coding gain in MIMO channels and are widely used [7]. Space-time trellis codes usually have time-invariant trellis structures just like convolutional codes. Thus, a product code based on convolutional codes is more suitable for integration with MIMO channels and poses an alternative to block product codes.

The type of proposed product code described in this chapter is called modified Convolutional Product Codes (CPC), considered as a different type of normal CPC [8]. The normal CPC depends on recursive systematic convolutional encoder, whereas the modified version of CPC will basically depend on non-recursive non-systematic convolutional encoder.

WiMAX system is a wireless communication system. It suffers from having a high *Bit Error Rate* (BER) at low *Signal to Noise Ratio* (SNR). Using modified version of CPC for WiMAX system reduces the BER at low SNR. Also using the modified version of CPC with WiMAX decreases the number of stages of its physical layer as described later.

2. CPC encoder

For a regular product code, the information bits are placed into a matrix. The rows and columns are encoded separately using linear block codes. This type of a product encoder is shown in Figure 1.

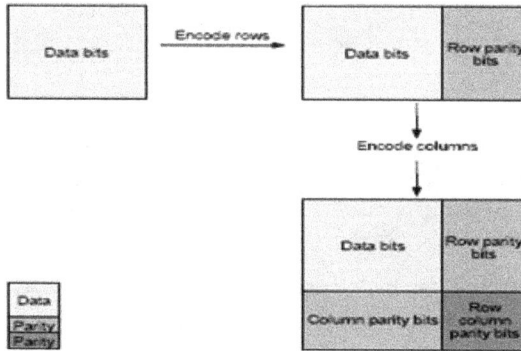

Fig. 1. Regular Product Code Encoding Procedure.

In CPC method, the information bits are placed into two dimensions (2D) matrix. The rows and the columns are encoded separately using recursive systematic convolutional encoders. Each row of the matrix is encoded using a convolutional code with generator polynomial (1, 5/7) octal and code rate (1/2) Figure 2. The same recursive systematic convolutional code with the same polynomial is used to encode each row. Once all rows have been encoded, the matrix is sent, if desired, to an interleaver. The original data matrix dimensions are (n × k), and the encoded data matrix dimensions will be (2n×k) for coding rate (1/2).

The coded rows matrix is then recoded column by column using the same or different recursive systematic convolutional encoder. Hence, the overall code rate is 1/4.

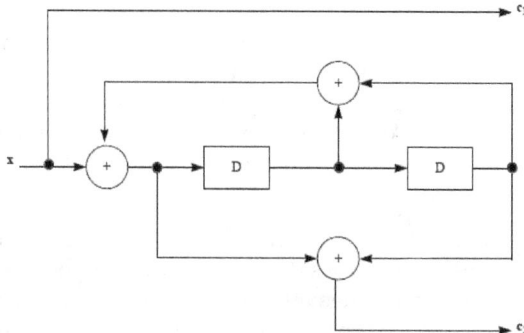

Fig. 2. CPC Convolutional Coding [1 , 5/7].

The general encoding procedure, which includes any type of convolutional encoder and interleaver, is illustrated in Figures 3 & 4.

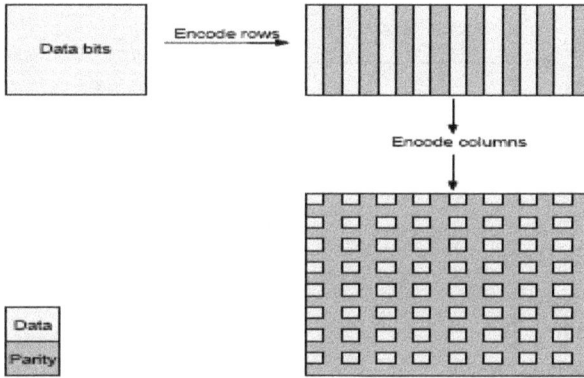

Fig. 3. CPC Encoding Procedure without an Interleaver.

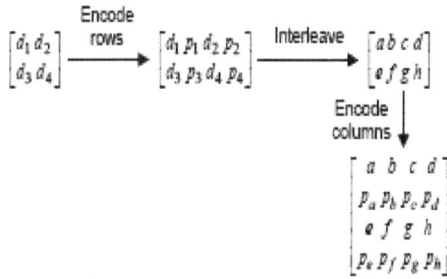

Fig. 4. Convolutional Product Code Encoder with any type of

Interleaver (d denotes data bits and p denotes parity bits).

3. CPC decoder

In the decoding process, the log-MAP soft decoding algorithm, [9] and [10], is used to iteratively decode the convolutional product code. Since columns were encoded last, each column is independently decoded one by one. The extrinsic information obtained from the columns is passed to the row decoder after being de-interleaved. Then, row decoding proceeds; rows are decoded one by one, and interleaved extrinsic information is passed to the column decoder. The CPC decoding procedure is depicted in Figure 5.

The decoding structure employed in this method is the same as that of serially-concatenated codes in Figure 6.6 [11].

4. Modified CPC encoder

In the modified version of CPC, the same technique is used for coding the message, except using nonrecursive nonsystematic convolutional encoder instead of recursive systematic convolutional encoders for coding both rows and columns. That means the both encoders of rows and columns will have coding rate (1/2), and generator polynomial (5,7) Octa Figure 7.

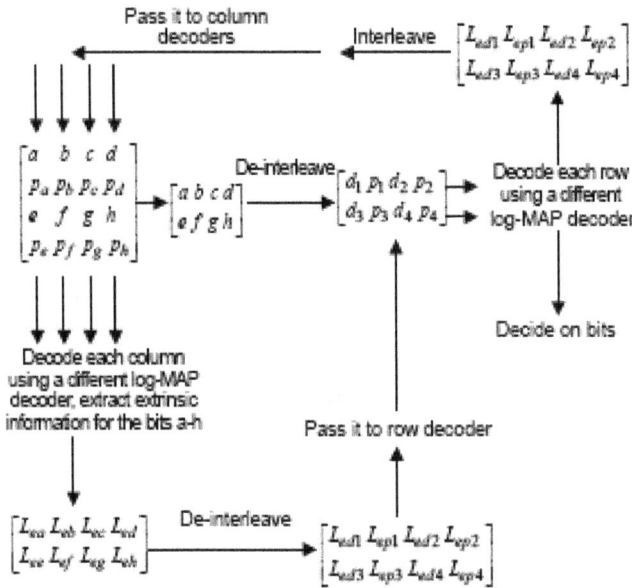

Fig. 5. Decoding Operation of the Convolutional Product Code.

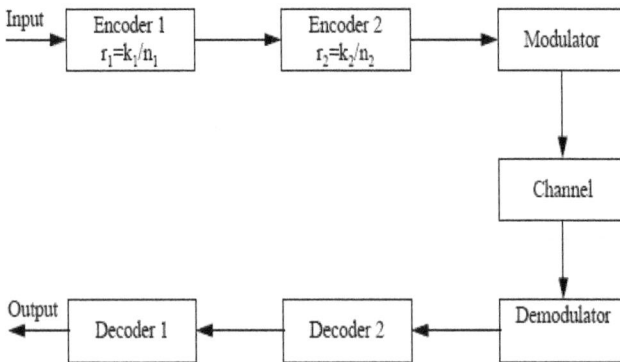

Fig. 6. Serial Encoding & Decoding Operations

The sequence of bits is fed into 2D matrix and fills it column by column. The size of this matrix depends only on the type of modulation used. For 16 QAM, the size of the matrix will be (nx4) and for 64 QAM the size of the matrix will be (nx6). These sizes simplify the process of mapping, as the symbol size in 16 QAM is 4 bits and in 64 QAM is 6 bits. So each row of those matrices will form one QAM symbol. The 'n' refers to the number of data subcarriers of OFDMA, 128 or 512.

The coding by modified CPC will be done in 2 stages

1.   Each column will be independently coded.
2.   Then each row of the resulting matrix will be coded by the same generator polynomials.

From Figure 7 since the generator polynomials used for coding both rows and columns are $(5,7)_{octal}$ with constraint length 3, not following the standard of WiMAX, each column is padded with two zeros for terminating its encoder. But each row is padded with two or three zeros according to the number of used subcarriers, 128 or 512, receptively to form the suitable size of the overall matrix. That matrix is then divided into smaller matrices with sizes (nx4) or (nx6) as described later.

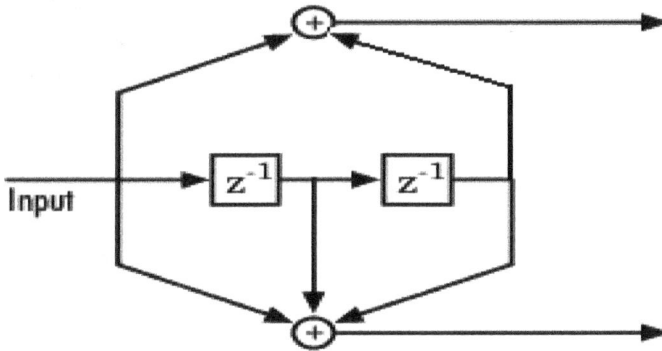

Fig. 7. Convolutional Coding [5,7].

After the coding process, the total number of bits will be more than the original message bits due to the increase in the overall code rate (1/4), and the addition of the zeros in both column and rows that used for the termination process. Therefore the following steps are done,

1.   Dividing the overall matrix produced from modified CPC into three matrices. Each one has a size (nx4) or (nx6) according to the type of QAM used as mentioned before. The reason for using three matrices only is to have a number of message bits equals to bits used in the convolutional code method, as a comparison between it and CPC is done.
2.   Applying symbol mapping for each one independently (16QAM or 64 QAM).
3.   Inserting the pilot and DC subcarriers for each matrix.
4.   Performing the IFFT on the three matrixes independently resulting in three OFDMA symbols.
5.   Applying (cyclic prefix) CP for each symbol.
6.   Sending each symbol independently.

The reason for using nonrecursive nonsystematic convolutional encoder instead of recursive systematic convolutional encoders is simplifying the termination of the encoder, as RSC contains a feedback and its termination will be more difficult. Also using the generator polynomials (5,7) leads to a little increase in the complexity of the system because of a few number of zeros will be added to terminate the two encoders.

5. Modified CPC decoder

At the receiver, the three OFDMA symbols are combined to form the original matrix which is decoded by Viterbi decoder. The Viterbi decoder uses the same generator polynomials (5,7) with hard decision for each row and for each column. The rows must be decoded first then the columns are done, because columns are encoded first Figure 6. To match the CPC method, the number of data bits will be reduced. For example in OFDMA (128-16QAM) and (128-64QAM) the number of data bits was 144 and 216 but in CPC method it becomes 136 and 204 bits receptively due to the number of zero bits added to terminate the two encoders.

6. Modified CPC minimum distance and its asymptotic performance

The Hamming weight of a binary codeword is defined as the number of '1's available in the codeword. The minimum distance of a linear code is the minimum Hamming weight of all the codewords. The minimum distance plays an important role in the code performance. As it gets larger, code performance improves, especially at high signal-to-noise ratio (SNR) values. The free distance of the component convolutional codes used in modified CPC with trellis termination will be called d_{free}. The minimum distance of the modified CPC in the case of no interleavers will be investigated.

No Interleaving

After the first stage of the modified CPC encoding operation (columns encoding), it is obvious that one of these columns should contain at least d_{free} number of '1's. This means that there are d_{free} rows containing at least a *single* '1' in the columns-encoded matrix. When rows are encoded, there exists at least d_{free} number of rows each containing at least d_{free} '1's. Hence, in total there are at least d^2_{min} '1's in the coded matrix. In summary, if no interleavers are used, the modified CPC minimum distance is d^2_{min}.

7. Advantage and disadvantage of CPC

CPC technique has mainly two main advantages that make it a motivating step for future considerations and improvements for practical systems.

1. Do not need another interleaver after channel coding because of converting into matrix (nx4) or (nx6) does almost the same job as the overall matrix will be filled column by column and will be read row by row after coding processes (block interleaver) since each row is used for making QAM symbol.
2. Reducing the BER at low SNR.
3. The product code we propose in CPC is constructed by using time invariant convolutional codes. Its component codes' trellis structure does not vary in time as in product codes constructed with Hamming, extended Hamming, BCH, and Reed Solomon block codes. The time invariant trellis structure of convolutional codes makes them more convenient for implementation
4. The number of states in CPC like a convolutional code can be set as desired.
5. Numerous practical techniques such as trellis coded modulation and puncturing can be simply utilized with convolutional codes as opposed to linear block codes.

6. Space-time trellis codes usually have time-invariant trellis structures just like convolutional codes. Thus, a product code based on convolutional codes is more suitable for integration with MIMO channels.
7. Increasing the free distance to be d $^2_{min}$.

But on the other hand it causes more delay for obtaining the original message because the code rate becomes 1/4 not 1/2 as in convolutional code. The performance of the system will be reduced and this is the price to be paid for the improvement obtained.

8. Results

This section contains comparisons between the modified CPC method and convolutional code, turbo code and LDPC code. Several results obtained at different types of the channels, modulation techniques (16QAM – 64QAM) and number of OFDM subcarriers (128 -512).

In this work, a matlab tool is used to simulate the physical layer of WiMAX and apply the mentioned coding methods.

8.1 AWGN channel

In this section the coded signal is transmitted through AWGN channel only. This can be done using matlab function **AWGN**. The syntax of this function is: $y = awgn (x, snr ,$ *'measured')* adds white Gaussian noise to the vector signal **x** to produce output signal **y** The scalar snr specifies the signal-to-noise ratio per sample, in dB. If x is complex, awgn adds complex noise. This syntax measures the power of x before adding noise.

We can derive the relationship between Es/N0 and SNR for complex input signals as follows:

$$E_s / N_0 \ (\text{dB}) = 10 \log_{10} \left((S \cdot T_{sym})/(N / B_n) \right)$$
$$= 10 \log_{10} \left((T_{sym} F_s) \cdot (S / N) \right)$$
$$= 10 \log_{10} \left(T_{sym} / T_{samp} \right) + SNR \ (\text{dB}) \tag{1}$$

Where

S = Input signal power, in watts

N = Noise power, in watts

Bn = Noise bandwidth, in Hertz

Fs = Sampling frequency, in Hertz = 1/Tsamp.

Tsamp = The period of each row of a frame-based matrix.

Tsym = The signal's symbol period.

A good rule of thumb for selecting the symbol period value is to set it to be what we model as the symbol period in the model. The value would depend upon what constitutes a symbol and what the oversampling applied to. From Figure 8 to Figure 13 BER versus different received SNR values are shown for the comparison between modified CPC and convolutional code, LDPC code and turbo code respectively. These comparisons are obtained for modulation type 16QAM and number of subcarriers equals 128 and 512 respectively.

The comparisons between modified CPC and convolutional code are shown in both Figure 8 and Figure 9. From Figure 8, it is shown that SNR will be improved by approximately 2 dB at BER equals 10 [-3] for modulation type 16QAM and number of subcarriers equals 128. Also, an improvement can be obtained when the number of subcarriers is increased to 512 as shown in Figure 9.

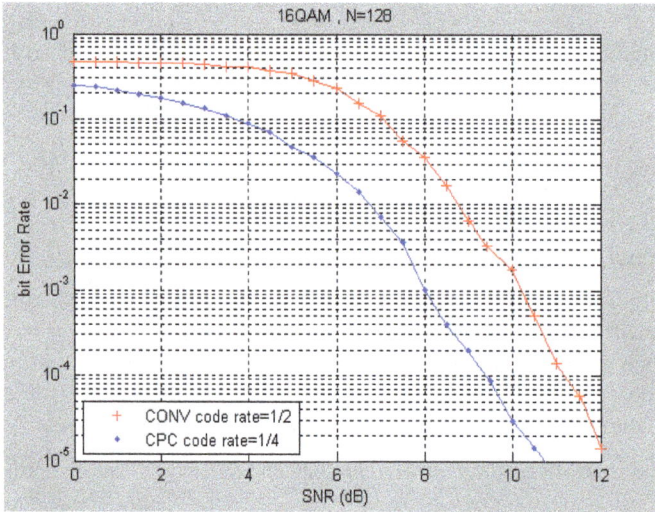

Fig. 8. BER Comparison between Conv code, CPC at 16 QAM, N=128.

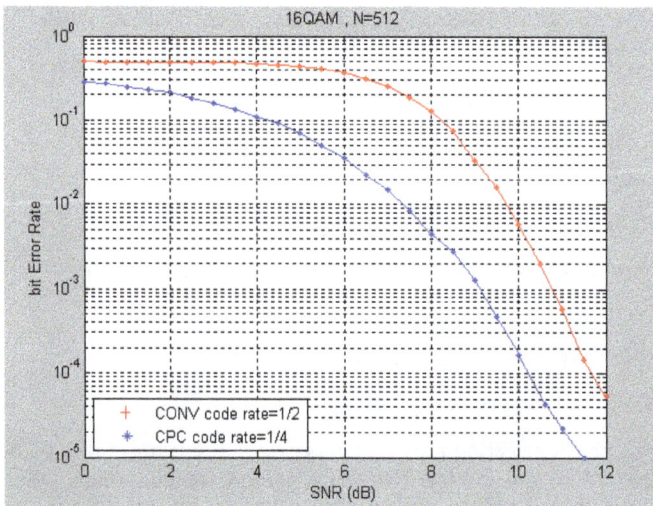

Fig. 9. BER Comparison between Conv code, CPC at 16QAM, N=512.

From Figure 10 and Figure 11, the results of comparisons between modified CPC and LDPC code are shown at different received SNR values. From these figures, we conclude that modified CPC gives good results at different SNR. Figure 10 shows that modified CPC coding technique gives better results than LDPC coding technique at 16 QAM and OFDM subcarriers equals 128. The improvement is more than 3 dB for 10 $^{-3}$. Also, there will be an improvement, when the number of subcarriers is increased to 512 as shown in Figure 11.

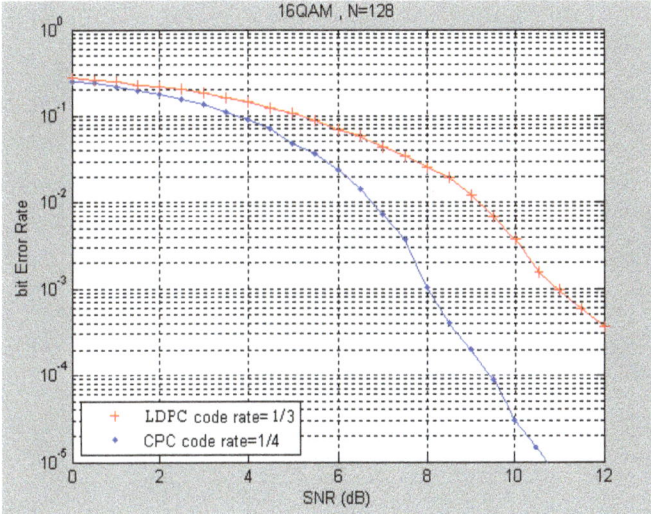

Fig. 10. BER Comparison between LDPC code, CPC at 16 QAM, N=128.

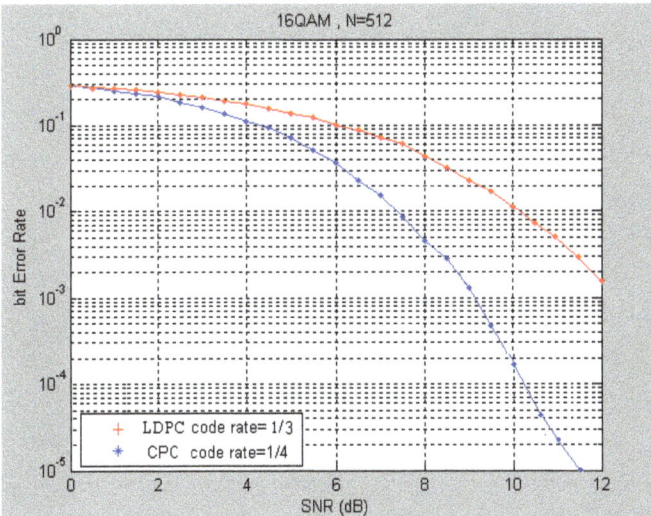

Fig. 11. BER Comparison between LDPC code, CPC at 16 QAM, N=512

From Figure 12 to Figure 13, the results produced from the comparisons between modified CPC and turbo code are shown at different received SNR values. As shown from these figures, modified CPC method gives good results compared to turbo coding. Figure 12 shows that using modified CPC method can give better results than turbo coding technique at 16QAM and OFDM subcarriers equals 128. This improvement is more than 3 dB for BER= 10 [-3]. Also other improvements can be obtained at different number of OFDM subcarriers (512) as shown in Figure 13.

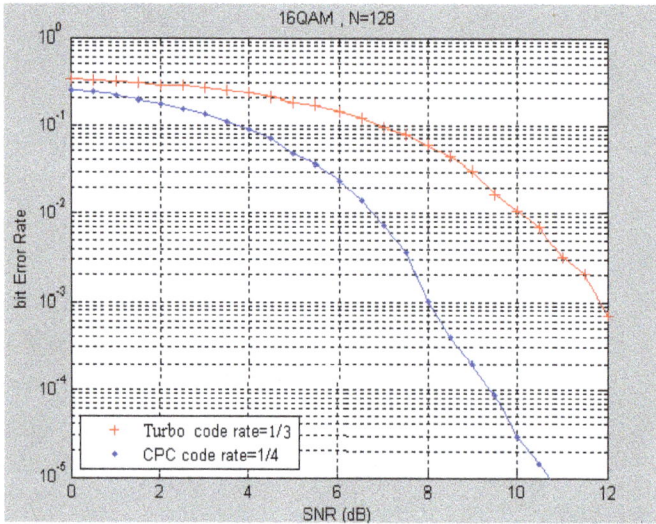

Fig. 12. BER Comparison between Turbo code, CPC at 16 QAM, N=128.

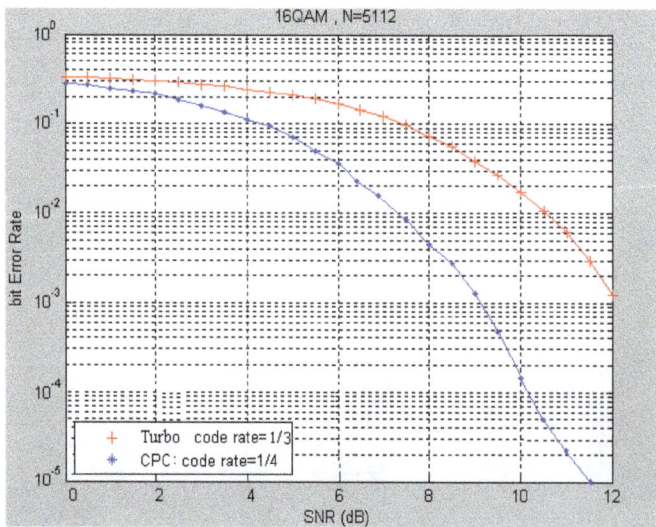

Fig. 13. BER Comparison between Turbo code, CPC 16 QAM, N=512.

The comparison for modulation type 64 QAM between modified CPC and convolutional code, LDPC code and turbo code are shown through Figure 14 to Figure 19. BER versus different received SNR values are shown in these figures. These comparisons are obtained and number of subcarriers equals 128 and 512 respectively. The comparisons between modified CPC and convolutional code are shown in both Figure 14 and Figure 15.

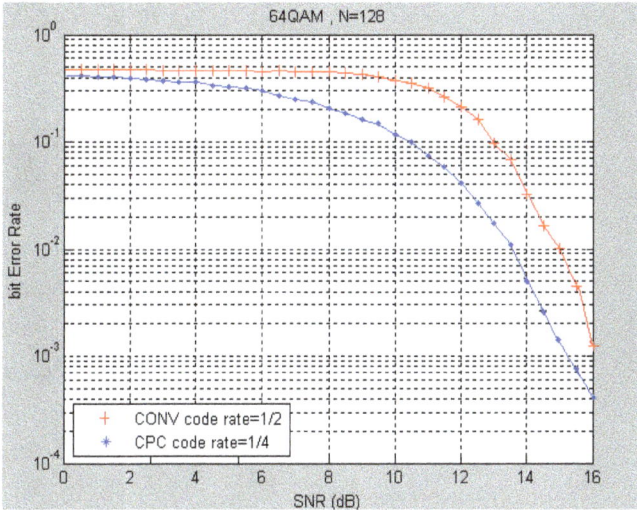

Fig. 14. BER Comparison between Conv code, CPC at 64 QAM, N=128.

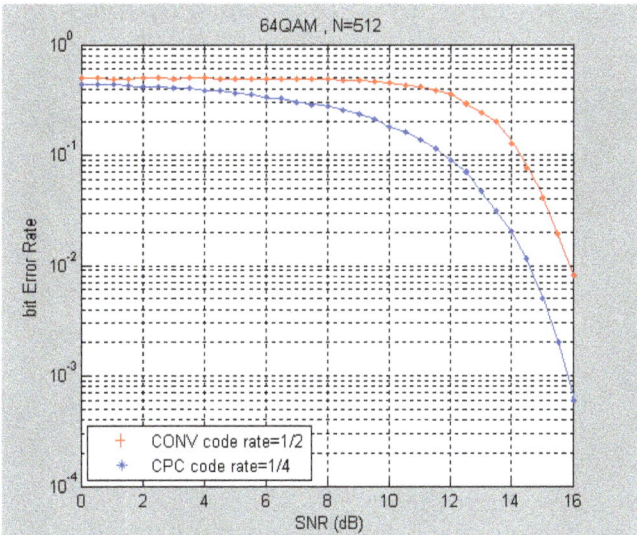

Fig. 15. BER Comparison between Conv code, CPC at 64 QAM, N=512.

From Figure 14, it is shown that SNR will be improved by approximately 1.5 dB at BER equals 10^{-2} for modulation type 64QAM and number of subcarriers equals 128. Also, an improvement can be obtained when the number of subcarriers is increased to 512 as shown in Figure 15. Figure 16 and Figure 17 show the results of comparisons between modified CPC and LDPC code are shown at different received SNR values. We conclude that modified CPC gives good results at different SNR. Figure 16 shows that modified CPC coding technique gives better results than LDPC coding technique at 64QAM and OFDM subcarriers equals 128. The improvement is more than 1.5 dB for 10 $^{-2}$. Also, there will be an improvement, when the number of subcarriers is increased to 512 as shown in Figure 17.

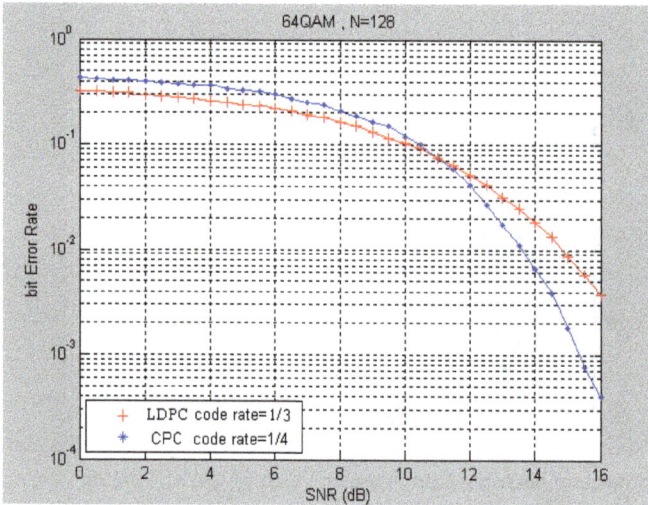

Fig. 16. BER Comparison between LDPC code, CPC at 64QAM, N=128.

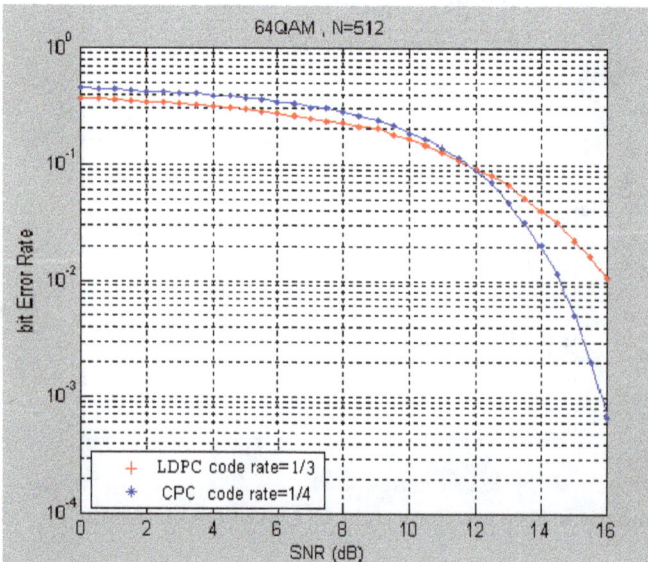

Fig. 17. BER Comparison between LDPC code, CPC at 64 QAM, N=512.

From Figure 18 to Figure 19, the results produced from the comparisons between modified CPC and turbo code are shown at different received SNR values. As shown from these figures, modified CPC method gives good results compared to turbo coding. Figure 18 shows that using modified CPC method can give better results than turbo coding technique at 64QAM and OFDM subcarriers equals 128. This improvement is about than 2 dB for 10^{-2}. Also other improvements can be obtained at different number of OFDM subcarriers (512) as shown in Figure 19.

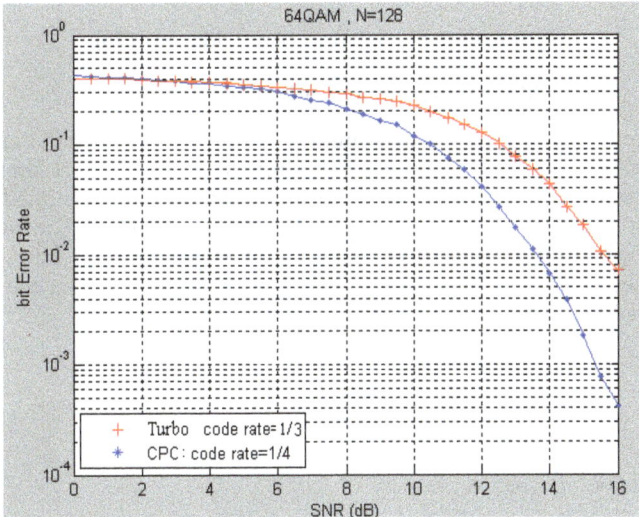

Fig. 18. BER Comparison between Turbo code, CPC at 64 QAM, N=128.

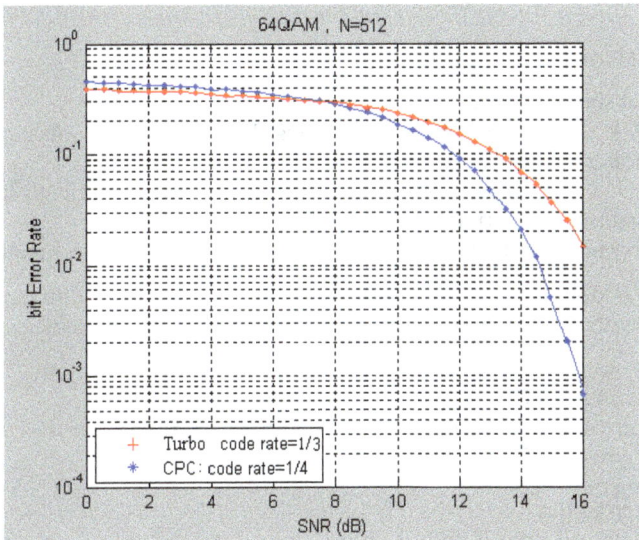

Fig. 19. BER Comparison between Turbo code, CPC 64 QAM, N=512.

8.2 AWGN plus fading channel

In this section, the transmitted signal is assumed to pass through time selectivity fading channel plus AWGN. This is done using matlab function **rayleighchan**. The syntax of this function is *chan = rayleighchan (Ts,Fd)* that constructs a frequency-flat ("single path") Rayleigh fading channel object.

Ts is the sample time of the input signal, in seconds. Fd is the maximum Doppler shift, in Hertz.

$$\text{Sample time} = 1/(\text{channel bandwidth} \times 28/25) \qquad (2)$$

$$\text{Maximum Doppler shift} = F_d = (v\, f_c)\, /\, C_0 \qquad (3)$$

Where f_c is carrier frequency, v is the maximum speed between transmitter and the receiver and C_0 is the speed of light. The Rayleigh multipath fading channel simulators of this toolbox use the band-limited discrete multipath channel model. It is assumed that the delay power profile and the Doppler spectrum of the channel are separable. The multipath fading channel is therefore modeled as a linear finite impulse-response (FIR) filter. Let S_i denotes the set of samples at the input to the channel. Then the samples y_i at the output of the channel are related to S_i through:

$$y_i = \sum_{n=-N_1}^{N_2} s_{i-n} g_n \qquad (4)$$

Where g_n is the set of tap weights given by:

$$g_n = \sum_{k=1}^{K} a_k \operatorname{sinc}\left[\frac{\tau_k}{T_s} - n\right], \quad -N_1 \le n \le N_2 \qquad (5)$$

In the equations above:

- T_s is the input sample period to the channel.
- τ_k ,where $1 \le k \le K$, is the set of path delays. K is the total number of paths in the multipath fading channel.
- a_k ,where $1 \le k \le K$, is the set of complex path gains of the multipath fading channel. These path gains are uncorrelated with each other.
- N_1 and N_2 are chosen so that $|\,g_n\,|$ is small when n is less than N_1 or greater than N_2

This simulation is done for different coding techniques that have different coding rates because we follow the standard in our simulation. The following parameters are used in our simulation:

1. Frequency band is 3.5 GHz.
2. Channel Bandwidth (1.25 MHz for IFFT size=128 and 5.00 MHz for IFFT size= 512).
3. Modulation types (16 QAM, 64 QAM).
4. Oversampling rate is 28/25.
5. Max speed 120 Kmph.
6. Convolutional code with rate equals (1/2), turbo code with rate equals (2/3) and LDPC code with rate equals (1/2).

From Figure 20 to Figure 25 BER versus different received SNR values are shown for the comparison between modified CPC and convolutional code, LDPC code and turbo code respectively through the fading channel. These comparisons are obtained for modulation type 16QAM and number of subcarriers equals 128 and 512 respectively.

The comparisons between modified CPC and convolutional code are shown in both Figures 20 and 21. In Figure 20, it is shown that SNR is improved by more than 4 dB at BER equals 10-2 for the number of subcarriers equals 128. An improvement is obtained if the number of subcarriers is increased to 512 as shown in Figure 21.

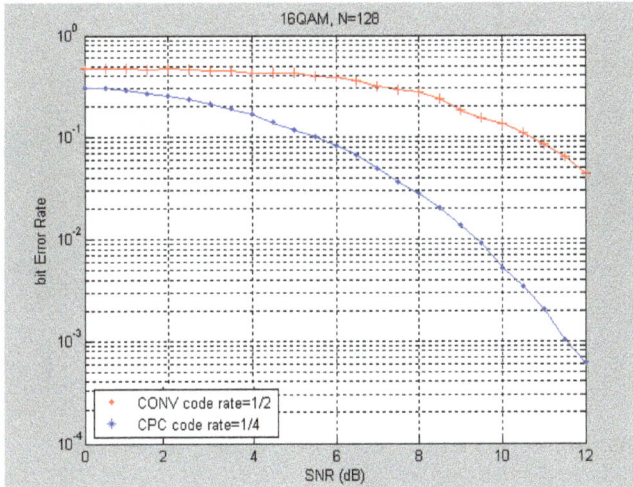

Fig. 20. BER Comparison between Conv code, CPC at 16QAM , N=128.

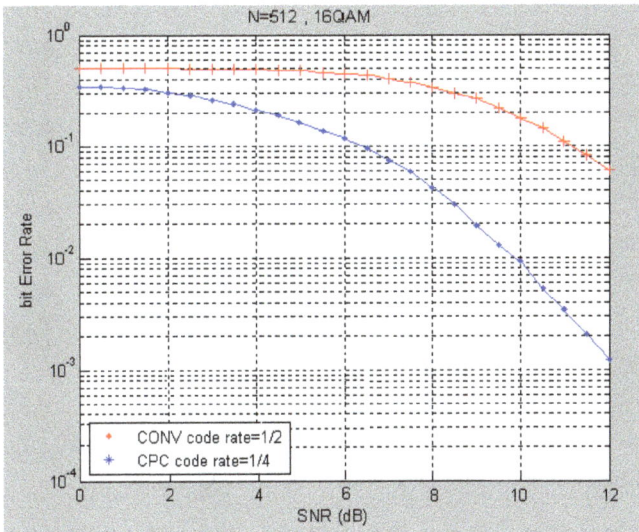

Fig. 21. BER Comparison between Conv code, CPC at 16 QAM, N=512.

The results of comparisons between CPC and LDPC code through the fading channel are obtained from Figure 22 to Figure 23. These comparisons are obtained for modulation 16QAQM at different SNR values. From Figure 22, it is shown that SNR is improved by about 2.5 dB at BER equals 10-2 for the number of subcarriers equals 128. Another improvement is also obtained at different number of OFDM subcarriers (512) as shown in Figure 23.

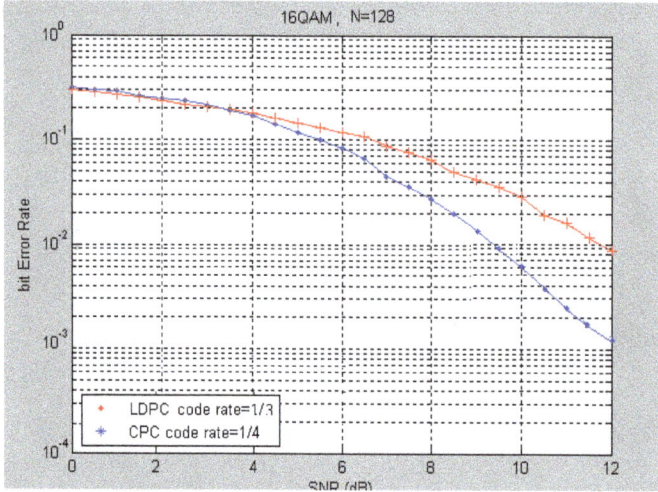

Fig. 22. BER Comparison between LDPC code, CPC at 16 QAM, N=128.

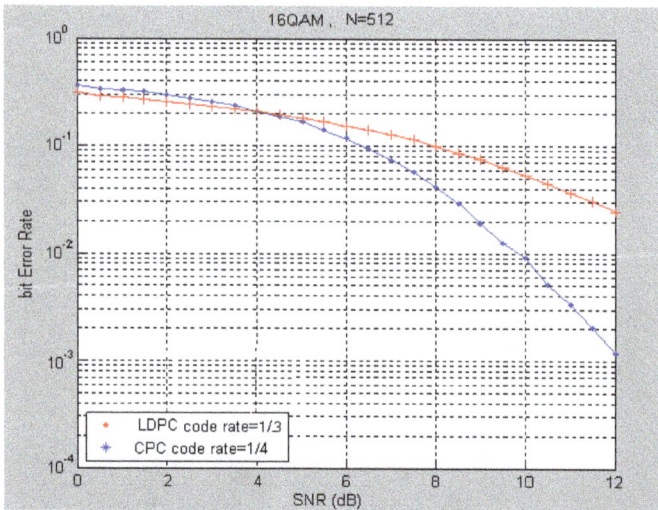

Fig. 23. BER Comparison between LDPC code, CPC at 16 QAM, N=512.

The results of comparisons between modified CPC and turbo code through the fading channel are shown from Figure 24 to Figure 25. These comparisons are done at different

SNR values for modulation type 16QAM. There is an improvement in SNR by more than 8 dB at BER equals 10^{-2} for 16QAM and number of OFDMA subcarriers equals 128, this is shown from Figure 24. Other improvements obtained at different number of OFDMA subcarriers (512) as shown from Figure 25.

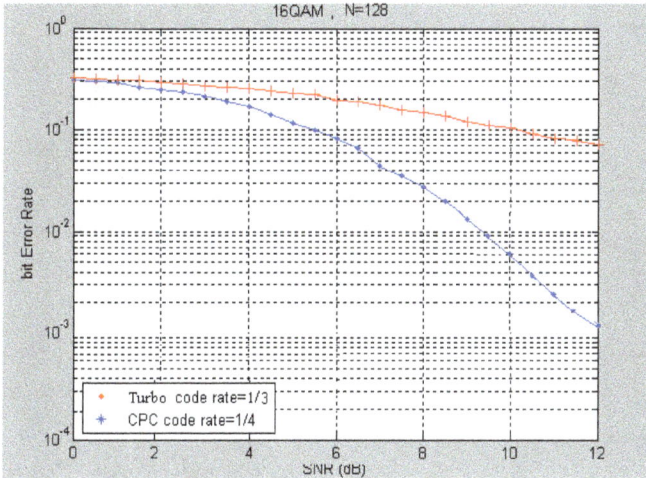

Fig. 24. BER Comparison between Turbo code, CPC at 16QAM , N=128.

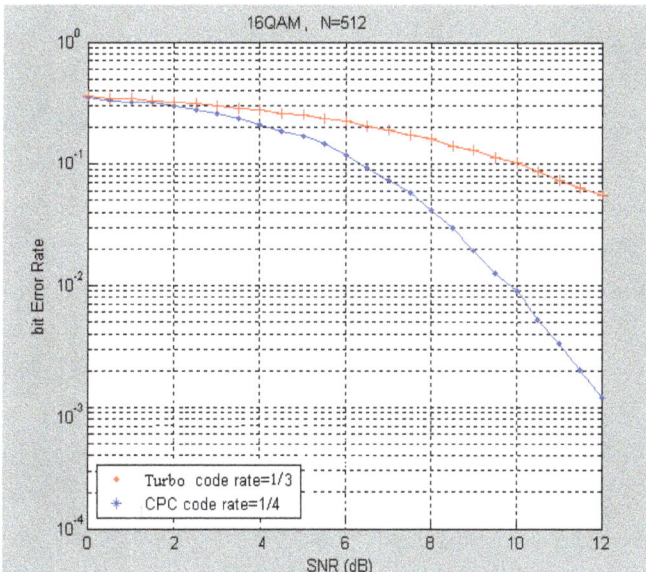

Fig. 25. BER Comparison between Turbo code, CPC at 16 QAM, N=512.

The comparison for modulation type 64QAM between modified CPC and convolutional code, LDPC code and turbo code are shown through Figure 26 to Figure 31 through the fading channel.. BER versus different received SNR values are shown in these figures. These comparisons are obtained and number of subcarriers equals 128 and 512 respectively. The comparisons between modified CPC and convolutional code through the fading channel are shown in both Figure 26 and Figure 27.

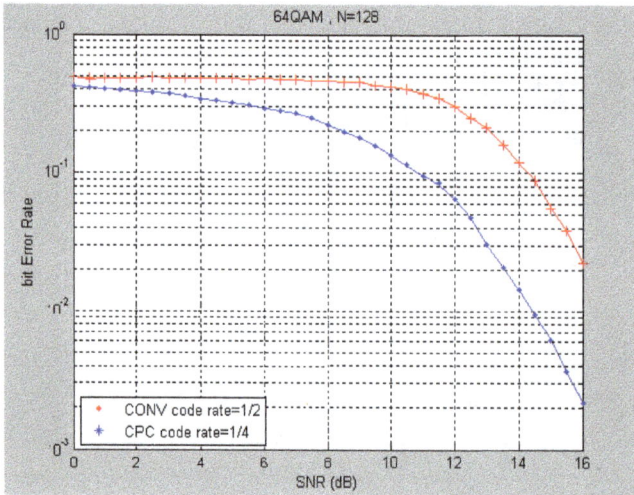

Fig. 26. BER Comparison between Conv code, CPC at 64QAM, N=128.

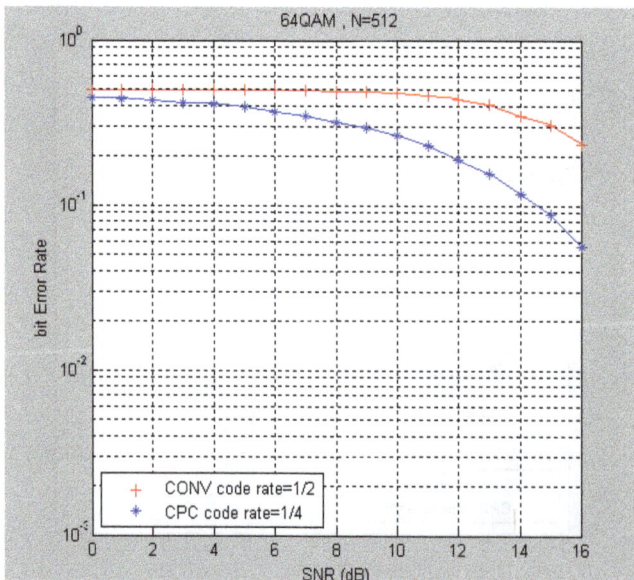

Fig. 27. BER Comparison between Conv. code, CPC at 64 QAM, N=512.

From Figure 26, it is shown that SNR will be improved by more than 2 dB at BER equals 10^{-2} for modulation type 64QAM and number of subcarriers equals 128. Also, an improvement can be obtained when the number of subcarriers is increased to 512 as shown in Figure 27. Figure 28 and Figure 29 show the results of comparisons between modified CPC and LDPC code through the fading channel are shown at different received SNR values. We conclude that modified CPC gives good results at different SNR. Figure 28 shows that modified CPC coding technique gives better results than LDPC coding technique at 64QAM and OFDM subcarriers equal 128. The improvement is more than 1.5 dB for 10^{-2}. Also, there will be an improvement, when the number of subcarriers is increased to 512 as shown in Figure 29.

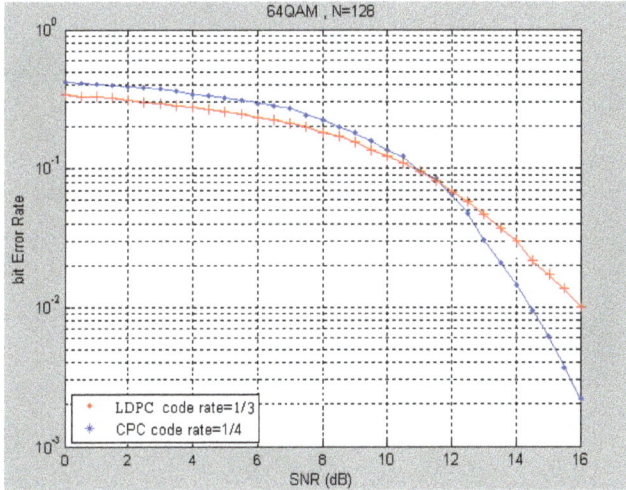

Fig. 28. BER Comparison between LDPC code, CPC at 64 QAM, N=128.

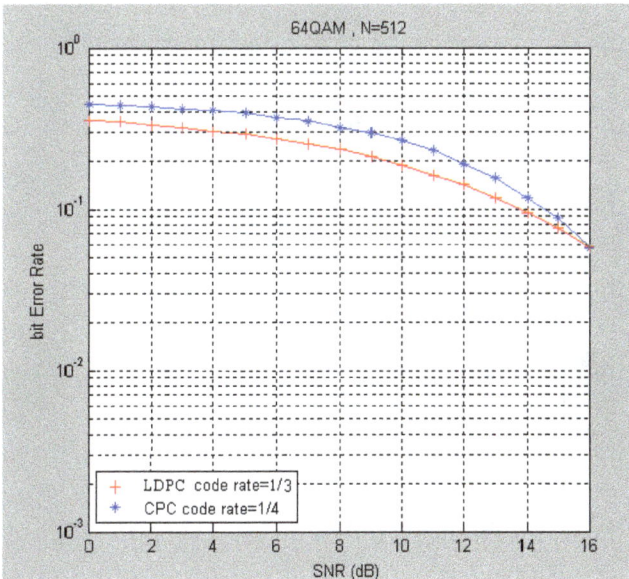

Fig. 29. BER Comparison between LDPC code, CPC at 64 QAM, N=512.

From Figure 30 to Figure 31, the results produced from the comparisons between modified CPC and turbo code are shown at different received SNR values for modulation 64 QAM. As shown from these figures, modified CPC method gives good results compared to turbo coding. Figure 30 shows that using modified CPC method can give better results than turbo coding technique at 64QAM and OFDM subcarriers equals 128. This improvement is about than 2 dB for 10 $^{-2}$. Also other improvements can be obtained at different number of OFDM subcarriers (512) as shown in Figure 31.

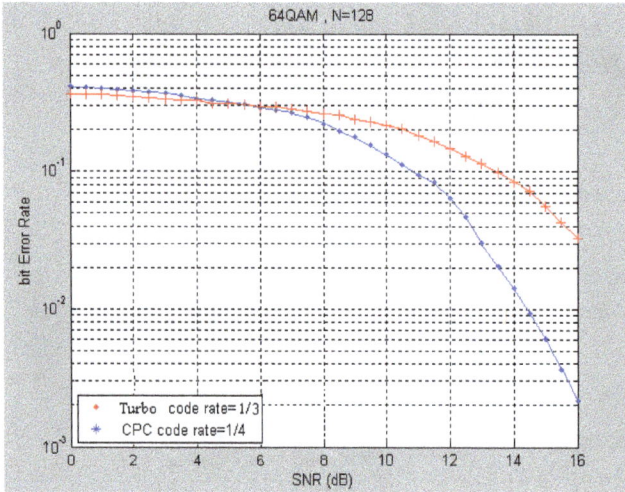

Fig. 30. BER Comparison between Turbo code, CPC at 64 QAM, N=128.

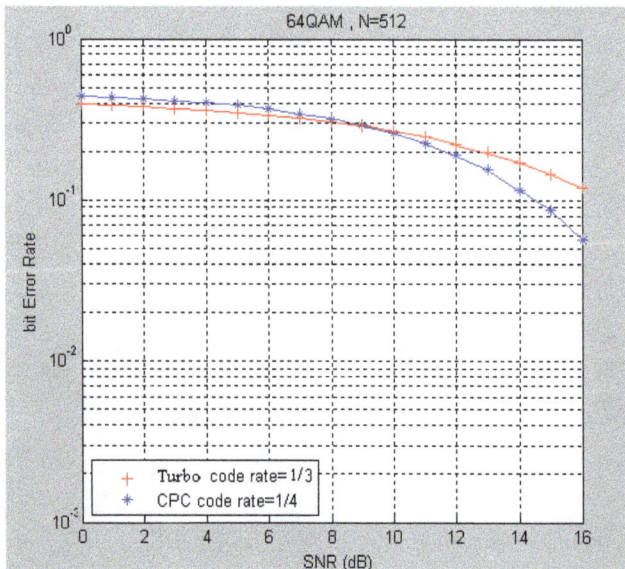

Fig. 31. BER Comparison between Turbo code, CPC at 64 QAM, N=512.

Due to the lower code rate of CPC (1/4), better results should be obtained comparing to the other coding techniques (Convolution – Turbo – LDPC) which has higher coding rate (1/2 or 1/3). But from the view of complexity, CPC is less complex than turbo or LDPC, so the type of coding can be used as an optional code instead of turbo or LDPC types. Another technique can be used with modified CPC to modify or increase its code rate, this technique called *puncture technique*.

9. Punctured CPC

To modify the rate of the coding process, a puncture technique is used. This technique enables to have a code rate equals 1/3. The modification is done by applying the puncture to the columns only, resulting in code rate of 2/3. So the overall code rate will be (2/3) x (1/2) = (1/3). Puncture enables to reduce the redundancy bits but on other hand it leads to increase the BER. From Figure 32 to Figure 42 the result of using CPC with puncture is shown, through AWGN plus fading channel, comparing with convolutional code, turbo code and LDPC code. From Figure 32 to Figure 34 BER versus different received SNR values are shown for the comparison between modified CPC, puncture CPC and convolutional code, LDPC code and turbo code respectively through the fading channel. These comparisons are obtained for modulation type 16QAM and number of subcarriers equals 128.

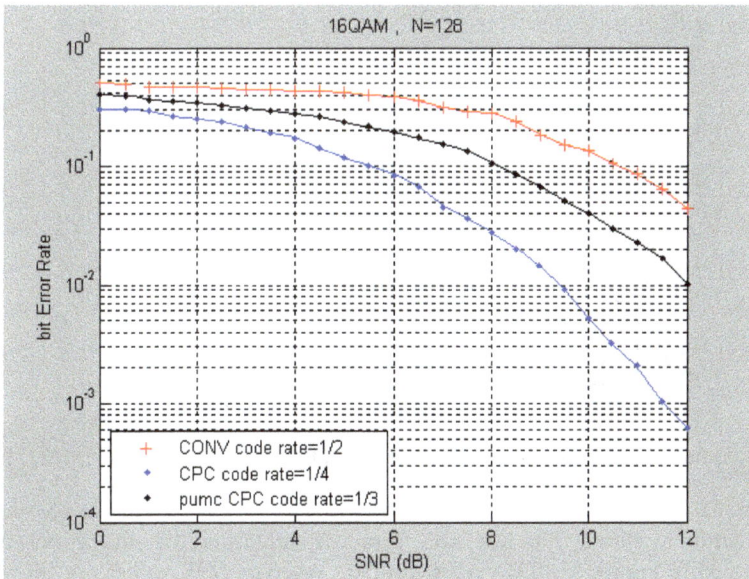

Fig. 32. BER Comparison between Conv code, CPC, punctured CPC at 16 QAM, N=128.

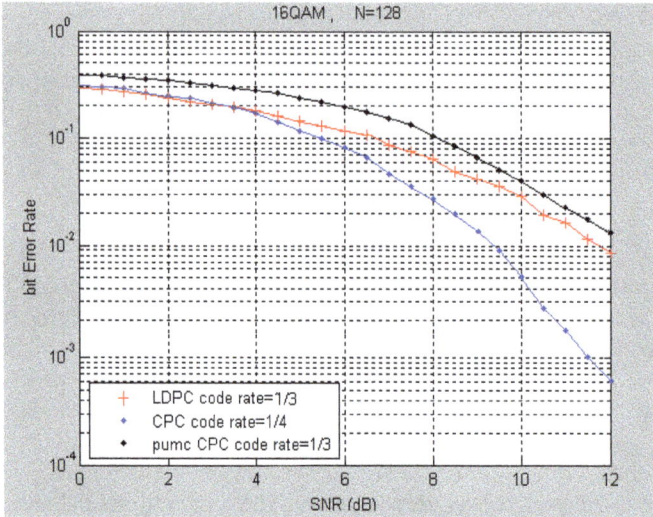

Fig. 33. BER Comparison between LDPC code, CPC, punctured CPC at 16 QAM, N=128.

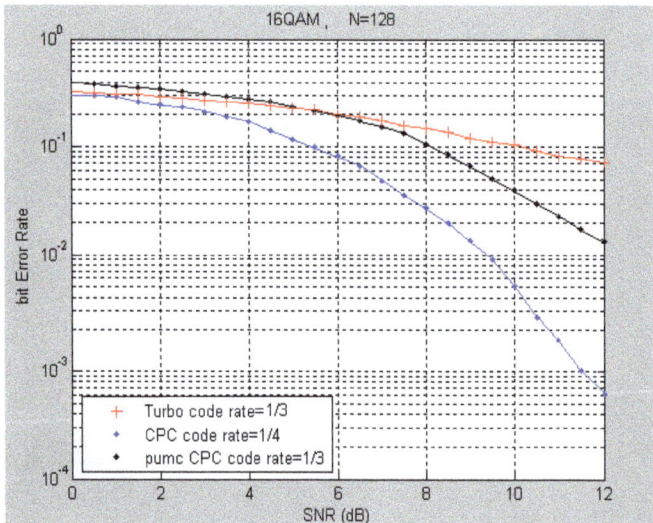

Fig. 34. BER Comparison between Turbo code, CPC, punctured CPC at 16 QAM, N=128.

From Figure 32 it is shown that the results obtained from LDPC code is approximately the same as puncture modified CPC, but LDPC is more complicated than the proposed method. From Figure 35 to Figure 37 BER versus different received SNR values are shown for the comparison between modified CPC, puncture CPC and convolutional code, LDPC code and turbo code respectively through the fading channel. These comparisons are obtained for modulation type 16QAM and number of subcarriers equals 512.

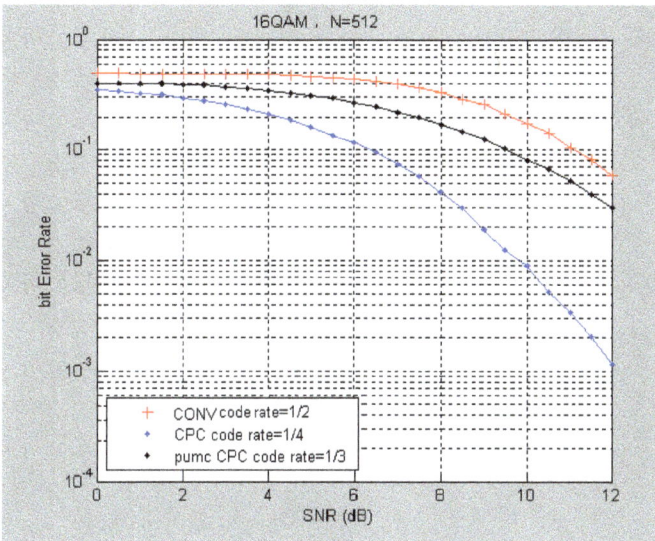

Fig. 35. BER Comparison between Conv code, CPC, punctured CPC at 16 QAM, N=512.

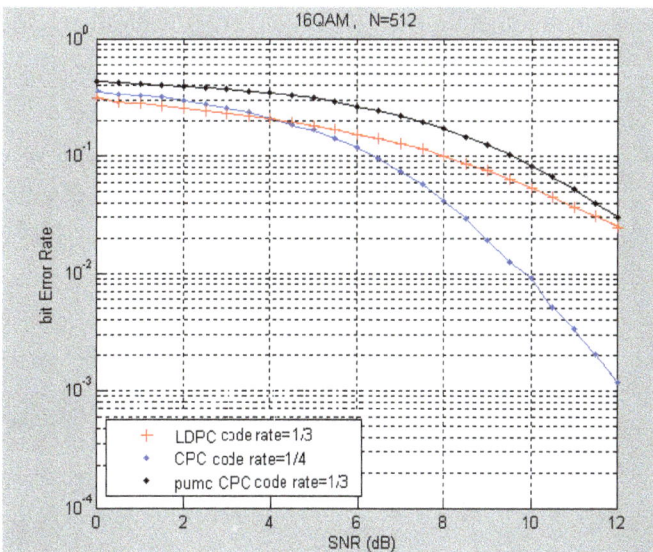

Fig. 36. BER Comparison between LDPC code, CPC, punctured CPC at 16 QAM, N=512.

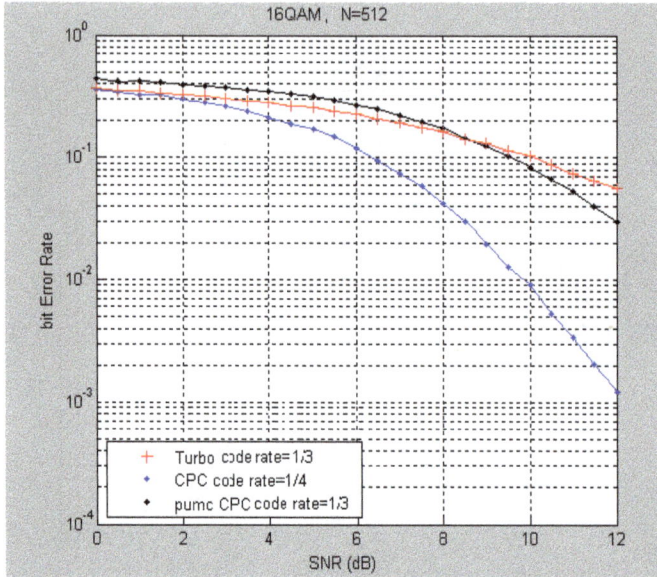

Fig. 37. BER Comparison between Turbo code, CPC, punctured CPC at 16 QAM, N=512

From Figure 36 it is shown that the results obtained from LDPC code is better than puncture modified CPC until SNR = 12 db, but LDPC is more complicated than the proposed method. From Figure 35 to Figure 37 BER versus different received SNR values are shown for the comparison between modified CPC, puncture CPC and convolutional code, LDPC code and turbo code respectively through the fading channel. These comparisons are obtained for modulation type 64QAM and number of subcarriers equals 128.

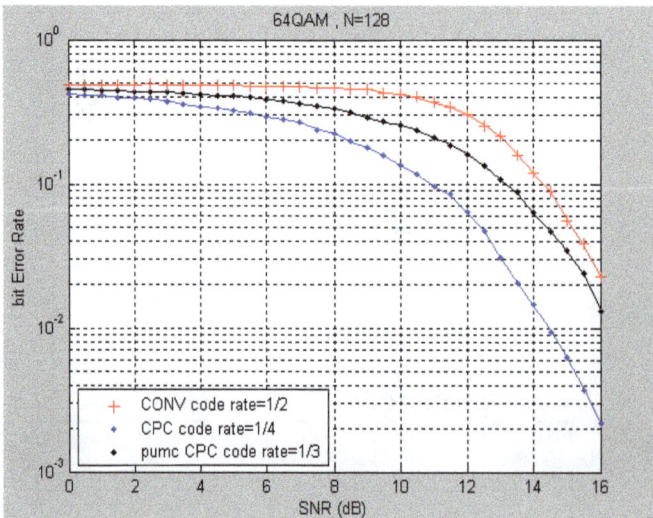

Fig. 38. BER Comparison between Conv code, CPC, punctured CPC at 64 QAM, N=128.

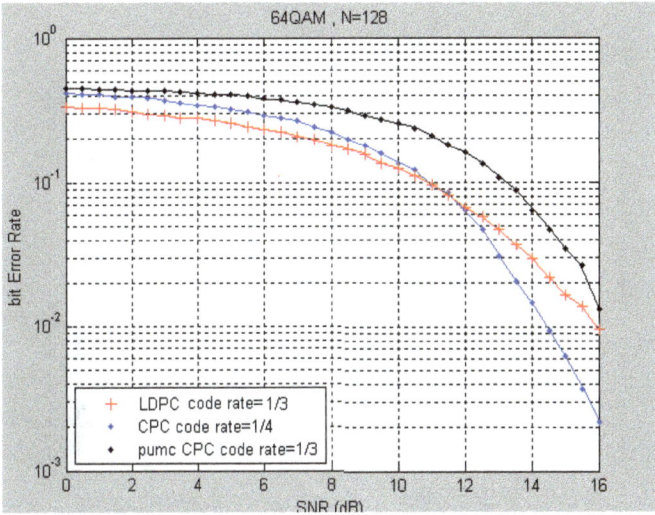

Fig. 39. BER Comparison between LDPC code, CPC, punctured CPC at 64 QAM, N=128.

From Figure 39 it is shown that the results obtained from LDPC code is better than t puncture modified CPC until SNR = 16db, but LDPC is more complicated than the proposed method. From Figure 35 to Figure 37 BER versus different received SNR values are shown for the comparison between modified CPC, puncture CPC and convolutional code, LDPC code and turbo code respectively through the fading channel. These comparisons are obtained for modulation type 64QAM and number of subcarriers equals 512.

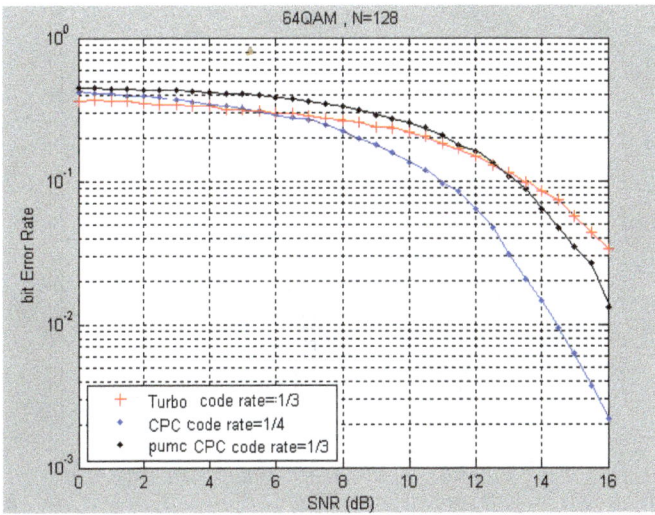

Fig. 40. BER Comparison between Turbo code, CPC, punctured CPC at 64 QAM, N=128.

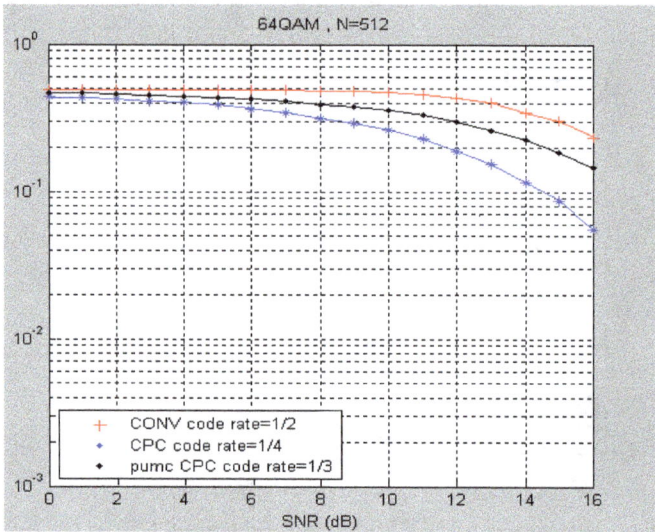

Fig. 41. BER Comparison between Conv code, CPC, punctured CPC at 64 QAM, N=512.

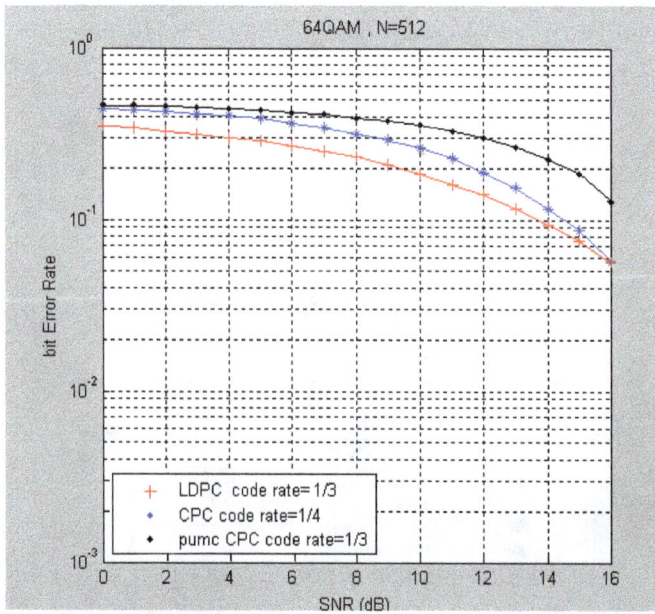

Fig. 42. BER Comparison between LDPC code, CPC, punctured CPC at 64 QAM, N=512.

From Figure 42 and Figure 43 it is shown that the results obtained from LDPC code and turbo code is better than puncture modified CPC, but LDPC and turbo code is more complicated than the proposed method.

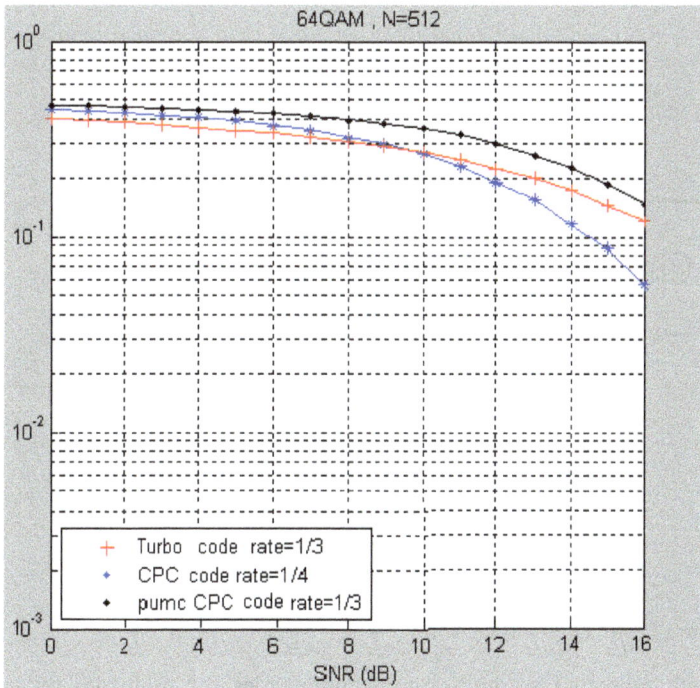

Fig. 43. BER Comparison between Turbo code, CPC, punctured CPC at 64 QAM, N=512.

10. Delay diversity scheme

In this section, both puncture and Delay Diversity Scheme (DDS) are together used to increase the efficiency of CPC system. To further improve the diversity of the channels a transmit diversity technique may be utilized.

Many transmit diversity techniques have been explored. One such technique is the transmit delay diversity Figure 6.44. In transmit delay diversity a transmitter utilizes two antennas that transmit the same signal, with the second antenna transmitting a delayed replica of that transmitted by the first antenna. By so doing, the second antenna creates diversity by establishing a second set of independent multipath elements that may be collected at the receiver.

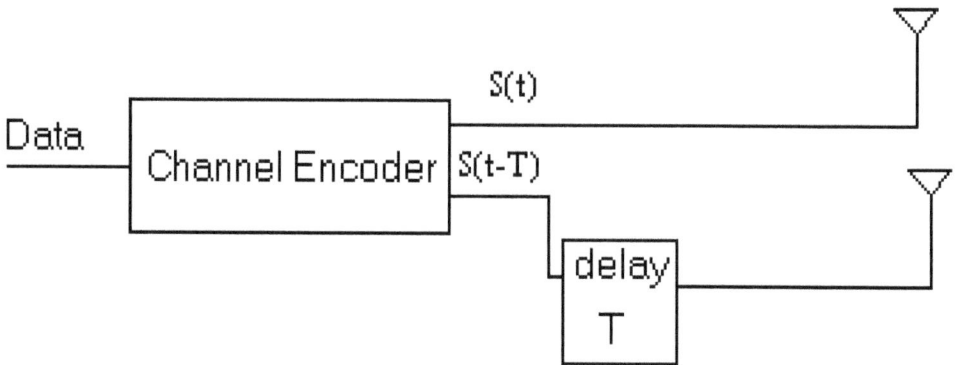

Fig. 44. Transmit Delay Diversity Scheme

If the multipath generated by the first transmitter fades, the multipath generated by the second transmitter may not, in which case an acceptable SNR will be maintained at the receiver. This technique is easy to implement, because only the composite TX0+TX1 channel is estimated at the receiver. The biggest drawback to transmit delay diversity is that it increases the effective delay spread of the channel, and can perform poorly when the multipath introduced by the second antenna falls upon, and interacts destructively with, the multipath of the first antenna, thereby reducing the overall level of diversity.

Our simulator for delay diversity technique is based on passing the same signal through the same path during two time intervals by using only one transmitted antenna Figure 6.45, not two transmitted antennas as in delay diversity technique. During these intervals the channel will have different fading and AWGN characteristics over the time.

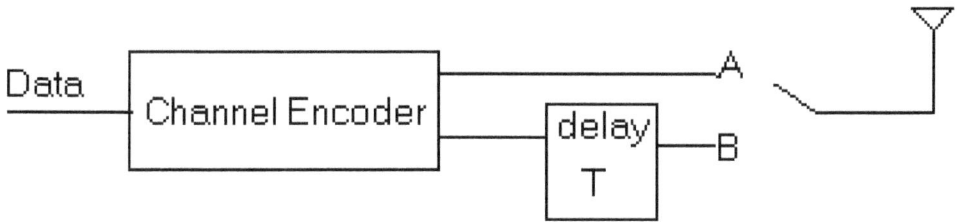

Fig. 45. Modified Transmit Delay Diversity Scheme

The receiver, by using one received antenna, chooses the best received signal according to its highest power.

11. Conclusions

In this chapter, we explained CPC method as a coding technique and our modification for it. Also the implementation of CPC in WiMAX system and the comparisons between its results and the results of other coding techniques such as convolutional, turbo and LDPC are investigated at different SNR for different number of subcarriers and at different types of modulation (16QAM – 64QAM).

12. References

[1] Nam Yul Yu, Young Kim, and Pil Joong Lee (2000). Iterative Decoding of Product Codes Composed of Extended Hamming Codes," 5th IEEE Symposium on Computers and Communications (ISCC 2000), France, pp. 732 -737, July.

[2] R. M. Pyndiah, (1998). Near-Optimum Decoding of Product Codes: Block Turbo Codes," IEEE Trans. Communication., vol. 46, no. 8, pp. 1003-1010, Aug.

[3] T. Shohon, Y. Soutome and H. Ogiwara. (1999). Simple Computation Method of Soft Value for Iterative Decoding of Product Code Composed of Linear Block Code," EIC Trans. Fundamentals, vol. E82-A, no. 10, pp. 2199–2203, Oct.

[4] Omar Aitsab and Ramesh Pyndiah. (1996). Performance of Reed Solomon Block Turbo Codes," Proc. IEEE LOBECOM' 96 Conf., London, U.K., vol. 1/3, pp. 121-125, Nov.

[5] David Rankin and T. Aaron Gulliver. (2001). Single Parity Check Product Codes," IEEE Trans. Commun., vol. 49, no. 8, pp. 1354-1362, Aug.

[6] Lin, S. and Costello. (1983). Error Control Coding: Fundamentals and Applications, Prentice Hall, Englewood Cliffs, New Jersey.

[7] V. Tarokh V, N. Seshadri, and A. R. Calderbank. (1998). Space-Time Codes for High Data Rate Wireless Communication: Performance Criterion and Code Construction," IEEE Trans. Inform. Theory, vol. 44, no. 2, pp. 744-765, Mar.

[8] Orhan Gazi and Ali Özgür Yılmaz (2006). Turbo Product Codes Based on Convolutional Codes," ETRI Journal, Volume 28, Number 4, Aug.

[9] Shannon, C. E. (1948). A mathematical theory of Communication," *Bell Syst. Tech. J., vol. 27, pp. 379–423, 623–656, July and Oct.*

[10] Proakis, J. G. and Salehi, M. (1993). Communication Systems Engineering, *Prentice Hall, New Jersey.*

[11] K.Miyauchi, S.Seki, and H. Ishio (1976). New Techniques for Generating and Detecting Multilevel Signal Formats," *IEEE Trans Communication, vol. COM-24, pp. 263-267, Feb.*

9

Design and Implementation of WiMAX Baseband System

Zhuo Sun, Xu Zhu, Rui Chen,
Zhuoyi Chen and Mingli Peng
Beijing University of Posts and Telecommunications,
China

1. Introduction

Design and implementation of a wireless communication system protocol stack on the hardware platform are challenging tasks, which are seldom mentioned in published results currently. In fact, the work aims at achieving the predetermined function and performance based on the specific hardware resource, which involves how to design the software architecture according to hardware, how to choice the suitable algorithms and program them optimally on programmable DSP or embedded processor, etc. Typically, there are two ultimate application modes for a communication protocol stack: dedicated ASIC or programmable DSP. However, before the protocol stack is formed into dedicated ASIC, the task of protocol or algorithms implementation and testing on the programmable DSP should be finished primarily. Therefore, the chapter will focus on the topic of development of the WiMAX PHY/MAC protocol on a multi-core DSP platform.

The contents of this chapter will be organized into three sections:

The first section, we will discuss the overall software and hardware architecture of a WiMAX TDD baseband system, and what are the most import considerations in this design phase.

The second section, we address on the topic of developing the WiMAX PHY protocol on the programmable multi-core DSP platform.

The third section, it is about the design and implementation of WiMAX MAC protocol on the embedded processor, on which embedded Linux OS is running.

2. Hardware platform for WiMAX system

2.1 Baseband system

2.1.1 Requirements

Wireless systems designers need to meet a number of critical requirements including processing speed, flexibility, and time-to-market, all of which ultimately drive the hardware platform choice.

- Processing throughput

WiMAX and LTE broadband wireless systems have significantly higher throughput and data rate requirements than W-CDMA and cdma2000 cellular systems. To support these high data rates, the underlying hardware platform must have significant capability of processing throughput. In addition, advanced signal processing techniques such as Turbo coding/decoding, and front-end functions including fast Fourier transform/inverse fast Fourier transform (FFT/IFFT),beam-forming, MIMO, crest factor reduction (CFR), and digital pre-distortion (DPD) are computationally intensive and require several billion multiply and accumulate operations per second.

- Flexibility

WiMAX is a relatively new market and is currently in the initial development and deployment stages. Similarly, 3GPP LTE is being defined and will go through numerous revisions before being finalized. While there are many competing mobile broadband technologies, such as WiMAX, LTE, and UMB, their common thread is OFDMA-MIMO (Parssinen et al., 2000). In this current scenario, having a flexible and reprogrammable product is necessary to provide a standards-agnostic or multi-protocol base station. Systems offering this flexibility can significantly reduce the capital expenditures and operating expenditures for wireless infrastructure OEMs and operators while alleviating risks posed by constantly evolving standards.

- Cost-Reduction Path

A valuable lesson learned from designing and deploying 3G systems is the importance of establishing a long-term cost-reduction strategy in the beginning. Evolving WiMAX and LTE standards are expected to stabilize. For OEM sand service providers to remain competitive in the marketplace, the cost of the final product eventually will be more important than flexibility. Choosing the right hardware platform also provides a seamless cost-reduction path for production volumes, saving millions of dollars in engineering costs that would otherwise be incurred by system redesign.

2.1.2 Generic hardware architecture

The goal of LTE and WiMAX is to provide a high-data-rate, low-latency and packet-optimized radio-access technology supporting flexible bandwidth deployments. In addition, new network architecture is designed with the goal to support packet-switched traffic with seamless mobility, quality of service and minimal latency.

Due to the demand of high performance of high data-rate, low-latency and reduced delays, the choice of LTE hardware platform is a challenging job. The general architecture of the mobile communications system is depicted as the figure below.

The architecture can be divided into three parts: radio frequency (RF), intermediate frequency (IF) and digital baseband. The mainstream design of receiver proceed as follows: sampling the analog signal in the intermediate frequency, then down-converting the digital signal to baseband, finally demodulating the digital baseband signal. The transmitter is just opposite of the receiver (Dohler et al., 2005). Moreover, there may be CFR (Crest Factor Reduction) and DPD (Digital Pre-Distortion) modules before converting to analog signal.

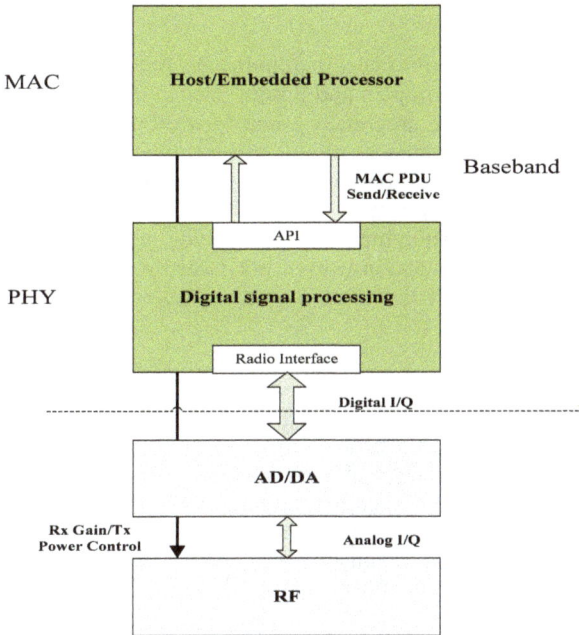

Fig. 1. The general hardware architecture of mobile communications system

The process in IF contains NCO (numerical controlled oscillator), CIC filter, half-band filter and FIR filter. All of the modules mentioned above have such a simple structure that they can be easily implemented in an FPGA. Because of the high data-rate of IF digital data, it is impossible to implemented in DSP. For example, first decimation filter in a digital wireless receiver, typically, is a CIC filter, operating at a sample rate of 50-100MHz. At these rates any DSP processor would find it extremely difficult to do anything. However, the CIC has an extremely simple structure, and implementing it in an FPGA would be easy. A sample rate of 100MHz should be achievable, and even the smallest FPGA will have a lot of resource available for further processing. In addition, the latest mobile communications system has employed MIMO technology, which means there will be two or more antennas. As the fact of introducing MIMO, the parallel sampling data will be processed simultaneously. Therefore, it is general to choose FPGA or ASIC as the processor in IF instead of DSP. For example, AD6654 is an IF to baseband receiver, with programmable decimating FIR filters, interpolating half-band filters and CIC filters built-in (Fathi, 2004).

Baseband is usually divided into two parts. One of them is digital signal processing, which is used for implementing PHY protocol. The other part is microprocessor used for implementing MAC and lower protocol (Goldfarb et al., 2000). The microprocessor is generally selected within ARM or Power PC processor. However, the choice of baseband digital signal processing solution is various. Generally there are four choices which are ASIC, DSP, FPGA and DSP+FPGA (Jusslia et al., 2001). Application-specific integrated circuit (ASIC), is an integrated circuit (IC) customized for a particular use, rather than intended for general-purpose use. Digital signal processors (DSP) are a specialized form of microprocessor, while FPGAs are a form of highly configurable hardware.

- ASIC

According to circuit functions and performance requirements, ASIC design needs to select circuit form, the device structure, process plan and design rules to minimize chip area, lower design cost and shorten the design cycle, and finally brings forward the correct and reasonable mask layout. Nevertheless, the disadvantages of full-custom design can include increased manufacturing and design time, increased non-recurring engineering costs, more complexity in the computer-aided design (CAD) system. Moreover, Due to the changing demand of mobile communications system, the equipment has to upgrade one day, but ASIC cannot upgrade flexibly. When the hardware platform does not meet the requirements, all of the equipment must be replaced. As a result, the cost of upgrading is very expensive.

- DSP

A digital signal processor (DSP) is a specialized microprocessor with an optimized architecture for the fast operational needs of digital signal processing. Digital signal processing algorithms typically require a large number of mathematical operations to be performed quickly and repetitively on a set of data. Signals (perhaps from audio or video sensors) are constantly converted from analog to digital, manipulated digitally, and then converted again to analog form. Many DSP applications have constraints on latency; that is, for the system to work, the DSP operation must be completed within some fixed time, and deferred (or batch) processing is not viable (Parssinen et al, 1999).

- Multi-core DSP

Multi-core processing is the technology or group of technologies that companies like Intel and IBM are betting will replace Instruction Level Parallelism and the clock rate ratchet: dual and quad core systems for desktop applications are already in volume production. As we have seen, in many cases the controlling factor in device performance has moved from the ability to complete computation to the ability to move data. Well designed multi-core architectures allow data stores from registers to main memory to be distributed throughout the system, in whatever way makes most sense for the application. In fact, in multi-core architectures the communications fabric can substitute for memory accesses by allowing direct communication between the processing elements. If matched to the task in hand, such an infrastructure can therefore intrinsically help to overcome any restrictions imposed by the need to move data.

- FPGA

The FPGA configuration is generally specified using a hardware description language (HDL), similar to that used for an application-specific integrated circuit (ASIC) (circuit diagrams were previously used to specify the configuration, as they were for ASICs, but this is increasingly rare). FPGAs can be used to implement any logical function that an ASIC could perform. The ability to update the functionality after shipping, partial re-configuration of the portion of the design and the low non-recurring engineering costs relative to an ASIC design (notwithstanding the generally higher unit cost) offer advantages for many applications.

- FPGA-DSP Co-Processing

FPGA and DSP represent two very different approaches to signal processing – each good at different things. There are many high sampling rate applications that an FPGA does easily, while the DSP could not. Equally, there are many complex software problems that the FPGA cannot address. Another advantage of co-processing is reconfigurable features of FPGA, which means that engineers can quickly build and modify the design architecture. Moreover, FPGA supports the integration of other components (such as Serial Rapid IO transceiver, PCI Express interfaces, glue logic and low-rate control task), which reduces overall system cost and power consumption. In addition, the integration of so many interfaces is valuable for scalability, which meets the changing demand of mobile communications system. As a result, the ideal system is often to split the work between FPGAs and DSPs.

2.1.3 Communication between MAC and PHY

In the following discussion, we adopt the picoChip PC7205 development platform as our baseband hardware platform, in which integrated one of the multi-core DSP processor (PC205) and one piece of FPGA. The PC205 process also includes an embedded ARM926EJ processor that could implement the MAC functions.

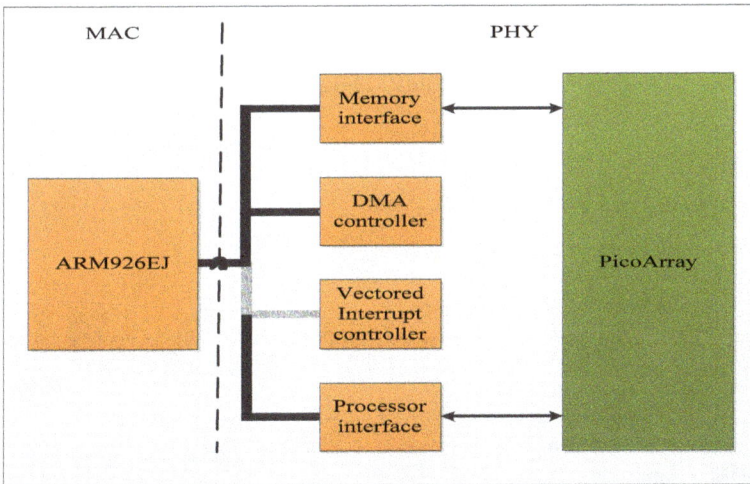

Fig. 2. PC205 block diagram

The PC205 microprocessor interface is designed for communications with a processor. No specific processor family is assumed and data can be exchanged over 8, 16 or 32-bit wide data bus. The processor interface is used for communication between MAC and PHY.

The Processor interface supports two basic transaction types

1. Single (GPR) – Reading or writing one word at a time.
2. Burst (DMA) – Reading or writing words at the same rate as the microprocessor proc clock.

GPR accesses allow access to the majority of memory mapped registers and services within the processor interface, GPR accesses can only be used for single read / write accesses. Typically, GPR is used for transmit control signals whose amount of data is small such as automatic power control (APC) signal between MAC and PHY.

DMA Accesses are primarily used for the efficient movement of data to and from the picoArray. Generally we use DMA to transmit the bulk data such as wireless frame between MAC and PHY. Fig.3 shows the DMA channel configured for write access. A FIFO buffers the data written from the microprocessor to the selected DMA channel. The FIFO output is connected to the picoBus and data is transferred to the internal array elements by using a get command from within the software.

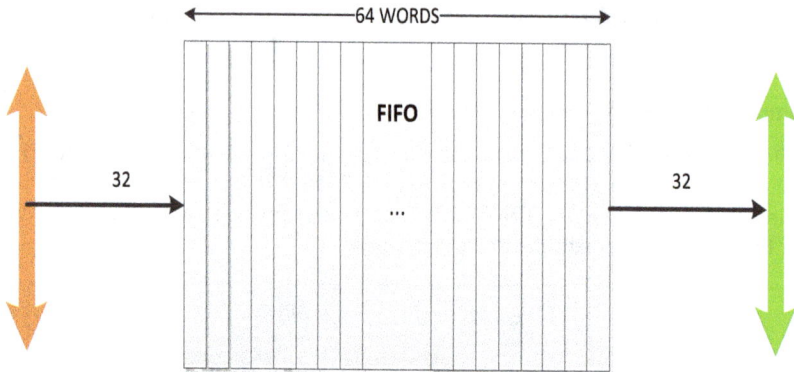

Fig. 3. DMA channel configured for write access

The PC205 support 3 DMA transfer mechanisms.

1.   Basic downlink Host processor to picoArray
2.   Basic uplink picoArray to Host processor – not practical
3.   HWIF_UL picoArray to Host processor

In the mode of Basic downlink, host processor initiates transfer. The process is as followed.

*   Open a transport session
*   Configure the transport session
*   Start the transport session
*   Write DMA data
*   Close the transport session

In the mode of basic uplink, host processor initiates transfer, but the processor has to pre-assumes data size, which is not practical.

In the mode of HWIF_UL, PicoArray uses handshake mechanism to indicate data size through GPR registers, and uses interrupt to initiate transfer through ITS register. The process is as followed.

*   Open and setup an interrupt clearing transport session used for clearing down interrupts.
*   Open and setup a HWIF_UL transport session for uplink DMA.
*   Poll for an event indicating the date available.
*   Read DMA data
*   Close the transport session

2.2 Connecting baseband to RF

2.2.1 Connecting to RF by using FPGA

Generally there are two kinds of interface signals between baseband and RF, which are data and control signals respectively. Data signal are usually 16 bits complex format. However, the control signals may adopt one communication protocol such as SPI, I2C and so on. Moreover, some of the protocols may be changed for the sake of implementation. The interface of DSP may not support all the protocols. As the result, it is the right way to introduce one piece of FPGA between the baseband and RF for flexibility and scalability. We can write the suitable protocols for almost all interfaces in FPGA.

In addition, for the sake of power consumption, more and more DSPs have chosen 1V and 1.8V as power supply of the core and interfaces respectively. But the other device may take 3.3V as the power supply of the interfaces. It is obvious that electrical characteristics don`t matched between the different interfaces. Nowadays, most of FPGA have more than one bank, and each bank can be supplied different power. We can use some of banks with 1.8V power as the interfaces with DSP, while the other banks with 3.3V power as the interfaces with some other device. As a result, it is easy to change the logic level.

2.2.2 Automatic gain control (AGC) and power control

Automatic gain control (AGC) Automatic gain control (AGC) is an adaptive function found in many electronic devices. In a digital communication receiver strong signals that fall outside the narrowband digital filter bandwidth, but inside the analog IF translator bandwidths, can overload or saturate the A/D converter. This results in the generation of in-band IMD products and can result in significant degradation of the desired signal. If large signal levels are detected at the A/D converter, the receiver gain may have to be re-distributed by reducing the pre-conversion analog gain and increasing the digital gain to maintain the desired signal output level. This will, however, reduce the desired signal-to-quantization noise ratio.

- Power Control

In the WiMAX system, there are two mechanisms for power control, which are open loop power control and closed loop power control respectively. It is necessary to have closed loop power control while open power control is optional. Closed loop power control means that the base station (BS) controls the transmission power of the mobile station (MS). The MS transmission power is controlled in order to avoid exceeding the BS`s total receiving power from an antenna. In the WiMAX standard, other uses of it are not defined (i.e., the uplink TPC algorithm is vendor specific).

- AGC and Power Control signal design

Usually RF device has the specific module for receiving gain and transmission power adjusting, which is controlled by voltage signals. Therefore, baseband just outputs direct current signals with variable amplitude to RF. Typically we can obtain the direct current signals with low rate DAC, but the interface between baseband and RF have to increases parallel lines used for transmitting the digital. Here we introduce a simple method to generate the direct current signals.

We can make use of Pulse Width Modulation (PWM) signal and a RC low-pass filter to generate the direct current signal. When the PWM signal duty ratio is 100 percent, the amplitude of direct current signal equals to the amplitude of PWM signal. When the PWM signal duty ratio is 50 percent, the amplitude of direct current signal equals to the half amplitude of PWM signal. The direct current signal is approximately linear with the duty cycle. In this case, it is necessary to use two digital signal lines for receiving gain and transmission power control.

2.2.3 Extern GPS synchronization signal

The IEEE 802.16 standard calls for the use of global positioning system (GPS) receivers to provide the precise time reference for synchronization of WiMAX networks. This operation is performed both during the startup and periodically in order to maintain the alignment with the external PPS pulse.

Briefly, the algorithm follows these steps. The controller of synchronization starts searching for the first PPS pulse while discarding the RX samples. Then it stalls the PHY while waiting for the PPS pulse and sends DL dummy complex samples. Once received the PPS pulse, after 100 ms, the controller starts passing the DL complex samples. For each frame period, the frame synchronization module receives the frame start indication and decides when a frame adjustment is required for maintaining the alignment with the external pulse.

3. Physical layer implementation

3.1 Introduction of PHY

Considering PHY implementation, the main functions of PHY layer may comprise API, control, transmit-path, receive-path and synchronization/radio interface for both BS and MS entities (LAN/MAN Standards Committee of the IEEE Computer Society et al., 2008). Fig.4depicts the relationship between these function blocks.

API:

The API provides an interface between the PHY and MAC. Its function is responsible for:

- The physical transfer of data to and from the MAC
- Buffering of data to and from the MAC
- Error checking and diagnostics on the data
- Interpreting data from the MAC and generating internal control data for other functions in the PHY
- Interpreting data from the PHY and parsing into data for the MAC

Control:

The control function is responsible for distributing control data originating in the API around the other functions of the PHY. Its most important task is to ensure that each of the data-path functions has access to the control data that it needs to process the data-path data it is currently working on. Secondly it orchestrates the collection of measurements from the PHY and provides routes for diagnostic information to be accessed by the MAC or other external processes.

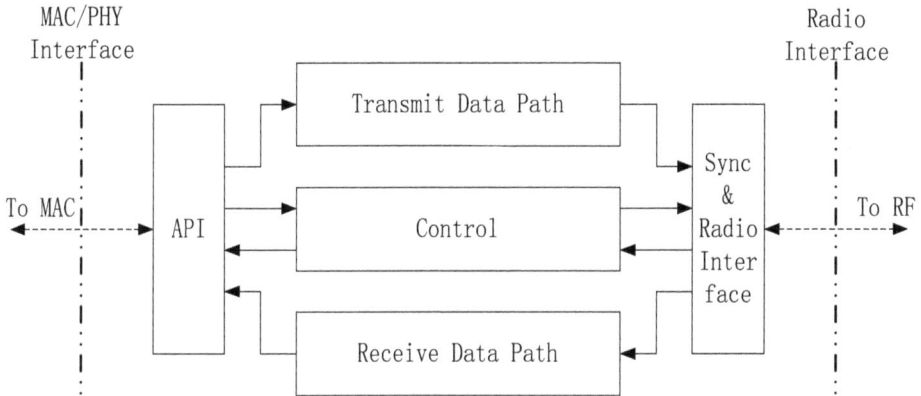

Fig. 4. Main function blocks of PHY Layer

Transmit-path:

This block provides the main data path for transmit in the PHY, which comprises the following stages:

- Unpacking of data into FEC blocks, Encoding, including randomization, FEC, interleaving, repetition---FEC block
- Data modulation: The conversion from bit stream to QPSK, 16QAM, 64QAM symbols---ConstPack block
- OFDMA zone processing, including pilot and preamble generation subcarrier permutation, subcarrier scrambling and zone boost---BurstZoneblock&AntEnc block
- Antenna processing, including IFFT, cyclic prefix, Peak-to Average reduction and transmitting filtering---FrondEnd block

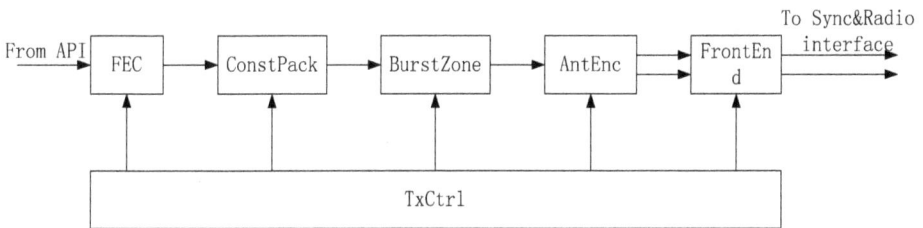

Fig. 5. BS transmit process

Receive-path:

This block provides the main data path for receive in the PHY, which including:

- Time and frequency synchronization process---MsRxAcq block &MsRxTf block
- Antenna processing with receiving filtering (Foschini, 1996), ALC, cyclic prefix removal and FFT---MsRxTf block

- OFDMA zone receive, with AAS/MIMO (Mugenet al., 2007) processing and buffering, subcarrier descrambling and depermutation, pilot extraction, CPE and frequency compensation, channel estimation and equalization, constellation demaping, channel state compensation and MRC---MsRxSym block &MsRxMap block
- Decoding: including derepetition, deinterleaving, depuncturing, H-ARQ (Lin and Yu, July 1982), FEC decoding, derandomising and repacking of user data into PDUs---MsRxBurstChain block

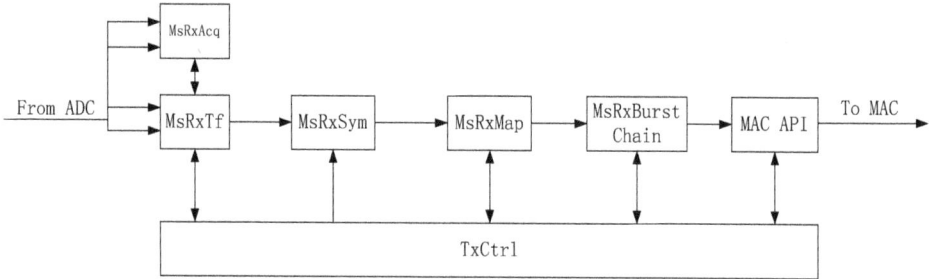

Fig. 6. MS receive process

Synchronization /radio interface:

This block is responsible for controlling the absolute and relative timing of uplink and downlink frames in the PHY and the radio. It is also responsible formultiplexing, demultiplexing and formatting data for the interface to the radio via the picoArray ADI (Asynchronous Data Interface) interface(PicoChip Company, 2008). For this reason realizing block may well be somewhat platform specific.

3.2 Link-level simulation

3.2.1 Simulation platform based on MATLAB

Before realizing the whole WiMAX PHY layer software on the picoArray DSP (PicoChip Company, 2008), a fixed-point link-level simulation is needed. First of all, it's important to make sure that the algorithms are correct and satisfy the performance demand in simulation environment. Because the development on DSP processor is time-consuming and expensive, the consequence of implementing a system that will never work in DSP processor can't be affordable. Secondly, it's very difficult to locate bugs and correct them in DSP processor. When the bugs have nothing to do with hardware, the bugs finding and correcting work can move back to simulation platform. This will save your development time and cost significantly. At last, there are varies wireless channel models in MATLAB, which are very useful for us to figure out how the system performs on different channel environment.

MATLAB simulation platforms are floating point in common situations. But this simulation platform does an extra job that it converts the calculation result from floating point to fixed-point result. That is to say, the simulation platform is a fixed-point platform which can be more approaching to fixed-point picoArray DSP processor. In this way, the performance between MATLAB simulation platform and DSP process on picoArray will be the same roughly. It makes the simulation more convincing.

3.2.2 Simulation platform development

The MATLAB simulation platform is built in the way that all blocks of the platform are map to the functions on picoArray respectively. So, the function blocks are transferred from MATLAB platform to DSP platform smoothly. Moreover, each block of the MATLAB fixed-point simulation can generate the corresponding result of this block which can be used as input of the followed block on the DSP platform directly. That is a very efficient way to verify the functions on the picoArray.

Matlab platform is built in accordance with the WiMAX physical layer protocol. The platform complies with the frame structure and resource distribution of WiMAX standard. It can generate any structure's sub-frame. Also it can provide fixed-point simulation for each sub-frame. Moreover, the platform can generate test vectors for each of the WiMAX PHY's module. The test vectors can be mapped to the AE level (picoChip processing unit), including the input control information and the input and output data of each AE.

3.3 PHY protocol implementation on picoArray DSP

In the following, the PHY protocol implementation on picoArray DSP, the chosen Multi-Core DSP for baseband application will be introduced in details.

3.3.1 PicoArray introduction

The picoArray multi-core DSP is based on a massively parallel architecture comprising large numbers of small independent processors. A DSP application is logically decomposed into a number of communicating sequential processes, each of which is assigned to a particular processor on the picoArray. The designing tools statically allocate processors and picoBus (PicoChip Company, 2008) resources for the system, so there is no need for an operating system. The static allocation of resources allows much of the system's runtime complexity to be moved back into the tool suite. It allows the hardware to be lightweight and hence allows a very high proportion of the power of the processor array to be used for the real DSP application.

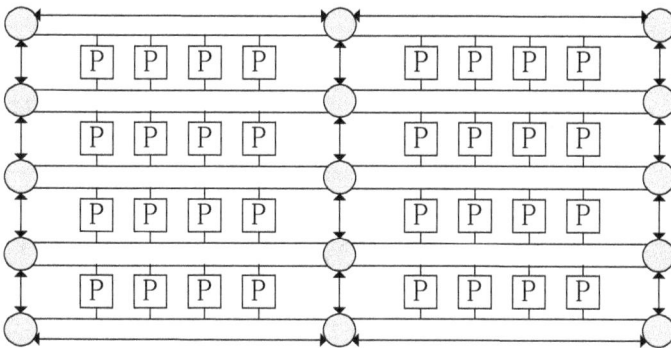

Fig. 7. A simplified representation of a picoArray

Fig.7 shows a simplified representation of the overall structure of the picoArray. Each box marked 'P' in the figure represents a single processor, referred to an Array Element(AE).

The processors are laid out in a grid, interconnected by a matrix of buses called the picoBus. Each AE is connected to two buses. The lines between the processors represent the picoBus, and the circles represent bus switches which connect buses together to provide routes between all AEs in the array. The communications between these processes, called signals, are then mapped on to physical segments of the picoBus between the assigned processors by suitable settings of the bus switches. The heavy red and blue lines illustrate two example connection paths between particular processors. Communication between AEs is time-multiplexed over the picoBus, which is a shared resource.

3.3.2 PHY implementation

A PicoArray process is composed of a number of AE which can work simultaneously. This structure is quite different from traditional single-core processor. As a result, developing work on picoArray will share nothing with that on traditional single-core processor. The major steps of developing work on picoArray are given as follows.

All the Instruction/Data memory that developers can utilized are in the core in traditional processors, so developers needn't care about how to arrange Instruction/Data memory for each functional block. But when doing developing work on picoArray, the first thing developers need to deal with is to select suitable AE for each functional block. There are some different categories of AE which have quite different abilities. Three types of AE are mostly used in our work: STAN2 (short for standard AE), MEM2 (short for memory AE) and CTRL2 (short for control AE).

Feature	CTRL2	MEM2	STAN2
Bus connections	4	2	2
Number of ports	32	12	10
Number of registers	15	15	15
Instruction/Data memory size options (bytes)	49152/16384 232768/32768 1 16384/49152 0 default 0	6656/2048 3 4608/4096 2 2560/6144 1 512/8192 0 default 0	512/256
Byte memory accesses	Yes	Yes	No

Table 1. AE types in PicoArray processor

For example, there are 196 STAN2 AEs, 50 MEM2 AEs and 2 CTRL AEs in one piece of picoArray PC205. The ability contrast is shown in the above table. When arranging the functional blocks into AEs, developers should select a good enough type of AE to implement the block according to the need of the blocks. Moreover, developers should optimize the code of the block to fulfill the limit of different types of AE. If one piece of picoArray can't hold all the blocks, another piece of picoArray processor should be introduced in to share the burden. The communication between the two pieces of picoArray processor is accomplished by IPI (Inter PicoArray Interface).

After developers have selected AEs for each block, memory access type should be selected. There are three types of data memory for data storage: data registers inside AE, data memory inside AE and SDRAM outside AE. The data registers are very fast access memories. When storing a small amount of variables, developers ought to use the registers as far as possible. There are only 15 registers in one AE as shown in table 1. If the size of data which needs to be stored exceeds the amount of unused registers, the data should be stored in the inside data memory. The access speed of this type of memory is a little slower than registers but is much faster than SDRAM outside of AEs. Last but not least, if the size of data exceeds the capacity of the memory inside AE, SDRAM is used to store it. In this situation, developers must arrange the SDRAM access area very carefully because the SDRAM is shared among different AEs which need it for their data storage. If some of the AEs use the same area in the SDRAM, fatal error will take place unexpectedly and is difficult to discover.

Then it comes to programming step. Firstly, functional code is created carefully based on the MATLAB simulation platform for each block. Then verify the code's syntax accuracy and logical accuracy with picoTools. The next step is taking the MATLAB simulation test vector as the input vector of each block. The function of block on every AE is verified by through comparison between the AE output and the MATLAB simulation result. The throughput matching is another important process when programming. The reason is if the throughputs among the blocks don't fit for each other, they can't work when connected together. If this problem happens, you should change your design to matching the throughputs demand. One principle is that the getting/reading port rate of the AE must be faster than the rate of the AE putting/writing data.

Last but not least, debug on picoArray. All the above coding and debug work is done with picoTools on development environment. It is easy to select the result of each block in the form of text document to be compared with MATLAB simulation result for verification. When it comes to the debug work on the picoArray, it is much more difficult to get the result of each block. As some hardware-related bugs can't be discovered on software environment, it's very important to get some methods for the debug work on picoArray. Fortunately, a probe mechanism is provided. You can configure the unused AE or unused SDRAM to get the output of the block which is needed to debug. The data in the 'probe' AE or SDRAM can be transmitted into text document to compare with MATLAB simulation result of the same block.

Some tips of developing on the picoArrays are given as follows:

First, the sum of the rates of signals connected to one single AE can't exceed 2 because there are two channels connecting one AE to picoBus. For example, if one AE has three signals names sig_A, sig_B and sig_C. The rates of the three signals are @2, @2 and @1 respectively. It is easy to find out that $1/2+1/2+1/1=2$, so there are no channel space for another signal to connect to this AE.

Secondly, for a process chain, the getting data rate of AE must be faster than the providing data rate of the previous AE. This is very important rule during the picoArray DSP design, because if this can't be satisfied, data flow would be blocked. Moreover, if one AE's data flow is blocked, the conjoint AE' data flow will be blocked too.

3.4 API architecture design

3.4.1 API introduction

The API interface is between the MAC and OFDMA PHY, which is defined to picoChip's IEEE 802.16e base-station PHY and optional lower MAC accelerator to perform CRC and HCS calculations. The API described in this document is based on the Wireless MAN-OFDMA PHY. The greatest feature of API is to process data and control information separately.

- The API addresses data and control plane functions. Data plane functions include the transfer of MAC PDUs in the uplink and downlink directions via the PHY service access point (SAP). Information required for uplink and downlink processing is sent separately within a frame configuration structure.
- Control-plane functions include determining the capabilities of the PHY, reconfiguration of the PHY and notification of error conditions, alarms and measurements gathered from the PHY or radio subsystem.
- The base-station PHY can be configured to perform CRC and HCS calculations to lessen the processor requirements for the MAC. Encryption and decryption is beyond the scope of this API document.
- A frame-sync interrupt is sent to indicate the start of every downlink sub-frame; this mechanism operates in parallel to this API.
- The response and indication primitives sent from the PHY to MAC can be masked at PHY configuration, allowing the MAC to select only the messages it is interested in.

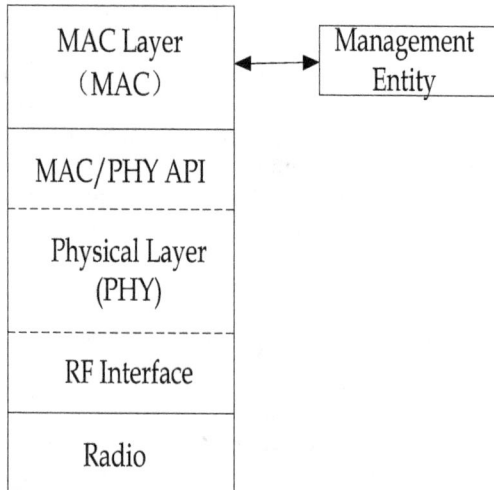

Fig. 8. 802.16e Protocol Stack

Fig.8 shows a modified version of the 802.16e reference model and the location of the API as described in this document (shown as MAC/PHY API). The 802.16e protocol stack is shown, together with the radio and management entity.

3.4.2 The block diagram of API

According to the function, the API is divided into two parts: communication with the MAC and the data processing chain of PHY layer. The first part includes getting messages from MAC, sending data to the SDRAM and regrouping the receive message to MAC. The messages from MAC include control-plane messages, which are used to perform configuration and reconfiguration of the PHY, and data-plane messages. Both messages have the same message format which includes message header, which descript message type and PHY entity, and message body. The receive messages to MAC are called response or indication messages. Then it comes to communication with PHY. The input signals of each module in the data processing chain can be classified into two types: control and data signal. Among them, all control signals come from the API; data signal is the output of the previous module. However, the beginning of the data also derives from the API. So, API separately process control and data information to produce the two signals. Their modules are Control system and Data processing, which are shown in Fig. 9.

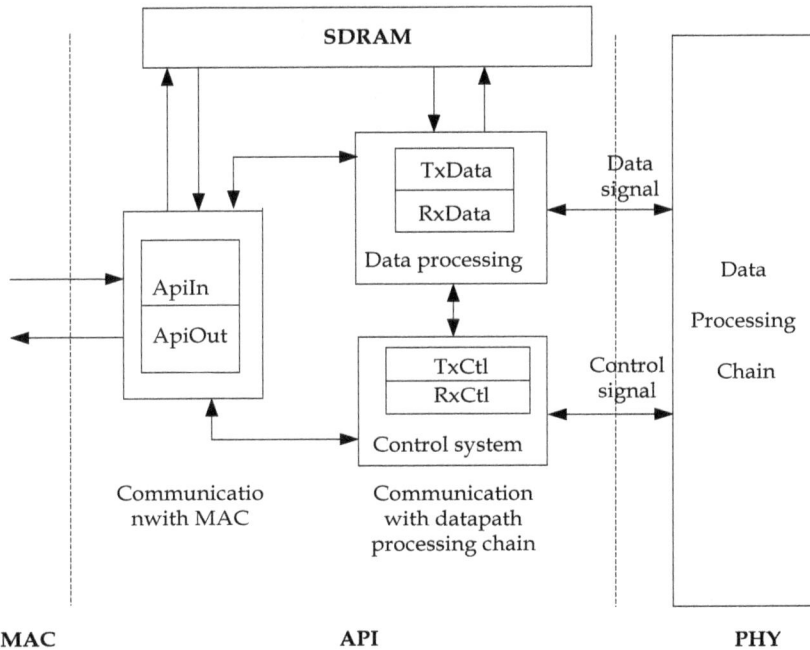

Fig. 9. The block diagram of API

4. MAC protocol implementation

4.1 Introduction of MAC functions

According to the OSI seven-layer network protocol，MAC lays between the PHY layer and the network layer, responsibility for the data convergence and resource scheduling. So there should be appropriate interfaces between the PHY and network layers. In order to highlight the implementation of the MAC layer, for the following description in the chapter, the network layer is designed simply. Its main functions and features are shown in Fig.10 (Du, 2010).

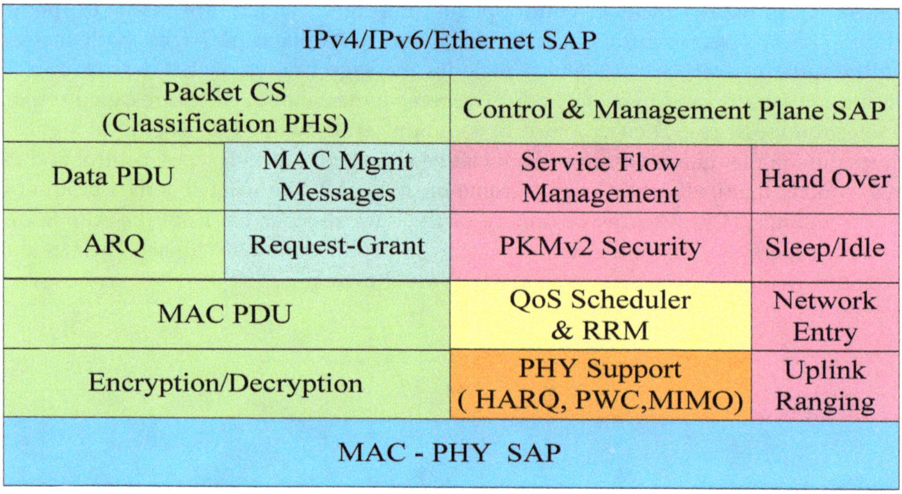

IPv4/IPv6/Ethernet SAP			
Packet CS (Classification PHS)		Control & Management Plane SAP	
Data PDU	MAC Mgmt Messages	Service Flow Management	Hand Over
ARQ	Request-Grant	PKMv2 Security	Sleep/Idle
MAC PDU		QoS Scheduler & RRM	Network Entry
Encryption/Decryption		PHY Support (HARQ, PWC,MIMO)	Uplink Ranging
MAC - PHY SAP			

Fig. 10. The structure of MAC functions

4.2 Implementation of MAC layer

The embedded ARM-Linux operation system is chosen as the development environment of the MAC layer. Considering the requirement of the processing time, the multi-threads techniques are designed so that the MAC layer can packet and parse messages in time. Besides that, the algorithms adopted by the system are needed optimizing to achieve the compromise between system performances and the complexity. Because the message process procedures are different at the different BS/MS state, the state machine is designed to track the state of the BS/MS. The implementation details are introduced in the following part.

4.2.1 State machine

This section will introduce the BS state machine and MS state machine briefly. For BS, the state machine of BS from the startup to normal and the state of the accessed MS to the current BS are designed. For MS, the corresponding state machine is also designed. The conditions of the state transfer are defined.

1.   State machine for BS

According to the functions of BS, the implementation of the BS state machine is divided into two modules. The first one is BS-State-Machine which manages the BS own state and state transition. The second one is MS-State-Machine which manages the states of MSs which have already accessed or attempted to access to the current cell. The BS-State-Machine is responsible for tracking the states of accessed MSs.

• BS-State-Machine module

The main functions of this state machine are to hold the current state of BS, parse the messages which are received from PHY layer, packet the messages which are sent to the PHY layer and transfer to another state. The BS state machine is designed as three states, namely STATE_CONFIG, STATE_NORMAL, STATE_NULL. The BS's initialization state is STATE_NULL. After the startup, the MAC layer entity of BS will configure the PHY layer with specific parameters and change the current state to STATE_CONFIG. The state will transfer to STATE_NORMAL when succeeding the configuration. The state of STATE_NORMAL indicates that the BS is working normally and any MS can try to access to it.

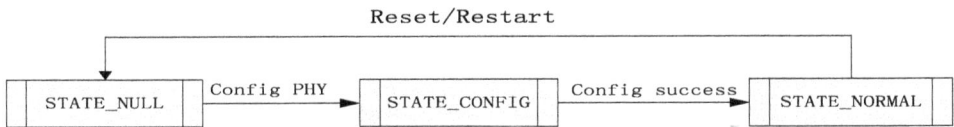

Reset/Restart

| STATE_NULL | Config PHY → | STATE_CONFIG | Config success | STATE_NORMAL |

Fig. 11. The state of BS-State-Machine

• BS-MS-State-Machine module

The main functions of BS-MS-State-Machine are to hold the current state of MSs, parse the messages received from the PHY layer, packet the messages, send to the PHY layer and transfer to other state. According to the possible state of MS, the BS-MS-State-Machine is defined as having nine states, which are depicted in Fig.12 and Fig.13.

After the success of the initial ranging, BS will create a corresponding state machine for the MS. The BS-MS-State-Machine transfers from STATE_NULL to STATE_CONNECTED. The MS will execute the process of registration. Once registration succeeds, the state will transfers to STATE_NORMAL. During the STATE_NORMAL, MSs can establish the connection with BS to transmit the traffic of video, voice and data services.

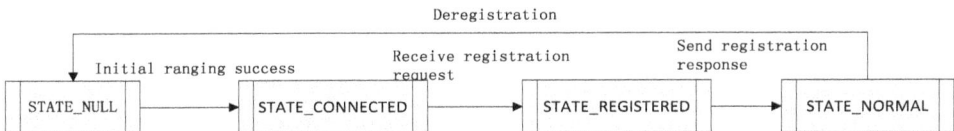

Deregistration

| STATE_NULL | Initial ranging success → | STATE_CONNECTED | Receive registration request → | STATE_REGISTERED | Send registration response → | STATE_NORMAL |

Fig. 12. The state of BS-MS-State-Machine

When the quality of the signals MS received is poor for a long time or the traffic capacity of the BS is saturated, the MS will consider to handover to another BS. Before the handover, the BS-MS-State-Machine will transfer to STATE_HOSCAN state to scan other BSs. After negotiating with target BSs, the MS will select the best one to process the handover confirm. Once receiving the allowance of the target BS, the BS-MS-State-Machine will transfer to STATE_HOPROCESS state to execute the handover. The target BS will initial a corresponding BS-MS-State-Machine and transfer to STATE_HOACCESSED state. The target BS changes into the serving BS. The target BS transfers the state to STATE_NORMAL and initial serving BS transfer the state to STATE_HOCOMPLETE. At this point, the scan and handover process is over.

Initial serving BS

| STATE_NORMAL | → | STATE_HOSCAN | → | STATE_HOPROCESS | → | STATE_HOCOMPLETE | → | STATE_NULL |

Neighbour scan — handover negotiate&confim — handover — Handover success

Target BS

| STATE_NULL | → | STATE_HOACCESSED | → | STATE_NORMAL |

Handover access — Handover success

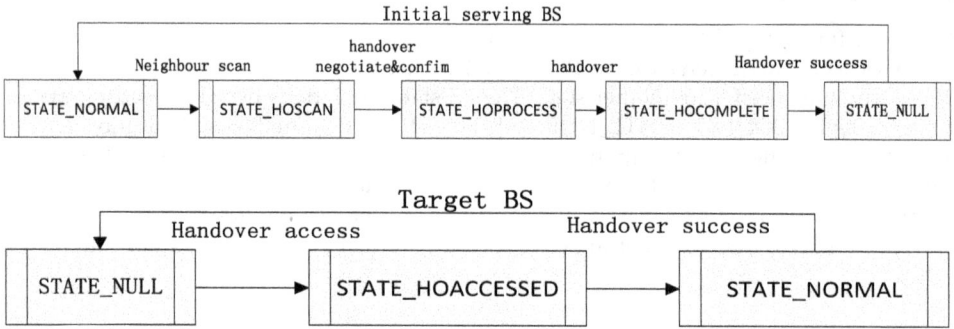

Fig. 13. The handover state of BS-MS-State-Machine

2. the design of MS state machine

The design principles of MS state machine are the same with those of BS. Similarly, the main functions of MS state machine are saving the current state, parsing the messages receiving from PHY layer, sending the packet messages to PHY layer and transferring the state.

• MS-State-Machine module

According to the behaviors of MS, the MS-State-Machine includes several states depicted as Fig.14. During the initial process of MS state machine, the state transfers from STATE_NULL to STATE_IDLE. When the MS receives the accessing indication from the upper layer, the MAC entity will process the PHY configuration and transfer the state to STATE_CONFIG. In order to synchronize with BS, the MS will scan the downlink channels to get downlink synchronization with the BS at the state of STATE_DL_SYN and achieve the uplink channel transmitting parameters to send the initial ranging request at the state of STATE_UL_SYN. After the synchronization, MS state machine transfers to STATE_RANGING and executes the process of registration at STATE_REGISTRATION. The state transfers to STATE_NORMAL after receiving the successful registration response from BS. The details are shown in Fig.14.

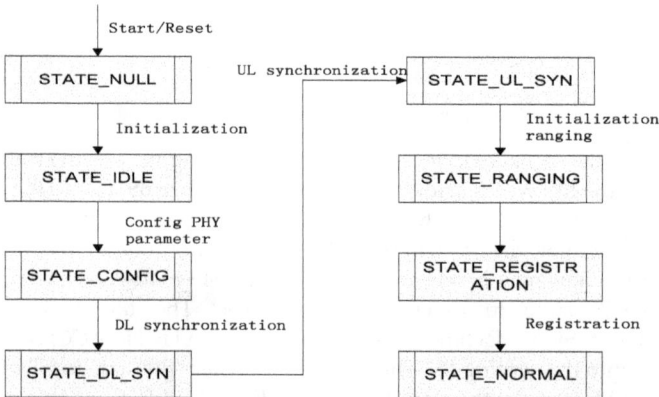

Start/Reset

| STATE_NULL | UL synchronization → | STATE_UL_SYN |

Initialization   Initialization ranging

| STATE_IDLE | | STATE_RANGING |

Config PHY parameter

| STATE_CONFIG | | STATE_REGISTR ATION |

DL synchronization   Registration

| STATE_DL_SYN | | STATE_NORMAL |

Fig. 14. The normal state setup of MSStaeMachine

When the quality of the MS received signals is poor for a long time or the traffic capacity of the BS is saturated, the MS will consider to handover and send the scan request to serving BS. After receiving the scan response, the MS state will transfer to STATE_Scanning_timerM. The MS will set the PHY configuration parameters at STATE_Scanning_PHYSyn according to the target BSs, synchronize with the target BS and send the scan result to the serving BS. When all the targets have been scanned, the state transfers to STATE_MOB_SCN_REP.

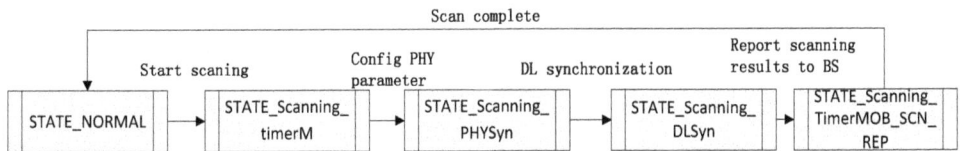

Fig. 15. MS scanning flow chart

Serving BS will consider the factors of signals strength and so on to select the best target BS to process the handover. The MS state will transfer to STATE_HO_Request. The MS will process the downlink synchronization with the target BS at the state of STATE_HO_PHY_SYN and execute the uplink synchronization by ranging mechanism at the state of STATE_HO_RANGING. Once having received the successful ranging response, the process of handover finishes. The MS sets the best target BS as the serving BS and transfers the state to STATE_NORMAL.

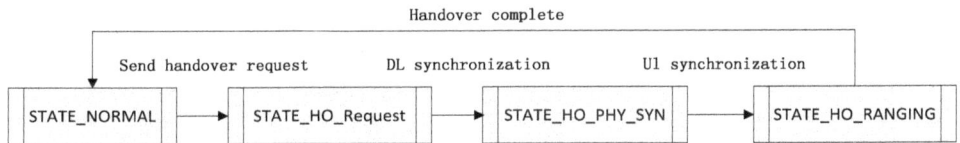

Fig. 16. MS handover flow chart

4.2.2 Multi-threading

Three threads for MAC protocol are designed, which are time-thread, MAC-thread and APP- thread to keep the system running. The functions performed by the three threads are slightly different at BS and MS sides. The following will make a brief introduction.

1.   three threads at BS side
Time-thread is mainly used for timing, polling the DMA channels to get the interruption to trigger the next operation. If the interruption type represents data arrival, the MAC layer will get the data from the PHY layer through the API for the further processing. If the interruption type represents frame beginning, then the frame number will increase by 1and the MAC entity will send the packet messages to the PHY entity through the API. Such a send-receive method is aimed to accommodate the requirements of limited time processing. The details can be referred to section 4.3. Because there are many timers in the MAC layer module, the time-thread will also be responsible for the timing and overtime processing.

MAC-thread is mainly responsible for processing the messages according to the current state, sending the API messages to PHY layer or getting the API messages from PHY layer.

APP-thread mainly executes the tasks of converging the uplink traffic data and sending the downlink traffic to the MAC-thread which will deal with PDUs forming. The triggers and relations of the three threads are shown in Fig.17

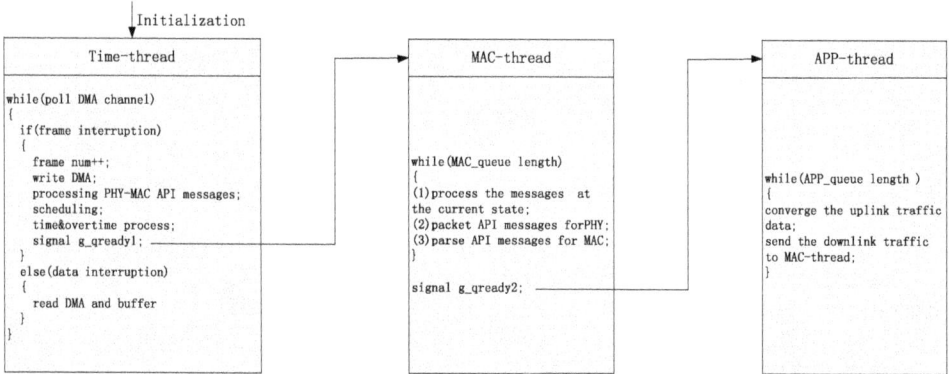

Fig. 17. The triggers and relations of the three threads at BS

2. three threads at MS side

The functions of the three threads at the MS side are similar with BS's. Once the frame interruption arrives, the time-thread will execute the overtime process. And then, the MAC-thread will process the API messages received from PHY layer according to current state. The MAC-thread will also packet the management messages and the uplink traffic data into PDUs that are sent to PHY layer. The APP-thread is mainly used to converge the downlink traffic data and send the uplink traffic data to the MAC-thread. The triggers and relations of the three threads are shown in Fig.18.

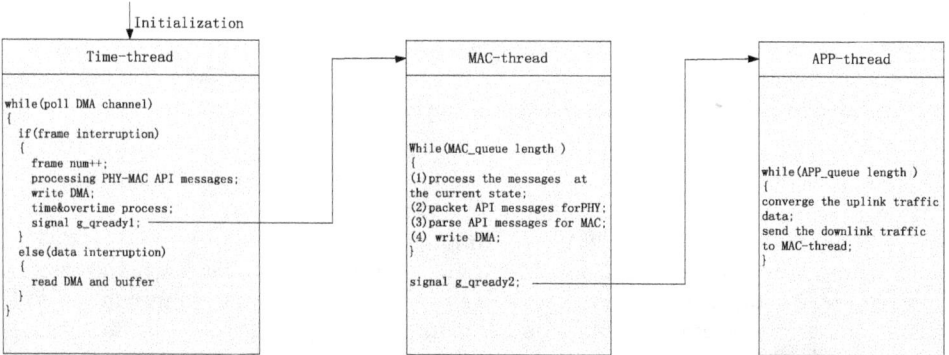

Fig. 18. The triggers and relations of the three threads at MS

4.3 Synchronization between MAC and PHY

The interaction between MAC layer and PHY layer is through API entity. The MAC layer parses the API messages received from the API entity and sends the packet API messages to the API entity. The API messages of txstart.request and rxstart.request at the BS side consist of DL-MAP and UL-MAP which are sent in the downlink subframe. The common place of DL-MAP and UL-MAP is that they all contain the frame number. If the frame number is the same, the DL-MAP indicates the current downlink subframe resource allocation and the UL-MAP indicates the current uplink subframe resource allocation. This is referred as minimum-time relevance. If the frame number in the UL-MAP is larger than that in the DL-MAP by1, the DL-MAP indicates the current downlink subframe resource allocation but the UL-MAP indicates the next uplink subframe resource allocation. This is referred as maximum time relevance. (Zeng, 2006).

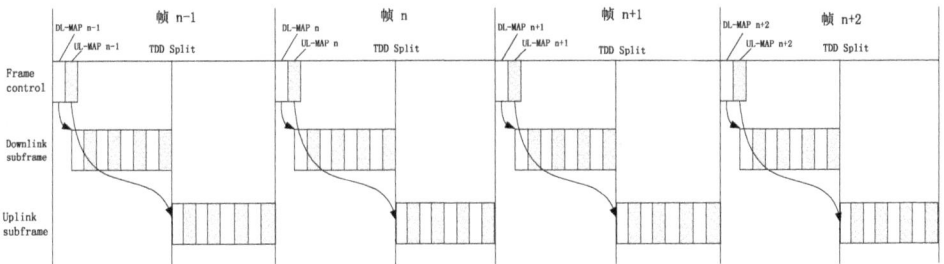

Fig. 19. Minimum time relevance of DL-MAP and UL-MAP

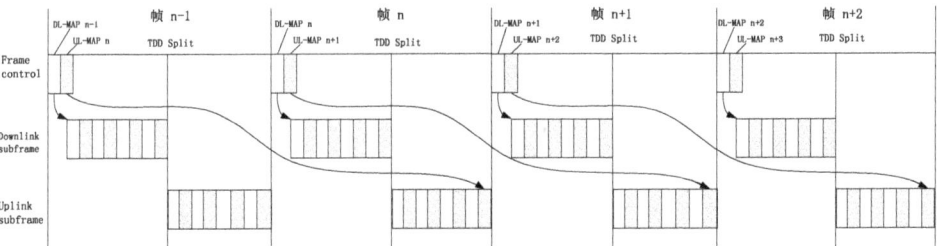

Fig. 20. Maximum time relevance of DL-MAP and UL-MAP

The API messages should be sent to the API entity before the transmission of the air interface. The processes of downlink and uplink transmission will be briefly introduced in the next section.

4.3.1 Downlink transmission

The construction and transmission of a downlink subframe at BS occurs as follows:

1. The BS MAC issues a TXSTART.request which includes the TXVECTOR describing the subframe structure according to the schedule. For a valid request, TXSTART.response returns the frame number(N) which it received from the MAC in the TXVECTOR structure.

2. The BS MAC issues one MACPDU.request for each downlink PHY burst(i.e. each burst described in TXVECTOR). It is not necessary for the BS MAC to wait for TXSTART.response before issuing MACPDU.request. Each MACPDU.request may contain multiple actual MAC PDUs. The MACPDU.request messages must be issued in the burst order specified by the TXVECTOR provided with the TXSTART.request and issued before transmission of the subframe is indicated by TXSTART.indication. An error is returned in MACPDU.response if a PDU is submitted too late.

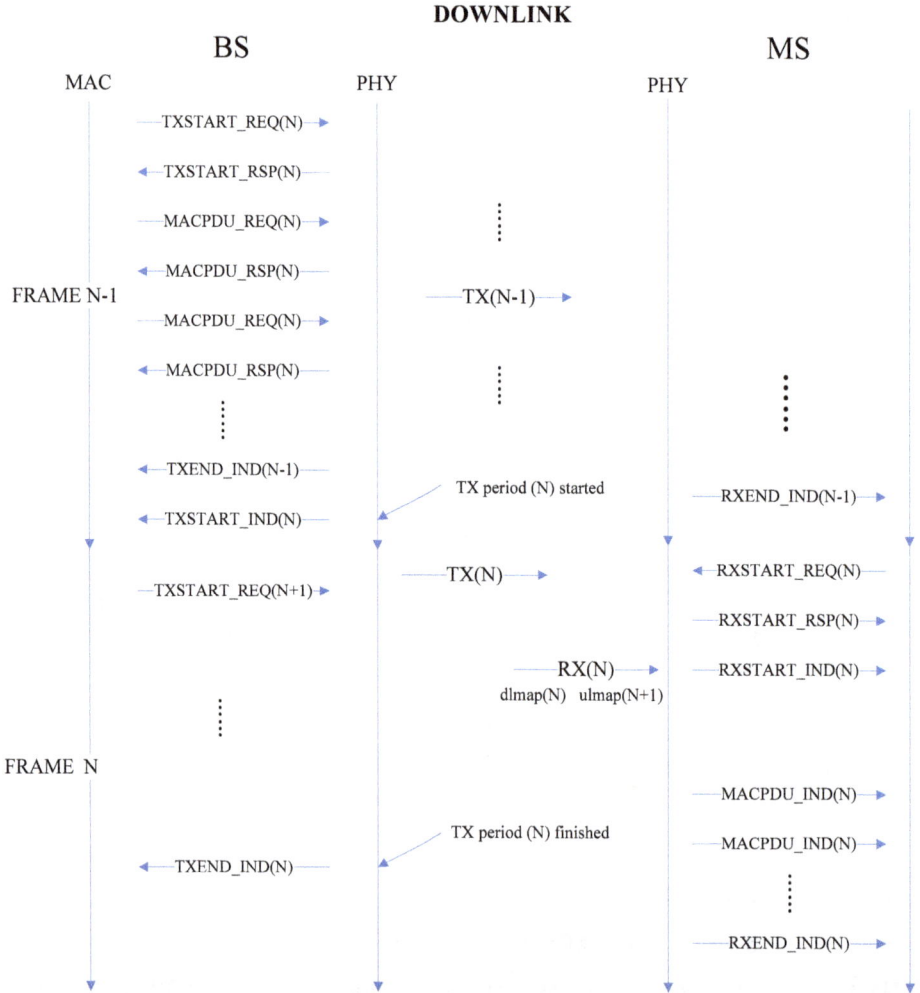

Fig. 21. API primitives for downlink transmission

3. A MACPDU.response is returned for each request indicating whether the PDU has been successfully queued for transmission. The TXEDN.indication message indicates whether the downlink subframe has been successfully transmitted or if transmission was aborted due to an error. If a TXSTART.request is not received then nothing is transmitted. Therefore to ensure continuous transmission, a TXSTART.request must be submitted for each and every downlink subframe (PicoChip Company, 2007).

The reception of a downlink subframe at MS occurs as follows:

1. The RXSTRAT.request message instructs the PHY to start reception of subframe(N). To allow the PHY to parse the DL-MAP the MAC must send a RXSTART.request message which includes the CIDs to decode. When reception of subframe(N) begins the RXSTRAT.indication message is send to MAC.

2. When the MS is registered with the BS and not in idle or sleep mode, it should issue a RXSTRAT.request for every subframe.

3. MAC PDUs received by the BS are transferred via the MACPDU.indication message. Each MACPDU.indication may contain multiple MACPDUs. Each received PDU is associated with the downlink frame number.

4. The first MACPDU.indicatiion for burst#0 will contain the FCH, the second for burst#1 will contain the DL-MAP. However, the PHY will parse the FCH and DL-MAP so it is not necessary for the MAC to send any downlink frame structure information to PHY.

5. The end of decoding for each downlink subframe is signaled via the RXEND.indication message. The downlink bursts for the MS may occur early in a subframe so RXEND.indication can be issued before the end of the downlink subframe (PicoChip Company, 2007).

4.3.2 Uownlink transmission

The construction and transmission of a uplink subframe at MS occurs as follows:

1. The MS MAC issues a TXSTART.request which includes the TXVECTOR describing the uplink subframe structure according to schedule. For a valid request, TXSTART. response returns the frame number (N) which it received from the MAC in the TXVECTOR structure.

2. The MS MAC issues one MACPUD.request for each uplink PHY burst described in TXVECTOR, there should be only one normal uplink burst allocated to the MS per frame. Each MACPDU.request may contain multiple actual MAC PDUs. The MACPDU.request message must be issued before transmission of the subframe is indicated by TXSTART.indicaation.An error is returned in MACPDU.response if a PDU is submitted late.

3. A MACPDU.response is returned for each request indication whether the PDUs have been successfully queued for transmission, The TXEDN.indication message indicates whether the uplink subframe has been successfully transmitted or if transmission was aborted due to an error. If a TXSTRAT.request is not received then nothing is transmitted. The MAC should only issue the TXSTART.request when it needs to transmit uplink data, perform ranging or send information on the ACK channel or fast-feedback channel (PicoChip Company, 2007).

UPLINK

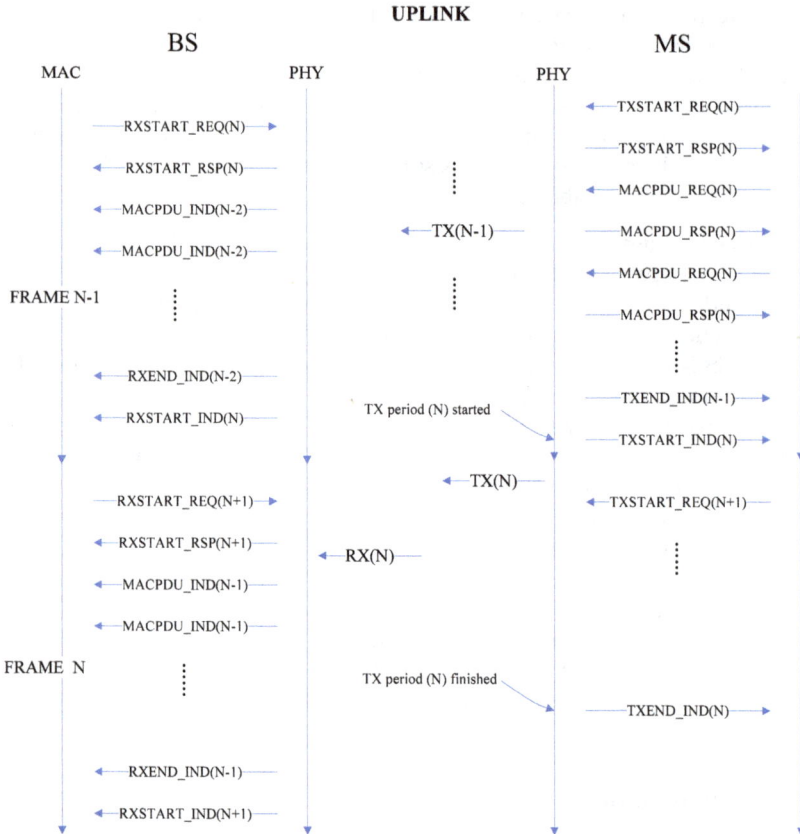

Fig. 22. API primitives for uplink transmission

The construction and transmission of an uplink subframe at BS occurs as follows:

1. The RXSTART.request message provides the RXVECTOR which describes the bursts on the uplink for frame N. The UL-MAP is transmitted at the start of downlink subframe N.

2. The RXSTRAT.request is issued against a particular downlink subframe and must be sent after the associated TXSTART.requset. A successful RXSTRAT.request returns the frame number N which is received from the MAC in the RXVECTOR structure. The RXSTRAT.request must also be issued before the start of downlink subframe N.

3. If a RXSTART.request is not received then the receiver is effectively disabled. Therefore to ensure continuous reception, a RXSTART.request must be submitted for each uplink subframe.

4. MAC PDUs received by the BS are transferred via the MACPDU.indication message. Each MACPDU.indication may contain multiple MAC PDUs. Each received PDU is associated with uplink frame number and the burst number specified in RXVECTOR. The end of decoding for each uplink subframe is signaled via the RXEND.indication(PicoChip Company, 2007).

5. Conclusions

In this chapter, the design and implementation of a WiMAX wireless baseband communication system are presented. Based on the discussion of the typical baseband hardware schemes, we adopt the Picochip multi-core DSP processor as the base of the baseband platform. The hardware, PHY protocol and MAC protocol are introduced in terms of design and implementation other than research aspect, the tradeoff between complexity and performances has been taken into account to meet the requirements. The PHY-MAC interface and API, link-level simulation and debugging method are also mentioned in this chapter, which may provide users better understanding of the development procedure.

6. References

Du, Y. (2010), The MAC layer, In: *IEEE 802.16m Broadband Wireless Technology and System Design*, pp. (181-220), Posts & Telecom Press, Peking, 978-7-115-22754-6.

Dohler, M. Lerau, C. &Hardouin, E. (2005). *UMTS FDD multi-Antenna Receiver Complexity Estimation*. France Telecom R&D, internal report. pp. 158-175.

Fathi, L. (2004). *Complexity of MPIC receiver for HSDPA R5*. France Telecom R&D, internal report. pp. 204-210.

Foschini. G.J.(1996).*Layered Space-Time Architecture for Wireless Communicationin a Fading Environment When Using Multi-Element Antennas* · Bell Labs Technical Journal, 1996, V01.1(2).pp.41-59.

Goldfarb, M. Palmer, W. & Murphy,T. (2000).*Analog baseband IC for use in direct conversion W-CDMA receivers*.Radio Frequency Integrated Circuits (RFIC) Symposium, 2000. Digest of Papers. 2000 IEEE, pp. 79-82.

Jan Mietzner, Robert Schober, Senior Member, Lutz Lampe, Wolfgang H. Gerstacker and Peter A. Hoeher.(2009). *IEEE COMMUNICATIONS SURVEYS & TUTORIALS, VOL. 11, NO. 2, SECOND QUARTER 2009, pp.* 87-105.

Jusslia, J. Ryynanen, J. &Kivekas,K. (2001).A 22-ma 3.0-db NF direct conversion receiver for 3g WCDMA.IEEE journal of solid-state Circuits, vol. 36, pp. 2025-2029.

LAN/MAN Standards Committee of the IEEE Computer Society and the IEEE Microwave Theory and Techniques Society.(December 2007). *Part 16: Air interface for Broadband Wireless Access System*, pp. (619-1082), P802.16Rev2/D2.

Parssinen, A. Jussila, J. &Ryynanen, J. (1999).*A 2-GHZ wide-band direct conversion receiver for WCDMA applications*.IEEE Journal of Solid-state Circuits, vol. 34, pp. 1893-1903.

Parssinen, A.Jussila, J.&Ryynanen, J. (2000).*A wide-band direct conversion receiver with on-chip A/D converters*.VLSI Circuits, 2000. Digest of Technical Papers. 2000 Symposium, pp. 32-33.

PicoChip Company. (July27,2007). BS MAC-PHY API Definition. pp. 11-13

PicoChip Company. (May21,2007). MS MAC-PHY API Definition. pp. 16-17

PicoChip Company. (September 17,2008).*Tools_Userdoc_Fullman_7.4.5*. pp. 123-138

S. Lin and P. Yu. (July 1982). *A Hybrid ARQ Scheme with Parity Retransmission for Error Control of Satellite Channel* , IEEE Trans on Communications. pp. 1701–1719.

Zeng, C. (2006), The support from MAC layer to PHY layer, In: The Principles and
     Applications of WIMAX/802.16, pp. (113-115), China Machine Press, Peking, 7-
     11120111-6.

10

Performance Analysis and Noise Immunity *WiMax* Radio Channel

Oleksii Strelnitskiy, Oleksandr Strelnitskiy, Oleksandra Dudka,
Oleksandr Tsopa and Vladimir Shokalo
*Kharkiv National University of Radio Electronics (KhNURE), Kharkiv,
Ukraine*

1. Introduction

Currently, *WiMAX* systems acquired widespread adoption, to help organize the network at *MAN (Metropolitan access network)* level. It is assumed that in the next decade, the performance of such systems can achieve 50 bits/second/Hz. This is due primarily to the fact that in *MAN* network large amounts of multimedia confidential information of high quality are transmitted. So it requires not only high speeds, but also the appropriate level of noise immunity. Therefore, the performance and noise immunity are the two main indicators of the radio system quality. The radio waves propagation (*RWP*) models in radio channel play a decisive role in their calculation. Several mechanisms of radio wave propagation are known, including the so-called street-wave channels (Fabricio, 2005) (Wei, 2007), that was already noted in the development of analog communication (Porrat, 2002). These channels are identified by the authors as wavelength channels formed by architectural buildings (WCAB) (Strelnitskiy, 2007). However, the existing mathematical models are extremely complicated, so they become ineffective in the calculation of large branching networks of urban and this fact makes it difficult to assess these characteristics. Performance is evaluated as the ratio of transmission rate to channel bandwidth and noise immunity is defined as the probability of a bit (BER) or packet errors (PER). The authors propose general approach to the WCAB representation in the form of microwave multipole.

Then the multipolar model of branched-line and outdoor radio channels is described. It allows us to calculate the attenuation of radio waves in the city streets. The reliability of the model is established by comparing the results of numerical and field experiments conducted by the authors. Performance and noise immunity of *WiMAX* communication channel estimates are given in the conclusion section.

2. Mathematical model of branched wavelengths, which are formed by architectural buildings

To form a mathematical model of *WCAB* we propose to adopt the following approach. Fig. 1 shows the city district fragment with the base station (BS) placed on the square.

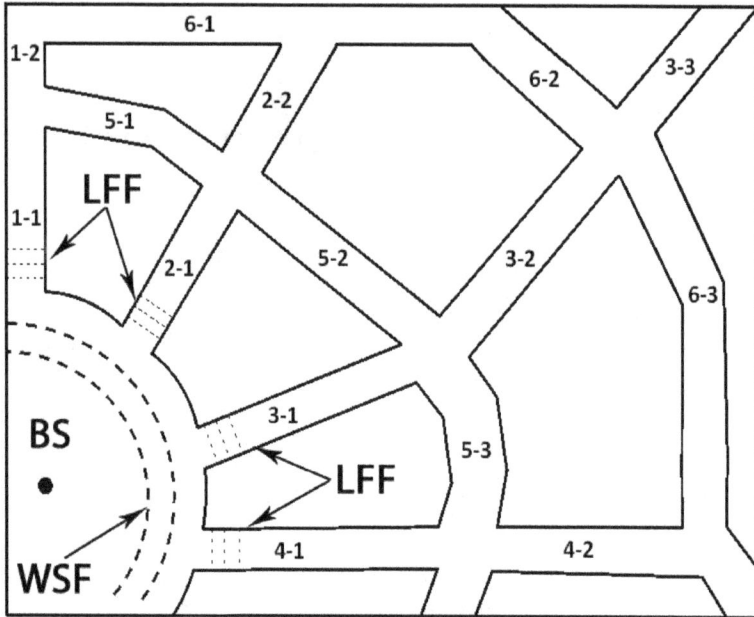

Fig. 1. Fragment of the city region

Numerals indicate segments of streets. For example, the designation 1-2 should be read: the second part of the first street. *BS* radiates waves of spherical front (*WSF*). Further we shall consider (based upon the Huygens principle) that in the radial streets (denoted by numbers 1-4 in Fig. 1) the radiated spherical wave is transformed into a series of waves with the locally flat front (*LFF*). A further approach is to use the following approximations. The street is represented by several continuous, homogeneous and smooth surfaces, which form the guide system with losses. Straight-line segment of length l of this guide system is replaced by an equivalent two-wire line segment. Wave resistance and wave factor β of this line are equal to the characteristic impedance and wave ratio of free space. Equivalence should be determined by equality of power transferred to the real and equivalent systems, i.e. largest attenuation. The attenuation is determined by losses in *RWP*, calculated by *RWP LAN-MAN* model (Strelnitskiy, 2008), phase – by value βl. As a result, segments of the line are easily represented by matrices of the quadripole scattering [S] (Gostev, 1997).

The properties of these segments are as follows: all lines have equivalent impedance equal to Z_0 because they spread a wave of T-type; street intersection (for example, 2, 3 c 5, 6 on Fig. 1) is a set of included equivalent line segments and in terms of circuit theory it is a power distribution system (*PDS*).

CPM with n equal divisions of channels, described by the matrix [S] of ideal multipoles, together with segments of lines with losses constitute a particular scheme, the calculation of which can be performed by cyclic algorithm. As a result, the considered scheme is equivalent to a multipole (Gostev, 1997) (Fig. 2), which can be used to determine the amplitude of the field at any point of *WCAB*.

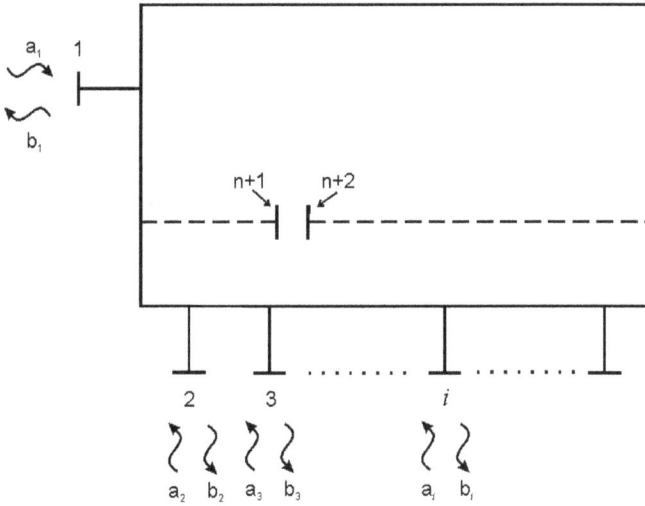

Fig. 2. Equivalent multipole

In this case, the problem is formulated as follows. Let there be a chain, equivalent to WCAB and containing the n external arms. It is required to determine the amplitude and phase of the field in a certain section of the circuit produced in accordance with these WCAB coordinates. In general, the circuit is excited with any number of arms (Fig. 2., where a_i, b_i – normalized amplitudes of the incident and reflected waves). For example in the case of Fig. 1 the number of excitation sources is 4.

The problem is solved by the method (Gostev, 1997). Let us isolate the circuit section in which you want to determine the amplitude and phase of the signal. Conventionally, we break the transmission line at this point (Fig. 2). Let us denote additional arms through n+1 and n+2, and the matrix of the resulting multipole - via $[S_{ij}](i, j = 1, 2, ..., n+1, n+2)$.

In (Gostev, 1997) it is shown that if an equivalent multipole is excited with the i-th arm, then the values of the normalized amplitudes of the reflected waves in the $(n+1)$ and $(n+2)$ arms will be written as:

$$b_{n+1} = \frac{S_{n+1,i}(1 - S_{n+1,n+2}) + S_{n+1,n+1}S_{n+2,i}}{(1 - S_{n+1,n+2})(1 - S_{n+2,n+1}) - S_{n+2,n+2}S_{n+1,n+1}} \cdot a_i, \tag{1}$$

$$b_{n+2} = \frac{S_{n+2,i}(1 - S_{n+2,n+1}) + S_{n+2,n+2}S_{n+1,i}}{(1 - S_{n+1,n+2})(1 - S_{n+2,n+1}) - S_{n+2,n+2}S_{n+1,n+1}} \cdot a_i. \tag{2}$$

In this case the resulting wave in the cross section

$$b_\Sigma = b_{n+1} + b_{n+2}. \tag{3}$$

If the circuit is excited with the arms, the resultant wave can be written as (Gostev, 1997):

$$b_{\Sigma} = \frac{\sum_{i=1}^{k}\left[S_{n+1,j}\left(1-S_{n+1,n+2}+S_{n+2,n+2}\right)+S_{n+2,i}\left(1-S_{n+2,n+1}+S_{n+1,n+1}\right)\right]}{\left(1-S_{n+1,n+2}\right)\left(1-S_{n+2,n+1}\right)-S_{n+2,n+2}S_{n+1,n+1}} \cdot a_i. \tag{4}$$

The S_{n+2}, S_{n+1} coefficients in the expressions (1, 2) are defined by cyclic algorithms given in (Gostev, 1997). For their use it is necessary to make the scheme which will be replaced by the multi-pole circuit. It is compiled on the basis of a multipole electrical circuit.

For example, in the case shown in the Fig. 1, we can make calculations from the block diagram shown in Fig. 3.

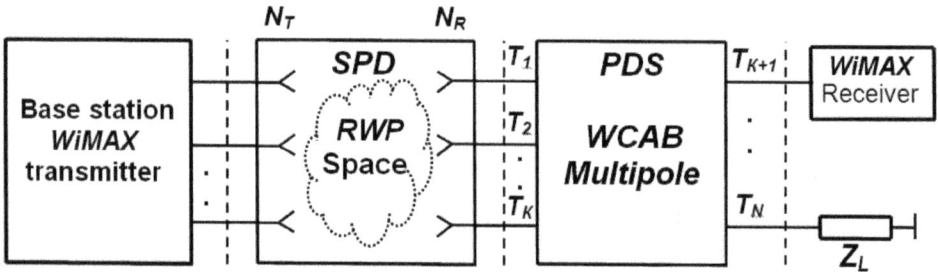

Fig. 3. WCAB block diagram

The scheme consists of a base station transmitter which is connected through its antenna, which has N_T emitters, with a spatial power distributor (SPD). On the SPD outputs there are the N_R receiving antennas, connected by the corresponding transitions to the equivalent line of PDS in the $T_1...T_K$ reference plane. Other reference planes are connected to the channel receiver, as well as loads of equivalent lines Z_L, equal to their characteristic impedance.

Amplitudes a_i in the $T_1...T_K$ reference plane depend on the relative position of the transmitting and receiving antennas and their radiation patterns. The coordinates of the receiving antennas in the cross sections of streets determine the positions of longitudinal sections of the streets along which the attenuation calculations are carried out. Let us represent the equivalent circuit for a part of urban area (Fig. 1). We assume that each segment of the street with length of r_i may be substituted by a segment of an ideal two-wire line, that is connected with the attenuator in cascade, its damping value α_i at RWP is equal to the damping on the street segment with length r_i.

The scattering matrix of quadripole equivalent to a cascading line connection and the attenuator is given by:

$$[S(r_i)] = \begin{bmatrix} 0 & \sqrt{\alpha\left(\dfrac{r_i}{r_0}\right)} \\ \sqrt{\alpha\left(\dfrac{r_i}{r_0}\right)} & 0 \end{bmatrix} e^{-i\beta r_i}. \tag{5}$$

The matrix is written on the assumption that the quadripole is consistent with the characteristic impedance of free space and reciprocal. Further let us assume that we need to determine signal strength along the street 2 (Fig. 1). To simplify the calculations, we assume that the wave processes occurring along the street 2, will be affected only by the adjacent streets 1, 3, 5 and 6. Then the electrical circuit of the equivalent multipole will have the form shown in Fig. 4, a.

It is easy to see that the diagram in Fig. 4, consists of the three basic elements: quadripole – cascade connection of attenuator and the ideal line segment - the ideal six-pole and the ideal eight-pole. We assign respectively, numbers 1, 2, 3 for the above basic elements and depict the equivalent circuit of the equivalent multipole (Fig. 4, b).

From the above equivalent circuit, it follows that the scattering matrix of the equivalent multipole can be obtained by applying the cyclic algorithms for cascade connection of quadripole 1 and six pole 2, and also quadripole 1 and eight-pole 3.

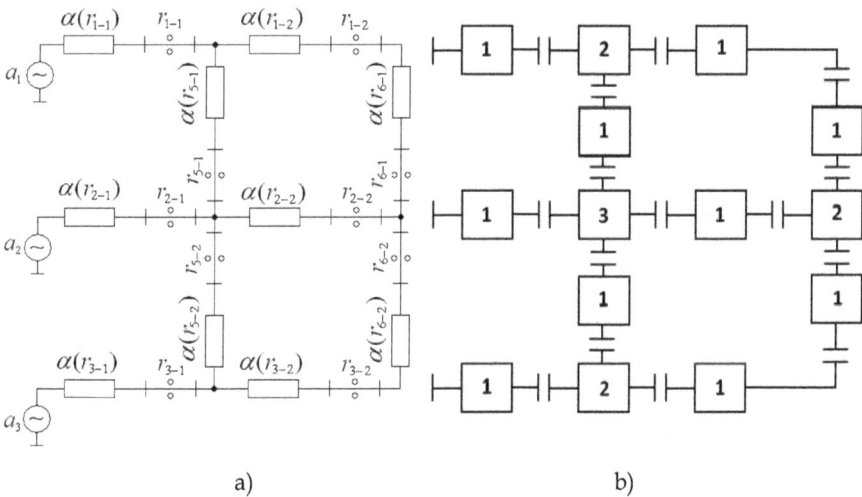

a) b)

Fig. 4. The electric circuit (a) and equivalent circuit (b) of WCAB multipole

Let's give the formula for calculating the scattering parameters of the quadripole 1 and eight-pole 2 (Gostev, 1997)

$$S_{11} = S_{11}^{(1)} + \frac{(S_{12}^{(1)})^2 S_{11}^{(2)}}{A}, \ S_{12} = \frac{S_{12}^{(1)} S_{12}^{(2)}}{A}, \ S_{13} = \frac{S_{12}^{(1)} S_{13}^{(2)}}{A}, \ S_{22} = S_{22}^{(2)} + \frac{S_{22}^{(1)}(S_{12}^{(2)})^2}{A}, \tag{6}$$

$$S_{33} = S_{33}^{(2)} + \frac{S_{22}^{(1)}(S_{13}^{(2)})^2}{A}, \ S_{23} = S_{23}^{(2)} + \frac{S_{22}^{(1)} S_{12}^{(2)} S_{13}^{(2)}}{A}, \ A = 1 - S_{22}^{(1)} S_{11}^{(2)} \tag{7}$$

where $[S_{ij}^{(1)}]$, $[S_{ij}^{(2)}]$ – scattering matrixes of quadripole and eight-pole.

The formulae describing connection of the quadripole 1 and eight-pole 3 (Gostev, 1997)

$$\hat{S}_{11} = S_{11}^M + \frac{4S_{12}^M S_{21}^M S_{11}}{1 - S_{11}(S_{22}^M + \sum_{N=2}^{N} S_{2,N+1})}, \hat{S}_{21} = \frac{S_{21}^M S_{21}^M}{1 - S_{11}(S_{22}^M + \sum_{N=2}^{N} S_{2,N+1})}. \tag{8}$$

Let is denote

$$S_{11}^M = S_{11}^{\ni}, \ S_{21}^M = S_{21}^{\ni}, \ 4S_{12}^M = S_{12}^{\ni}, \ S_{22}^M + \sum_{N=2}^{N} S_{2,N+1} = S_{22}^{\ni}. \tag{9}$$

Considering notation (9) expression (10) can be written this way

$$\hat{S}_{11} = S_{11}^{\ni} + \frac{S_{12}^{\ni} S_{21}^{\ni} S_{11}}{1 - S_{11} S_{22}^{\ni}}, \ \hat{S}_{21} = \frac{S_{21}^{\ni} S_{21}}{1 - S_{11} S_{22}^{\ni}}. \tag{10}$$

The scattering matrix of an equivalent quadripole of i-row of the schema will be:

$$\left[S^{\ni(i)}\right] = \begin{bmatrix} S_{11}^{(i)} & N_i S_{12}^{(i)} \\ S_{21}^{(i)} & S_{22}^{(i)} + \sum_{N_i=2}^{N_i} S_{2,N_i+1} \end{bmatrix}, \tag{11}$$

where $S_{i,j}^{(i)}$ – are the scattering coefficients of the divider or quadripole of i-row; N –amount of inputs of i- row element.

The above formulae (1) - (11) constitute a WCAB mathematical model.

3. Model of street wave channels formed by architectural buildings when WiMAX system works in the city

In this section, the general WCAB model developed in Section 2 is refined for the case of outdoor radio channels taking in consideration the characteristics of WiMAX antenna systems and RWP canyon model.

This section also describes the attenuation of radio waves along the street radio channels of the central district of Kharkov. The measurements were made at 3.5 GHz with the WiMAX base station and a created mobile laboratory. A comprehensive analysis of the results is completed - the mechanism of formation of field distribution along the streets is elucidated. Comparative results of calculations and experiments are presented. Practical suitability of the created model in the problems of forecasting of attenuation in outdoor WCAB is proved.

3.1 Experimental studies of attenuation in the street wave channels formed by architectural buildings when WiMAX system works in the city

The design of digital wireless communication systems is based largely on the design of the radio channel. The accurate model of radio channel as we know from (Hata, 1980), is always based on the experiment.

For the case of digital information transmission system (DITS) with WiMAX-technology there appeared a number of articles (Fabricio, 2005); they highlight some issues of radio wave propagation in urban environments. In the end, for example, modulation types are revealed which are peculiar to one or another level of signal/noise (*S/N*) ratio at the reception point. However, the experimental results described in the mentioned works are of particular nature. They cannot be used to construct a general *RWP* model of *WiMAX* wireless channels, both because of the limited number of experimental studies, and because of the lack of their systematization on any grounds. In particular, the mechanism of wave propagation along urban wave channels formed by the architectural building is not studied.

Increased knowledge of the laws of propagation of *WCAB* wireless networks with *WiMAX* technology, especially in city streets, is important in connection with putting into operation mobile *WiMAX* systems at the present time and requires conducting extensive experimental work. Some of the experiments are done within the present study.

The purpose of the work in this subsection was to conduct experiments and analyze their results in the propagation of *DITS* signals with *WiMAX* technology along the street *WCAB* in a large industrial city (Kharkov). The map of the part of Kharkov where the investigations were made is shown in Fig. 5.

Fig. 5. The Map of the study area in Kharkov (BS - base station location)

For measurements a mobile laboratory was created, its general form is shown in Fig. 6, and its structure in Fig. 7.

Fig. 6. General view of the mobile laboratory

The mobile laboratory is equipped as follows: *WiMAX Breeze-Max 3500* (*Alvarion*) subscriber station, «Asus» notebook, NovAtel SS-11 *GPS* receiver, voltage transformer VT – 12V/220V and storage battery SB–12V. To measure the signal/noise ratio (*S/N*) and signal level (S) we used special software interface that was provided by «Alternet». *WiMAX* base station (BS) was placed at the altitude of $h_{BC} = 80$ m. (Gasprom building, Fig. 6, right picture).

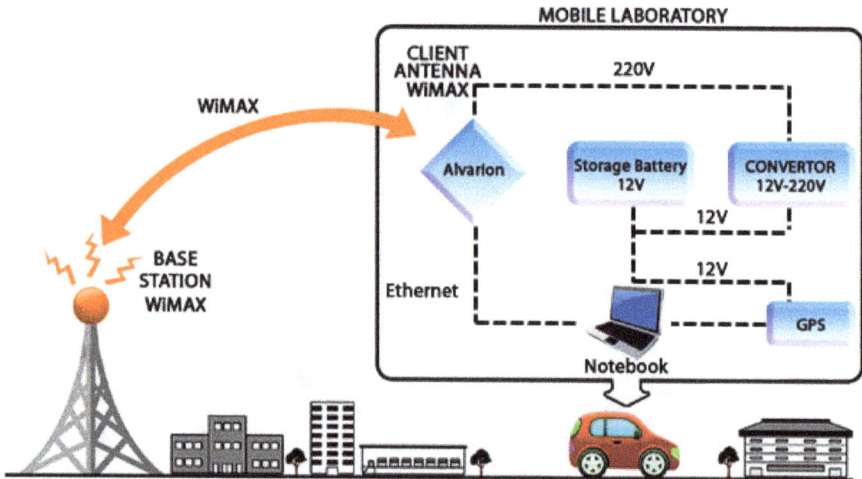

Fig. 7. The structure of the mobile laboratory

In the BS four quadrant antennas is used. One of the sectors of the polar pattern (PP) (Fig. 8, a) serves the area shown in the map (Fig. 5), the direction of maximum radiation is almost identical with the direction (orientation) of Lenin Ave.

F, dB

a)   b)

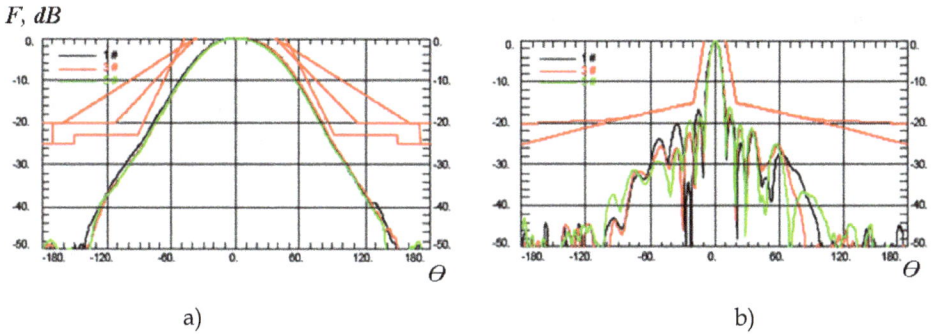

Fig. 8. Polar patterns: a) for BS antennas; б) for subscriber station (SS) antennas

Let us describe the experiments conducted, their results are shown in Fig. 9-13. The first experiment is to measure the radiation pattern along the maximum base station (along Lenin Ave). The results of the experiments in the form of dependency of $S(r)$ and $S/N(r)$ are shown in Fig. 9 a, b. In the experiments, the maximum distance r was equal to 4 km, which corresponded to the maximum range of confident communication. In the figure dots

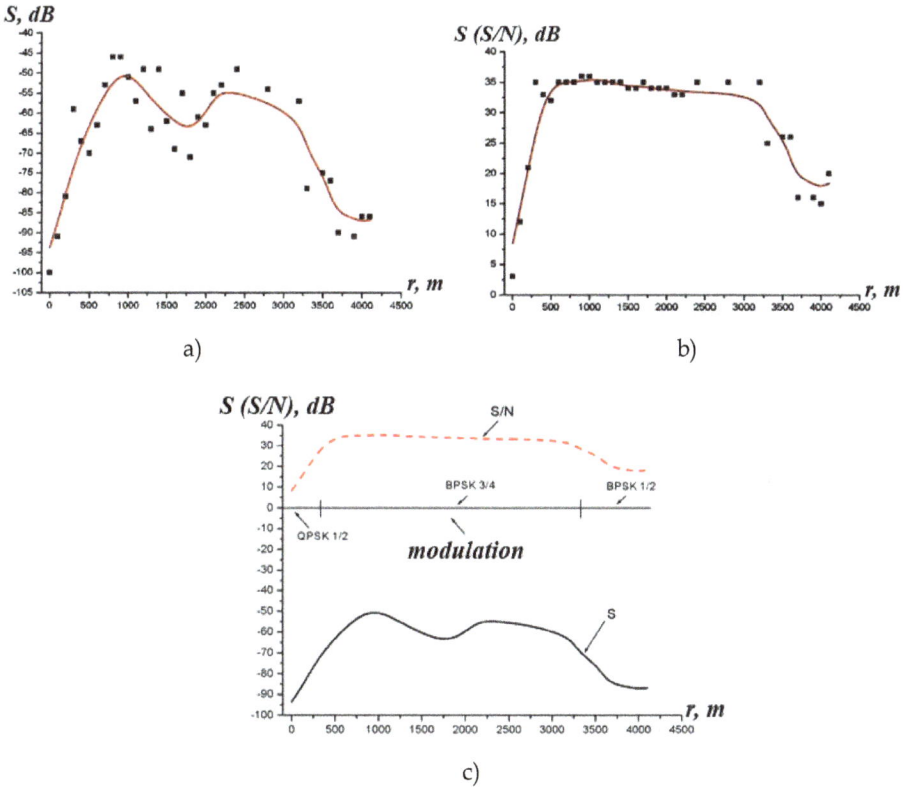

a)   b)

c)

Fig. 9. The measurement data on Lenin Avenue

represented the data of single measurements at the reference distance $r_0 = 100\ m$ and with $100\ m$ step. At each point of measurement the aperture of subscriber station (SS) antenna, fixed on a tripod (Fig. 6), was placed perpendicular to the direction of maximum reception at the height of 1.5 m above the street cover. Initial measurements were averaged and smoothed using the «Origin 6.1» software. The processed results as solid curves are shown in Fig. 9 a, b. The same curves are shown in conjunction in Fig. 9, c. The fig shows the pattern of change of modulation type along the route. On the distances axis segments with one or another kind of modulation (QPSK 1/2, BPSK 3/4, and BPSK 1/2) are shown.

It is known that the WiMAX equipment is adaptive and allows you to maintain a constant transmission rate (or S/N level) of digital stream with a decrease in the signal (Fabricio, 2005).

From the presented data it follows that when S <-65 dB the adjustment does work and the S/N ratio decreases with the distance at the same rate as the signal level. This result significantly refines the capabilities of WiMAX for adaptation, since in (Balvinder, 2006) it is shown that the lower limit of adaptation is the -75 dB signal level.

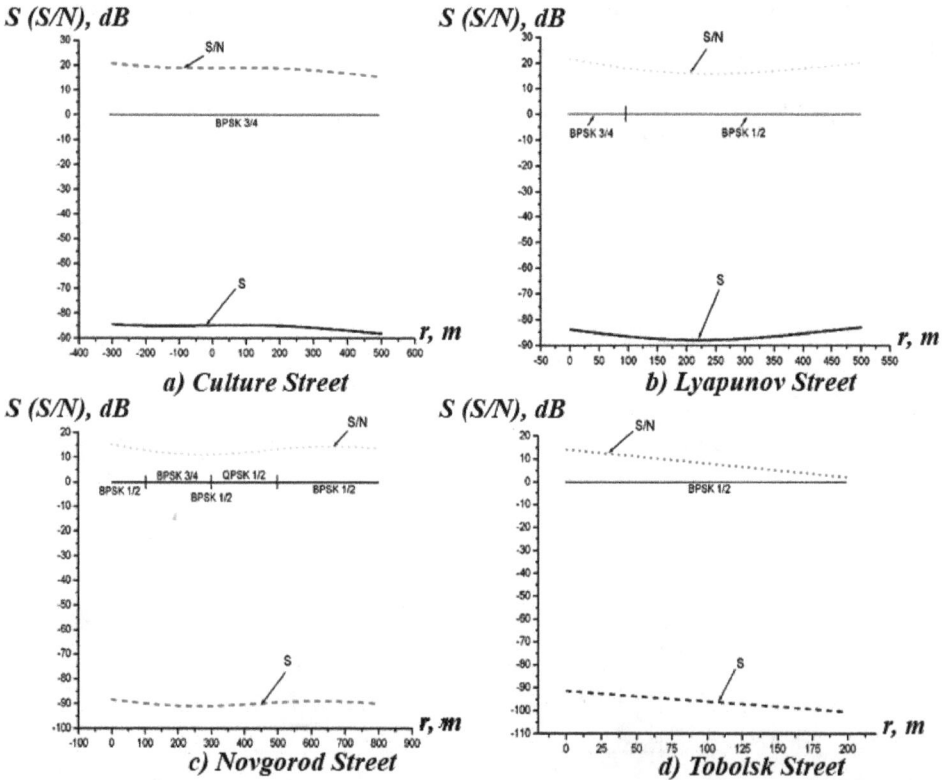

Fig. 10. Measuring the level of S, S/N values and modulation types outside the main lobe of PP of base station antenna

At the same time the data received are well correlated with the results presented in (Porrat, 2002) (Fig. 10, c and Fig. 11) in changes of modulation types on the track. Analyzing them together with the data of the experiment, we can conclude that in the tested part of Kharkov the transfer of information with *WiMAX* wireless communication system is carried out with the rate of 1-2 Mbps.

The distances shown in Fig. 11, a-d, were measured from H points in the direction of arrows (Fig. 5). In this case the main maximum PP of subscriber station (Fig. 8, b) was set along the street axis (approximately at the angle of $90°$ to the direction of maximum BS radiation). For this reason the signal level decreased by *30 dB* compared to its level on Lenin Ave, which corresponds to PP value at $\Theta = 90°$.

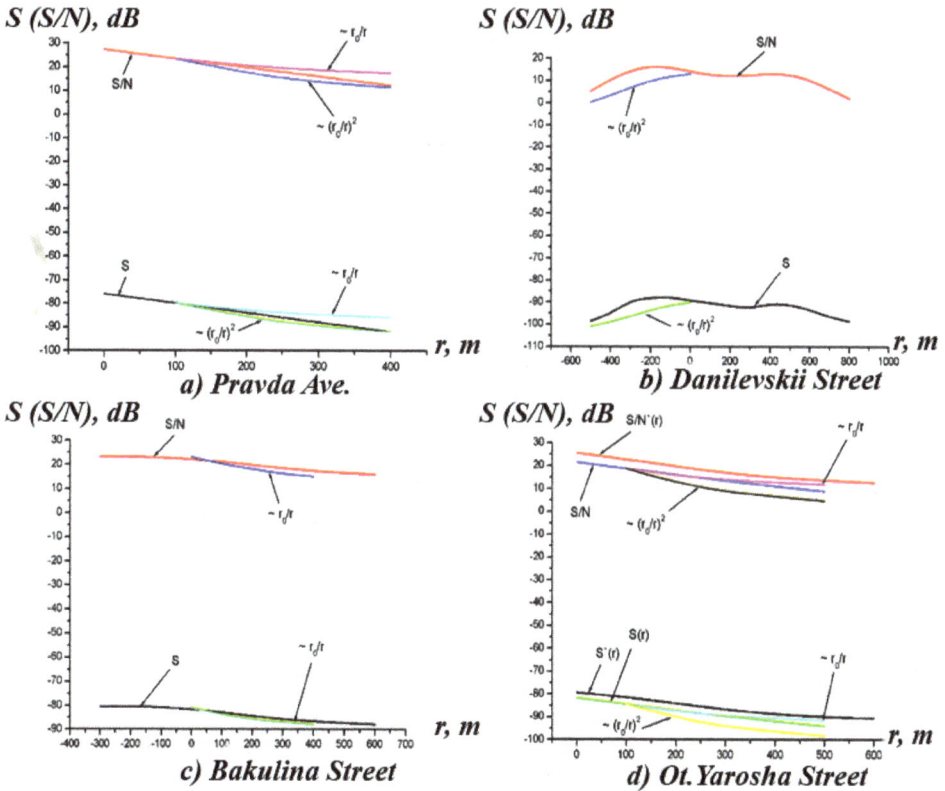

Fig. 11. Approximation of the measured distributions of fields are known functions

According to *WiMAX* radio access technology on the base station sector antennas with wide PP are used (Fig. 8, a), but subscriber stations have embedded antenna with narrow PP and low level of back lobe reception (Fig. 8, b).

This feature of the *WiMAX* apparatus allows us to offer a new method of experimental evidence for the existence of the wave channels and comparison of the signal levels S and signal/noise S/N ratio, created at the receiver due to different mechanisms of propagation.

Let us consider Fig. 12. Here BS is located on the longitudinal pattern of the street. Then, under the assumption that there exist wave channels, at the cross street, formed by ensembles of buildings D1 and D2, two streams of energy should appear (indicated by arrows in Fig. 12). These streams are running waves moving toward each other. They interfere, forming a mixed wave.

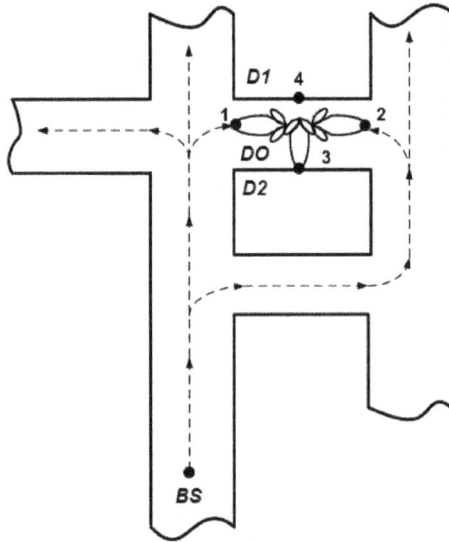

Fig. 12. *WCAB* structure

From the above description, we get the following method of experimental proof of the existence of the wave channel. The polar pattern of the receiving antenna is sent to a maximum PP to point 1 (Fig. 12), thus recording the flow of energy moving from point 1 to point 2. Then PP maximum goes to point 2 and a reverse flow of energy is recorded. The presence of both flows indicates the existence of a wave channel. Orienting the PP maximum to point 3 (points on the walls of houses ensembles), we can detect the intensity of the signal formed by the diffraction of radio wave propagation (from BS through the roofs of the D2 ensemble of houses).

The novelty of the proposed method in comparison with known works (for example, (Volkov, 2005)) is that using the antenna with a narrow PP one can detect the direction of energy flow along the streets and separate the contributions of different *RWP* mechanisms to the received signal level.

The experimental studies by the proposed method were performed in one of the four sectors of BS.

The results of measuring the levels of signal S and signal/noise S/N are shown in Fig. 10. Here curves 1 and 3 – shows the dependence of the ratio S/N and S signal levels, respectively, along the street when moving from point 1 to point 2, and curves 2 and 4 – are the same curves, only measured when the vehicle was moving in the opposite direction (i.e. the aperture of the receiving antenna was rotated at $180°$).

Based on the above reasoning, we can easily conclude that in this case the *RWP* mechanism by *WCAB* acts. As before, the measurements were made at the height of the receiving antenna $h_{AC} = 1,5 M$ over the street surface. Further experiments showed that the intensity of the signals at points 1 and 3 differ in –(10÷15) dB, i.e. contribution of diffraction mechanism to the intensity of the signal is more than an order of a magnitude smaller than the *RWP* mechanism by *WCAB*

The established fact of the interference of counter propagating waves in the street channels in the presence of diffraction field component can explain the pattern of change in signal attenuation along the streets. Fig. 11 shows the measured signal and signal/noise levels for a number of streets in conjunction with the curves of decrease of *S* and *S/N* under the laws of $(r_0 / r)^2$ or r_0 / r. It is easy to see that the experimental curves fall off more slowly than $(r_0 / r)^2$ (Fig. 11, a, b, c) or even than r_0 / r (Fig. 11, в). These dependencies as it is known from (Grudinskaya, 1967), are characteristic for Fresnel and Relay zones at the *RWP* over the reflecting surface.

The reducing of the extent of decrease of the power flux density of the signal *P* in the experiment compared with the above case we explain as follows. For simplicity, we assume that the phasing of the two interfering flows in a street corridor, and the diffraction component of the field is such that the vector sum can be replaced by algebraic one. Then the expressions for the damping power of the street channel from the normalized distance k can be easily written as:

$$\alpha\left(\frac{r_0}{r}\right) = \frac{PD(r)}{PD_{max}} = \left[\frac{1}{k^n} + \frac{M(l)}{[l-(k-1)]^n} + L(r)\right]. \tag{12}$$

In (12) it is indicated: PD_{max} – is the maximum power flux density at the point 1, when $r = r_0$ and the energy moves in the direction of point 2 (Fig. 12); k = 1, 2...l, where l – number of sections the street of length $l \cdot r_0$ was divided into; $M(l) = PD_{max}(l) / PD_{max}(r_0)$, where $PD_{max}(l)$ – maximum power flux density at point 2 when energy moves in the direction of point 1 (Fig. 14); $L(r) = PD_d(r) / PD_{max}$, where $PD_d(r)$ – power flux density in the street due to diffraction of the radio channel.

In Fig. 13 calculations of value $\alpha(r)$ at $l = 10$, M = 0,1 , $r_0 = 100$ m are given assuming that $L(r) = const = 0,1$, and $n = 2$.

Curve $\alpha(r)$ at the decreasing site is well described by the function $(r_0 / r)^{1.5}$. Depending on the values of $M \in [0,1]$ and l the pattern of measuring the field distribution along a particular street can be described either by the inverse power function (Fig. 11,a), or by polynomial of *n* power (Fig. 10,a).

We obtained experimental field distributions along the street *WCAB* that well match with the data in (Porrat, 2002). Here they also conducted measurement of radio waves attenuation in street channels (Ottava), only at *900 MHz* frequency and with the help of nondirectional antennas.

$\alpha(r)$

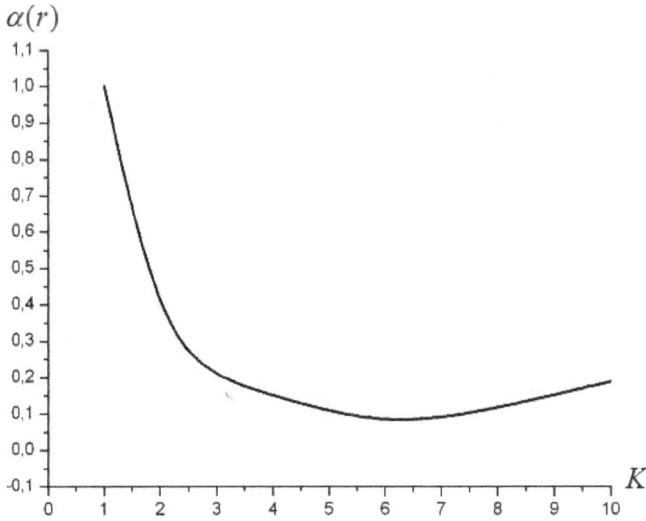

Fig. 13. Calculation results for formula (8)

Another proof of the validity of the results of the experiment is the data of repeated measurements shows in Fig. 11 d. These experiments were conducted one week after the first experiments. The qualitative nature of the curves is identical in both cases.

Thus, for *WCAB* of different frequencies of microwave range interference is inherent leading to the formation of mixed waves. The mathematical description of these waves is well developed in the theory of microwave circuits which is recommended for the calculation of street *WCAB* without diffractive component of the field. Another proof of the feasibility of the approach investigated to the *WCAB* analysis is presented in (Waganov, 1982).

3.2 Theoretical research of attenuations in the street wave channels formed by architectural buildings at the example of *WiMAX* system

When applying the *WCAB* model worked out in section 2, in the case of functioning of *WiMAX* systems, we must know the architectural features of the area where the measurements were done.

For the analysis area (Fig. 5) the following data were obtained. Height above the sea level for most of the analysis area varies smoothly from 135 to 145 m, which allows us to characterize the underlying surface as slightly undulating. The area is characterized by high building density, which makes it possible to approximate the lateral surface of wavelengths to a solid wall. Studies have shown that the material of the walls of most buildings in this case is brick. All the streets have asphalt as the underlying surface. Electrical parameters of brick can vary greatly enough and, according to the paper (Volkov, 2005), for this frequency range they are: specific permittivity $\varepsilon_1=2..15$, conductivity $\sigma=0,002..0,01$. The conductivity of asphalt is in the same range as for bricks ($\sigma_2=0,002..0,01$), and permittivity $\varepsilon_2=2..5$. Data as for the *WCAB* length and width are given in Table. 1.

Street name	Segment length, m	Width, m
Lenin Ave	4000	50-55
Bakulina	600	25-30
Danilevskii	1000	30-35
Lenin	1200	30-35
Lyapunov	550	30-50
Samokisha	200	25-50
Culture	600	30
Galana	500	30-40

Table 1. WCAB lenght and width

In the case of street branched radio channels general WCAB model, created in section 3.1, must be supplemented by the calculated damping ratios for RWP taking into consideration characteristics of WiMAX antenna system. Thus, the damping on the straight segment of the street should be calculated by the formula

$$\alpha\left(r_i / r_0\right) = (D_1 - \Delta D_1) + (D_2 - \Delta D_2) + \frac{PD(r_i / r_0)}{PD_{max}}, \; [dB], \qquad (13)$$

where D_1 – maximum directional antenna factor (DAF) of base station (14 dB), D_2 – maximum DAF of client adapter antenna (16,5 dB); $PD(r_i / r_0) / PD_{max}$ – relative power flux density, which was calculated using the models described in RWP LAN-MAN (Strelnitskiy, 2008); $\Delta D_1, \Delta D_2$ – amendments that allow change of DAF in a given direction, calculated from the mentioned considerations.

Antenna parameters significantly affect the nature and level of the signal and noise. The complexity of the problems of determining the signal strength and signal/noise level is that you need to know not only the maximum DAF D_{max}, but also DAF in a particular direction in azimuth θ and the corner of the place $\varphi - D(\theta, \varphi)$.

The applied methods of reducing antenna extraneous emission, and side lobe suppression leads to the complication of the analytical description of the antenna as a whole (Wheeler, 1947). In addition, such description often presents a trade secret of manufacturing companies and antenna technical data contain only its simplified polar pattern and basic characteristics. You should also take into account the fact that under real conditions of installation (the roof of a building, an antenna mast), due to the influence of the earth's surface and surrounding objects, the shape of a real PP is different from that calculated. Therefore, considering a large amount of analyzed data based on the information provided by the manufacturer of antennas, in the polar coordinate system we made polar patterns of the base station antenna (Fig. 14, a) and PP of the customer WiMAX antennas (Fig. 14, b).

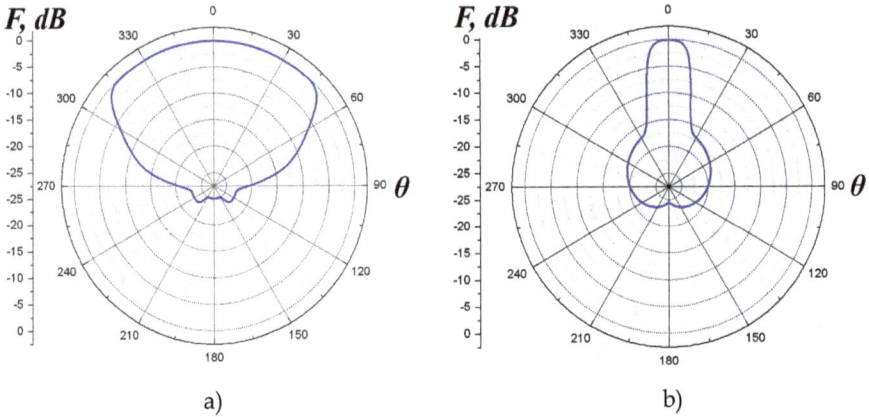

Fig. 14. PP of the base station (a) and customer $WiMAX$ antenna (b)

Mutual arrangement and orientation of antennas in the calculation of communication systems can be quite different. Therefore, the actual antenna gain at the base station in the direction of the client adapter interacting with it is defined by the angles that define the direction from the antenna of the transmitter to the receiver antenna, and vice versa, – in horizontal and vertical planes θ_{TR}, θ_{RT}, φ_{TR}, φ_{RT} respectively.

In Fig 15,a relative position of antenna polar patterns in the horizontal plane is given and the same thing is given in a plane passing perpendicular to the plane $\lambda, 0, \xi$ through the studied antennas in Fig. 15,b.

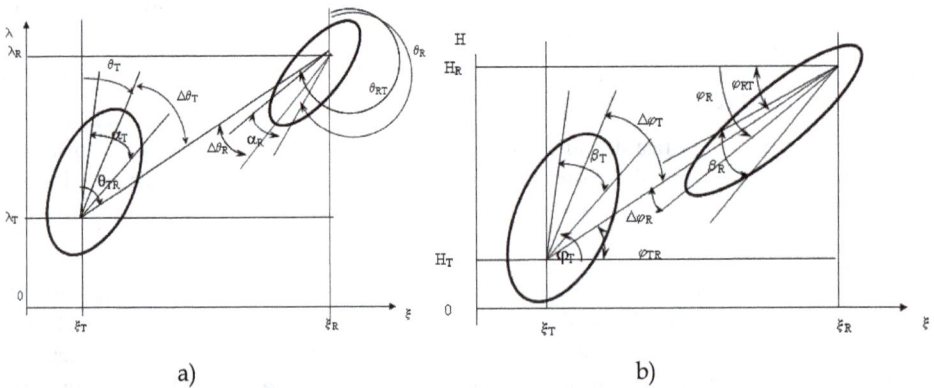

Fig. 15. Parameters of relative positions of antennas in the horizontal (a) and in the vertical (b) planes

In figures the following geographic coordinates are marked: λ_T, ξ_T – latitude and longitude of the location of the transmitter antenna, λ_R, ξ_R – latitude and longitude of the location of the receiver antenna; α_T, β_T – width of the transmitter PP in the horizontal and vertical planes respectively; α_R, β_R – the same for PP of the receiver; θ_T, θ_R – azimuth of maximum radiation and reception of the transmitting and receiving antennas, respectively; φ_T, φ_R – elevation angles of maximum radiation and reception of the transmitting and receiving antennas, respectively; r_i – distance between the transmitter and receiver; H_T, H_R – height of transmitting and receiving antennas, respectively.

Corners θ_{TR} and φ_{TR} are obtained from the geometrical problem in Fig.15:

$$\theta_{TR} = arccos\frac{\sin\lambda_R - \sin\lambda_T(\sin\lambda_T \cdot \sin\lambda_R + \cos\ \lambda_T \cdot \cos\lambda_R \cdot \cos(\xi_R - \xi_T))}{\cos\lambda_T \cdot \sin(arccos(\sin\lambda_T \cdot \sin\lambda_R + \cos\lambda_T \cdot \cos\lambda_R \cdot \cos(\xi_R - \xi_T)))}, \tag{14}$$

$$\varphi_{TR} = arctg\frac{H_R - H_T}{r_i}. \tag{15}$$

To determine θ_{RT} and φ_{RT} it is necessary to change indexes T to R and R to T in (14).

The correction that takes into account DAF changes in a given direction of PP, can be found from:

$$\Delta D = \sqrt{D_{\Delta\theta}^2 + D_{\Delta\varphi}^2}, \tag{16}$$

where $D_{\Delta\theta}$ and $D_{\Delta\varphi}$ – coefficients of directional antennas in both horizontal and vertical planes, respectively.

DAF relative to field strength can be calculated by the formula:

$$D_{\Delta\theta}, D_{\Delta\varphi} = 20 \cdot lg(\tau) \ (dB), \tag{17}$$

where τ – value, taking into account the reduction in antenna gain in direction $\Delta\theta, \Delta\varphi$ compared to the maximum gain.

Corners $\Delta\theta, \Delta\varphi$ can be calculated by the formulas:

$$\Delta\theta_{T,R} = \left|\theta_{T(R)} - \theta_{TR(RT)}\right|,$$
$$\Delta\varphi_{T,R} = \left|\varphi_{T(R)} - \varphi_{TR(RT)}\right|. \tag{18}$$

The width of base station antenna PP: in the horizontal plane – $90°$, in the vertical plane – $8°$. The width of customer adapter antenna PP: in the horizontal plane – $20°$, in the vertical plane – $20°$.

Formulas for calculation τ for different antenna types have the form:

- customer WiMAX adapter antennas:

$$\tau = \frac{4b^2 \cdot \cos^2\gamma}{\left(4b^2 - 1\right)\cos^2\gamma + 1}, \tag{19}$$

$$b^2 = \frac{1}{2} \cdot \frac{1 - \cos^2\alpha}{1 - \left(\sqrt{2}\cos\alpha - 1\right)^2}, \qquad 0° \leq \alpha \leq 65°, \ -90° \leq \gamma \leq 90°, \tag{20}$$

where γ – corner $\Delta\theta_T, \Delta\theta_R, \Delta\varphi_T, \Delta\varphi_R$ depending on the particular antenna (transmitting - T, receiving - R) and plane (horizontal θ or vertical φ); α – PP width in horizontal (α_T, α_R) or vertical (β_T, β_R) planes;

- *WiMAX* base station antennas:

$$\tau = \frac{(1-a)\cos\gamma + \sqrt{(1-a)^2 \cdot \cos^2\gamma + 4a}}{2}, \tag{21}$$

where $0 \le a \le 1$; at $a = 0$ $-90° \le \gamma \le 90°$; $a = 1$ $-180° \le \gamma \le 180°$.

Corner γ is calculated by the formula: $\gamma = arctg\left(\dfrac{H_T - H_R}{r_i}\right)$ Fig. 16.

Here are some examples of calculations on the proposed model by means of *Microwave Office* application package.

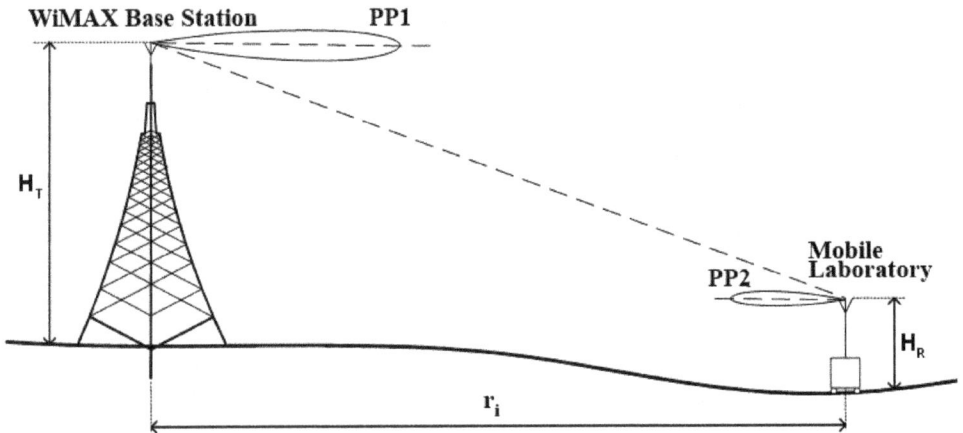

Fig. 16. For the calculation of the angle γ

The example of a streets connection scheme used to calculate the attenuation along Lenin Avenue in the *Microwave Office* application package, is shown in Fig. 17 ($l = r_i$ – line segment length r_i; $AT(m,n)$ – attenuators (m – street number, n – attenuator serial number for the street with m number); PI – power indicator). The scheme is activated by three sources of locally plane waves from the streets: Romain Rolland, Galana and Lenin Ave.

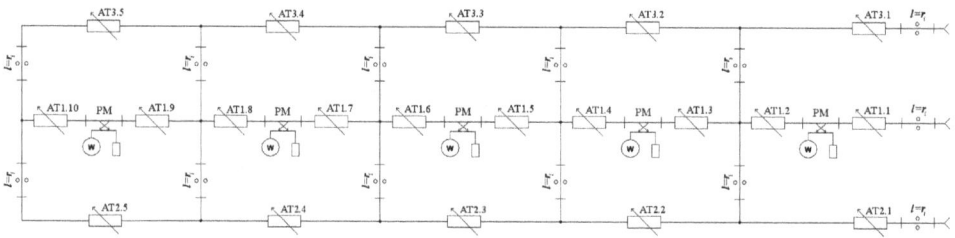

Fig. 17. The scheme for calculating the attenuation along Lenin Avenue ($m=1$ – Lenin Ave, $m=2$ – Romain Rolland street, $m=3$ – Galana street)

For signal level S assessment in r_i points on Lenin Ave. the attenuator AT(1,1) was attributed with the decay $\alpha(r_i / r_0)$, calculated by the formula (9). Summand $PD(r_i / r_0) / PD_{max}$ in formula (9) is calculated using ratio (10) for a power flux density and the expression (11) for the field strength at the receiver.

$$PD_{max} = \frac{1}{2} \cdot \frac{|E|^2}{120\pi}. \tag{22}$$

$$\dot{E}(r) = \frac{\sqrt{P \cdot Z_0 \cdot D_2}}{2\pi} \cdot \left[\frac{1}{r} \cdot \left(1 + R_B \cdot e^{-k \cdot r \cdot i} + R_\Gamma \cdot e^{-k \cdot r \cdot i} + R_\Gamma \cdot e^{-k \cdot r \cdot i} \right) \cdot e^{-k \cdot r \cdot i} \right], \tag{23}$$

where P – emitting power at the point of transfer; Z_0 – impedance of free space; D_2 – directional antenna factor; R_B, R_Γ – reflection coefficients for vertical and horizontal polarizations. Similarly, attenuation for all other attenuators AT(2,n) and AT(3,n) were calculated. At the same time power meter in the lines that imitate the streets 2 and 3, were excluded and were attributed to the attenuator loss distribution over the entire length of the segments. Comparative data on the results of the calculation according to the method described and experimental results are shown in Fig. 18, 20, 22. Fig. 19, 21 shows a connection diagram of the streets used to calculate the attenuation along the streets in the *Microwave Office* application package.

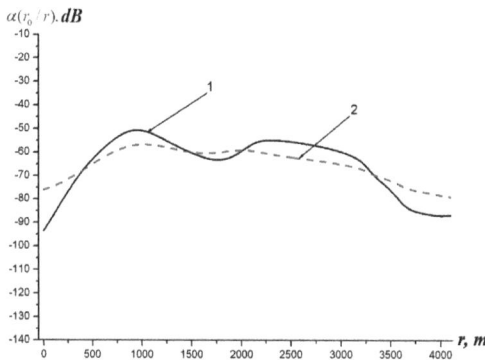

Fig. 18. Attenuation along the Lenin Ave (1 – experiment, 2 – calculation)

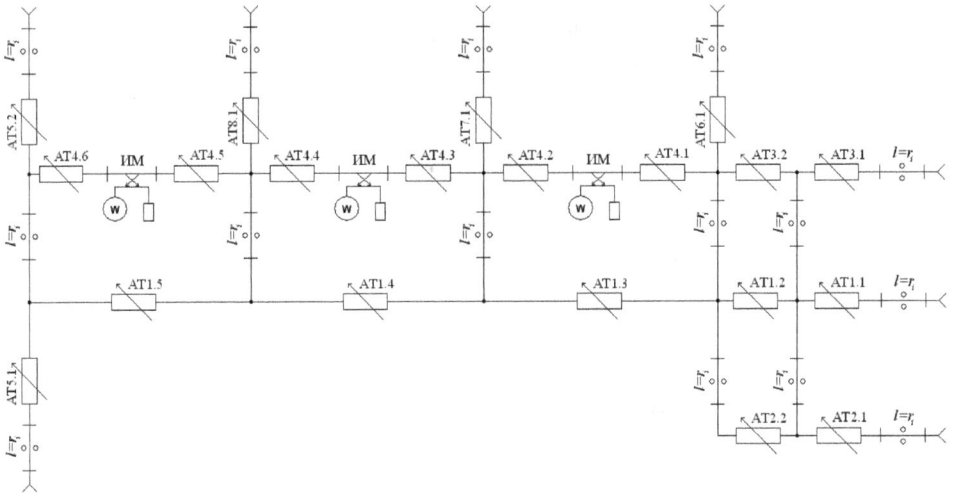

Fig. 19. The scheme for calculating the attenuation along Lenin street (m=1 – Lenin Ave., street (m=1 – Lenin Ave., m=2 – Romain Rolland street, m=3 – Yaroslav Galan street, m=4 – Lenin street, *m=5* – Novgorod street, *m=6* – Culture street, *m=7* – Baculina street, *m=8* – Ak. Lyapunov street)

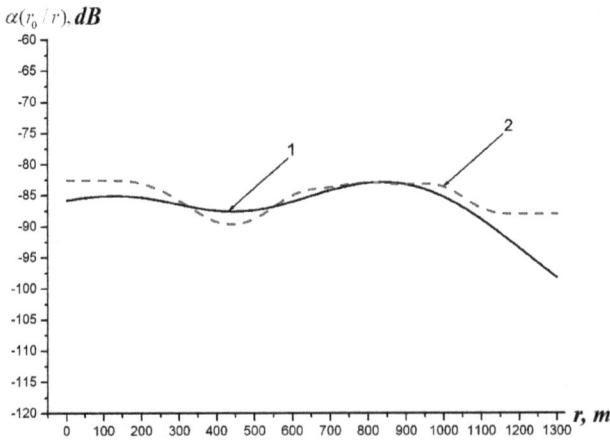

Fig. 20. Attenuation along Lenin street (1 – experiment, 2 – calculation)

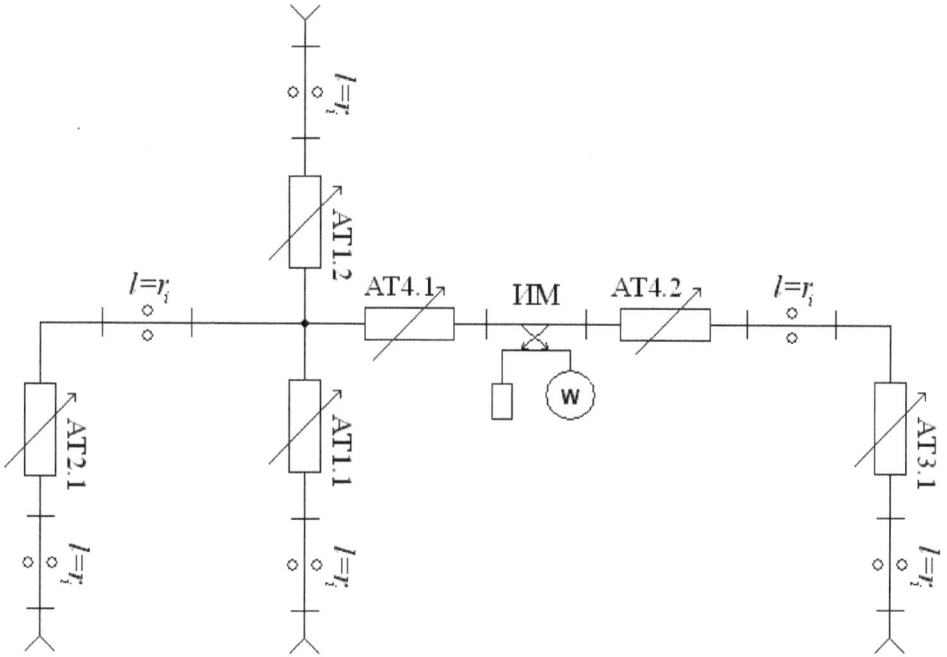

Fig. 21. The scheme for calculating the attenuation along the Danilevskii street (m=1 – Lenin Ave., m=2 – Romain Rolland street, m=3 – Yaroslav Galan street, m=4 – Danilevskii street)

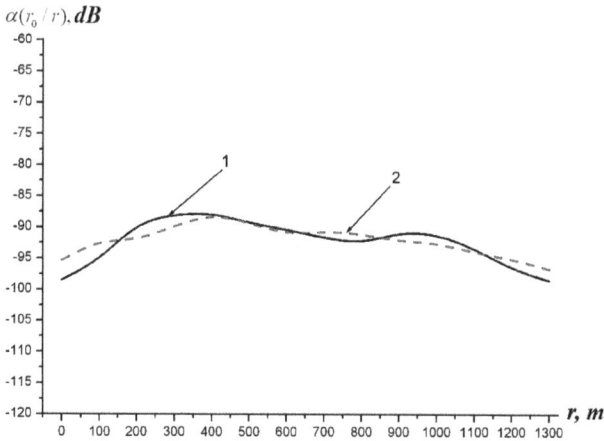

Fig. 22. Attenuation along the Danilevskii street (1 – experiment, 2 – calculation)

It is easily seen that the theoretical curves agree well with experimental data.

4. Assessment of performance and noise immunity of the *WiMAX* system in the city

Let us estimate the performance and security of *WiMAX* channel in the particular example of its operation in the telemedicine system (Fig. 23). It is expedient to consider this example because in constructing the telemedicine system branched channels of street *WCAB* are used.

In this case, the base station and special ambulance are supplied with *WiMAX* equipment. Medical team transfers the data about the patient via *Wi-Fi*. Information comes from the base station to the telemedicine center and providing consultations to the medical team.

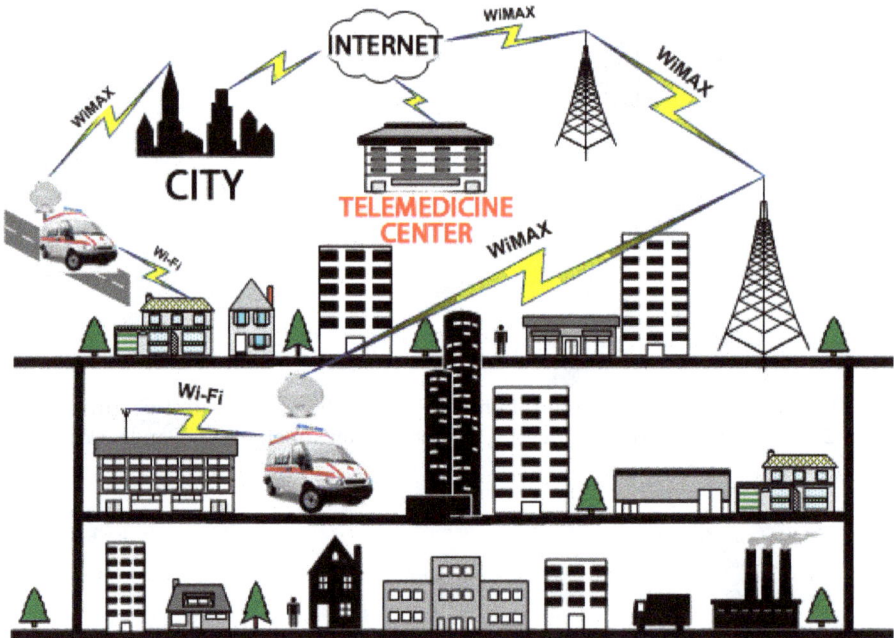

Fig. 23. Configuration of telemedicine network based on *Wi-Fi* and *WiMAX* technologies

Let us define the system efficiency, assuming that the telemedicine system serves population on Lenin Ave. The experimental and theoretical S/N values are shown in Fig. 5.14 (curves 1 and 2). From the comparison of these two curves it follows that the proposed model can be applied to calculate the performance of branched *WCAB* of *MAN* level (Strelnitskiy, 2009).

The rate of information transmission in *WiMAX* channel can change significantly (Fig. 24) depending on their bandwidth, which varies according to (IEEE Standard, 2004) from $1,7MHz$ (curve 1) to $3,5\ 7MHz$ (curve 2).

Comparing our results with the standards, we conclude that in the absence of interference in branched *WiMAX* channels, multimedia data can be transferred with high quality, which is very important during, for example, medical operations (Strelnitskiy, 2008).

Let us plot the graphs of the packet errors probability versus interference level for networks such as *MAN* (Fig. 25).

Analyzing the results in Fig. 25 with the recommendations of the video transmission standards, we conclude that the presence of noise value less than $P_J / P_S = 0,4$ in branched *WiMAX* channels, multimedia data can be transferred with high quality at a distance of *3 km,* that is very important in the construction of telemedicine networks in big cities.

The model examined in this chapter can be used not only to assess the performance and noise immunity of *WiMAX* radio channel on *WCAB* conditions, but also to forecast its physical level security [Strelnitskiy, 2011].

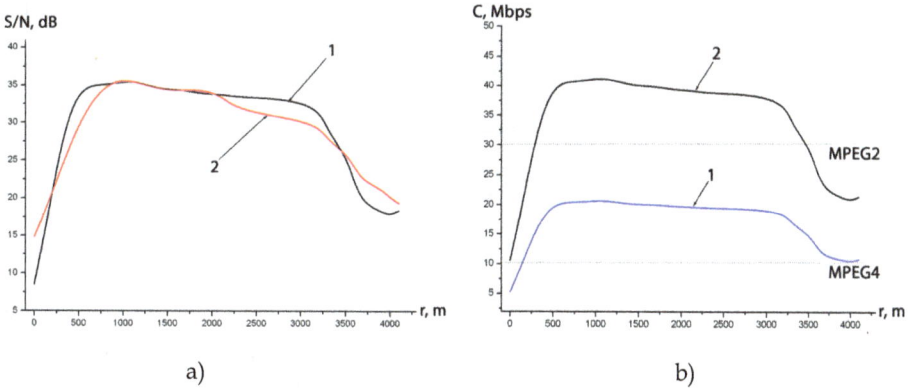

a)   b)

Fig. 24. Dependency of S / N to distance (a) and transmission rate (b) for Lenin Ave.

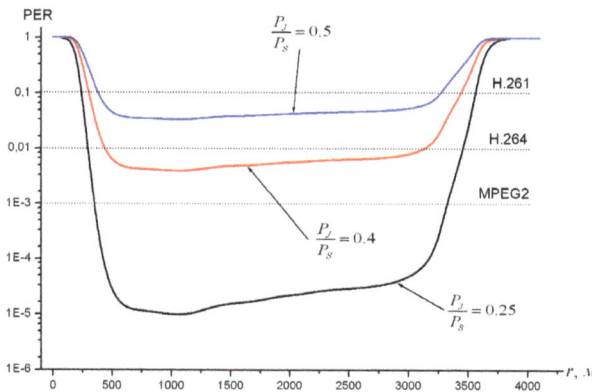

Fig. 25. The dependence of the probability of packet errors on the distance for different values of noise for Lenin Ave. (P_J – interference power, P_S – signal power)

5. Conclusion

1. The approach is developed to create a simplified *WCAB* model, which is based on the idea of representation branched wavelengths in the form of two-wire lines segments connected to one multipole and equivalent in the level of transmission power of street wavelengths. We propose a model to calculate the S-parameters of such multipole-based on using known cyclic algorithms and on registered losses equal to the *WRP* attenuation along *WCAB*.

2. Version of a mobile laboratory on the basis of *WiMAX* subscriber station was created and the *WRP* patterns of street *WCAB* were measured. From the comparison of the data we obtained in the signal/noise ratio with the data of *WiMAX* standard we discovered that the speed of information transmission by the *WiMAX* system operating in the city is less than 2 Mbps at a distance of 4 km.

3. New data on the possibilities of adapting the *WiMAX* system to maintain constant transmission speeds were obtained. It is shown that the adaptation of this system is realized if the signal level is above -65 dB (the lower limit declared before was -75 dB (Balvinder, 2006).

4. Using the property of high antenna directivity of *WiMAX* client adapter we offer a new method of detection the street wavelengths and reception of the diffraction component of the field. Using this methodology, we proved the dominant existence of *WCAB* in the central area of the city of Kharkov and showed that the level of the diffraction component does not exceed - 10 dB at the chosen measurement conditions ($h_{БС} = 80 м$, $h_{AC} = 1,5 м$ and low-rise building of analysis area).

5. It is shown that the patterns of distribution of the field along the city streets are largely predetermined by the interference of waves. A verbal description of this process is given and by comparing with the results (Porrat, 2002) a conclusion is made that it is valid for the cases of measurements in different cities and different frequencies of microwave range.

6. It is concluded that *WRP* dependencies along the street channels identified in the analysis of experimental results are characteristic of microwave circuits, which suggests the possibility of using their well-developed theory to create a *WCAB* model.

7. The ability to predict the attenuation in branched outdoor radio channels using the proposed *WCAB* model was proved.

8. The formula was derived for the calculation of the channel performance and the functional dependence of attenuation on the track was found, which allow to determine the performance of communication systems not worse than previously known mathematical models, but with much less time-consuming. Good agreement between calculated and experimental data gives the right to recommend the above model to calculate the performance and speed of information transmission in wireless access systems of *MAN* level.

9. The assessment of communication system noise immunity on *MAN* level has been given by the wave propagation in the channels formed by buildings. Comparing these results with the standards of video transmission, it was concluded that the presence of noise of less than $P_J / P_S = 0,4$ in branched *WiMAX* channels multimedia data with high quality can be transferred at a distance of 3 km, which is very important in the construction of telemedicine networks in major cities.

6. Acknowledgment

These studies were supported by the grant from the State Foundation for Fundamental Research, Ministry of Science and Education of Ukraine (#F25/217-2008).

7. References

Balvinder, B.; Eline, R. J., and Franca-Neto, L. M. (2006), RF System and Circuit Challenges for *WiMAX*. *Intel® Technology Journal*, Vol.08, Issue 03, (August 2004). pp. 189-201, ISSN: 1535-864X

Fabricio, L. F., and Cardieri, P., (2005), Coverage Prediction and Performance Evaluation of Wireless Metropolitan Area Networks based on IEEE 802.16. *Journal* of communication and information systems, Vol.20. No.3, (2005), pp. 132-140, ISSN: 1980–6604

Gostev, V.I., Konin, V.V., and Matsepura, A.L., (1997), Linear multichannel microwave devices, Radio amator Publishing house, Kiev: 315 p. (in Russian)

Hata, M., (1980), Empirical formula for propagation loss in land mobile radio service, *IEEE Transactions on Vehicular Technology*, Vol. 29, No. 3, (Aug 80), pp. 317-325.

IEEE Standard, (2004), Standard for Local and metropolitan area networks. Part 16: Air Interface for Fixed Broadband Wireless Access Systems. IEEE P802.16-REVd/D5-2004, 915 p.

Porrat, D. (2002). PhD Thesis: Radio Propagation in Hallways and Streets for UHF Communications.

Strelnitskiy, O.O.; Strelnitskiy, O.E.; Tsopa, O.I., and Shokalo, V.M., (2011), Prediction Model of Energy Security for the Systems of Subscriber Radio Access with Branched Street and Corridor Communications Channels. *International journal «Radioelectronics and Communications Systems»*, Allerton Press, Inc., Springer, Vol. 54, No. 2, (2011), pp. 61-68, ISSN: 0735-2727.

Strelnitskiy, O.O.; Tsopa, O.I. and Shokalo, V.M., (2009), Approximate Model for Estimation of Efficiency and Noise Immunity of Branched Street and Corridor *Wi-Fi* and *WiMAX* Communication Channels. *International journal «Telecommunication and Radio Engineering»*, Begell House, Vol. 68(17), (2009), pp. 1511-1528, ISSN: 0040-2508

Strelnitskiy, O.E.; Tsopa, O.O.; Tsopa, O.I., and Shokalo, V.M. (2008), The variant of quality increasing of video information transmission via *WIMAX* fixed connection radio channel. *Proceeding of IX International Conf. Modern problems of Radio Engineering, Telecommunications and Computer Science /TCSET'2008/*, Lviv-Slavsko, Ukraine, (February 2008), pp. 388-389, ISBN: 978-966-553-678-9

Strelnitskiy, A.A., Strelnitskiy, A.E., Tsopa, O.I., and Shokalo, V.M., (2008), Version of the model of the wideband signal attenuation in radio link when calculating the local communication networks protection, *Scientific Journal Information Protection*, Vol. 3(39), pp. 38-43 (in Russian).

Strelnytskiy, A.A.; Strelnytskiy, A.E.; Tsopa, O.I., and Shokalo, V.M., (2007), The Model of the Multiterminal Network for Attenuation Calculation of the Radio Waves in the Wave Channels of the Architectural Buildings (KNURE - WCAB Model). *Microwave & Telecommunication Technology, CriMiCo2007.* 17th International Crimean Conference Publication, (September 2007), pp. 213-214, ISBN: 978-966-335-012-7

Volkov, L.N., Nemirovsky, M.S., and Shinakov Y.S., (2005), Digital communication systems: basic methods and characteristics, Eco-Trends, Moscow: 392 p. (in Russian)

Waganov, R.B., and Katseneleybaum, B.Z. (1982), Fundamentals of diffraction theory. Science, Moscow: 272 p. (in Russian) Grudinskaya, T.P., (1967), Wave propagation, High school, Moscow: 244 p. (in Russian)

Wei, Z., (2007), Capacity analysis for multi-hop *WiMAX* relay. *Proceeding of the International Conference on Wireless Broadband and Ultra Wideband Communication*, (Mar. 2007), University of Technology Sydney, Sydney, pp. 1-4

11

On PAPR Reduction Techniques in Mobile WiMAX

Imran Baig and Varun Jeoti
Universiti Teknologi PETRONAS,
Malaysia

1. Introduction

The mobile Worldwide Interoperability for Microwave Access (Mobile WiMAX) air interface adopts orthogonal frequency division multiple access (OFDMA) as multiple access technique for its uplink (UL) and downlink (DL) to improve the multipath performance. All OFDMA based networks including mobile WiMAX experience the problem of high peak-to-average power ratio (PAPR). The literature is replete with a large number of PAPR reduction techniques. Among them, schemes like constellation shaping, phase optimization, nonlinear companding transforms, tone reservation (TR) and tone injection (TI), clipping and filtering, partial transmit sequence (PTS), precoding based techniques, selective mapping (SLM), precoding based selective mapping (PSLM) and phase modulation transform are popular. The precoding based techniques, however, show great promise as they are simple linear techniques to implement without the need of any complex optimizations. This chapter reviews these PAPR reduction techniques and presents a Zadoff-Chu matrix transform (ZCMT) based precoding technique for PAPR reduction in mobile WiMAX systems. The mobile WiMAX systems employing random-interleaved OFDMA uplink system has been used for determining the improvement in PAPR performance of the technique. It has been further used in selective mapping (SLM) based ZCMT precoded random-interleaved OFDMA uplink system. PAPR of these systems are analyzed with the root-raised-cosine (RRC) pulse shaping to keep out-of-band radiation low and to meet the transmission spectrum mask requirement. Simulation results show that the proposed systems have low PAPR than the Walsh-Hadamard transform (WHT) precoded random-interleaved OFDMA uplink systems and the conventional random-interleaved OFDMA uplink systems. The symbol-error-rate (SER) performance of these uplink systems is also better than the conventional random-interleaved OFDMA uplink systems and at par with WHT based random-interleaved OFDMA uplink systems. The good improvement in PAPR offered by the presented systems significantly reduces the cost and the complexity of the transmitter.

This chapter is organized as follows: Section 2 describes the background of the random-interleaved OFDMA uplink systems and SLM based random-interleaved OFDMA uplink systems, while in section 3, we present a detailed literature review. In section 4, we present our proposed system models with improved PAPR, section 5 presents the computer simulation results and section 6 concludes the chapter.

2. Background

The mobile Worldwide Interoperability for Microwave Access (Mobile WiMAX) is a broadband wireless solution that enables the convergence of mobile and fixed broadband networks through a common wide area radio-access (RA) technology and flexible network architecture. Since January 2007, the IEEE 802.16 Working Group (WG) has been developing a new amendment of the IEEE 802.16 standard i.e. IEEE 802.16m as an advanced air interface to meet the requirements of ITU-R/IMT-Advanced for 4G systems. The mobile WiMAX air interface adopts orthogonal frequency division multiple access (OFDMA) as multiple access technique for its uplink (UL) and downlink (DL) to improve the multipath performance. The scalable OFDMA (SOFDMA) is introduced in the IEEE 802.16e amendment to support scalable channel bandwidth.

OFDMA is a multiple access version of the orthogonal frequency division multiplexing (OFDM) systems. OFDMA system splits the high speed data stream into a number of parallel low data rate streams and these low rates data streams are transmitted simultaneously over a number of orthogonal subcarriers. The key difference between OFDM and OFDMA is that instead of being allocated all of the available subcarriers, the base station assigns a subset of carriers to each user in order to accommodate several transmissions at the same time. An inherent gain of the OFDMA based systems is its ability to exploit the multiuser diversity through subchannel allocation. Additionally, OFDMA has the advantage of simple decoding at the receiver side due to the absence of inter-carrier-interference (ICI). Other benefits of OFDMA include better granularity and improved link budget in the uplink communications (Knopp & Humblet, 1995; Tse, 1997).

There are two different approaches to do subcarrier mapping in OFDMA systems, localized subcarrier mapping and distributed subcarrier mapping. The distributed subcarrier mapping can be further divided in to two modes, interleaved mode and random interleaved mode. Fig.1 shows the subcarrier mapping in interleaved mode, where the subcarriers are mapped equidistant to each other's. Fig.2 explains the subcarrier mapping in random-interleaved mode, where the subcarriers are mapped randomly based on some permutation algorithm to each other's. Fig.3 further explains the concept of localized subcarrier mapping, where the subcarrier mapping is done in adjacent.

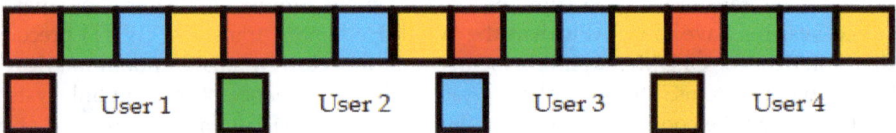

Fig. 1. Interleaved OFDMA

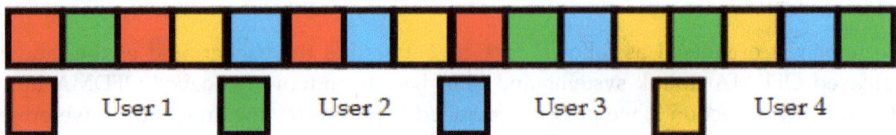

Fig. 2. Random-Interleaved OFDMA

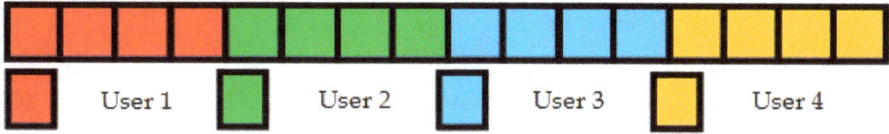

Fig. 3. Localized OFDMA

OFDMA is widely adopted in the various communication standards like WiMAX, mobile broadband wireless access (MBWA), evolved UMTS terrestrial radio access (E-UTRA) and ultra mobile broadband (UMB). OFDMA is also a strong candidate for the wireless regional area networks (WRAN) and the long term evaluation advanced (LTE-Advanced).

However, OFDMA has some drawbacks, among others; the peak-to-average power ratio (PAPR) is still one of the major drawbacks in the transmitted OFDMA signal (Wang & Chen, 2004). Therefore, for zero distortion of the OFDMA signal, the high-power-amplifier (HPA) must not only operate in its linear region but also with sufficient back-off. Thus, HPA with a large dynamic range is required for OFDMA systems. These amplifiers are very expensive and are major cost component of the OFDMA systems.

Thus, if we reduce the PAPR it not only means that we are reducing the cost of OFDMA systems and reducing the complexity of analog-to-digital (A/D) and digital-to-analog (D/A) converters, but also increasing the transmit power, thus, for same range improving received signal-noise-ratio (SNR), or for the same SNR improving range. Fig.4 illustrates the block diagram of the OFDMA uplink systems. In OFDMA uplink systems the baseband modulated symbols are passed through serial-to-parallel (S/P) converter which generates complex vector of size M. We can write the complex vector of size M as follows:-

$$X = [X_0, X_1, X_2 \dots X_{M-1}]^T \tag{1}$$

Then the subcarrier mapping of these constellations symbols can be done on in one of the subcarrier mapping mode: interleaved mode, random-interleaved mode or in localized mode respectively. After the subcarrier mapping, we get frequency domain samples: $\{\hat{Y}_l : l = 0,1,2,\dots,N-1\}$. Mathematically, the subcarrier mapping in interleaved mode can be done as follows:-

$$\hat{Y}_l = \begin{cases} X_{\frac{l}{Q}} & , l = Q.k \quad 0 \le k \le M-1 \\ 0 & otherwise \end{cases} \tag{2}$$

where N : System subcarriers, M : User subcarriers, Q : Subchannels/Users, $(Q=N/M)$, $0 \le l \le N$-1 and $N=Q.M$. The subcarrier mapping in random-interleaved mode can be done mathematically as follows:-

$$\hat{Y}_l = \begin{cases} X_{\frac{l}{\hat{Q}}} & , l = \hat{Q}.k \quad 0 \le k \le M-1 \\ 0 & otherwise \end{cases} \tag{3}$$

where, $0 \le l \le N$-1 and $N=Q.M$, and $0 \le \hat{Q} \le Q$. The subcarrier mapping in localized mode can be done mathematically as follows:-

$$\hat{Y}_l = \begin{cases} X_l & 0 \le l \le M-1 \\ 0 & M \le l \le N-1 \end{cases} \qquad (4)$$

The kth subcarrier of each group is assigned to the kth user with index set $\{(k), (Q+k), ..., ((M-1)Q+k)\}$. Suppose the kth user is assigned to subchannel k then the complex baseband ZCMT precoded interleaved OFDMA uplink signal for k^{th} user can be written as follows:-

$$x_n^{(k)} = \sum_{l=0}^{L-1} \hat{Y}_l^{(k)} . e^{j2\pi \frac{(lQ+k)}{N} n}, \ n = 0,1...N-1 \qquad (5)$$

The kth subcarrier of each group is assigned to the kth user with index set: $\{(\gamma_{q,1}),(Q+\gamma_{q,2}),...,(M-1)Q+\gamma_{q,M-1})\}$, where $\{(\gamma_{q,1}),(\gamma_{q,2}),...,(\gamma_{q,M-1})\}$ are independent and identically distributed random variables with uniform distribution on $(q=0,1,2,...,Q-1)$. Suppose the kth user is assigned to sub-channel k then the complex baseband random-interleaved OFDMA signal for k^{th} user with N system subcarriers and M user subcarriers can be written as follows:-

$$x_n^{(k)} = \sum_{l=0}^{L-1} (\hat{Y}_l^{(k)} . e^{j2\pi \frac{(lQ+\gamma_{q,k})}{N} n}), \ n = 0,1...N-1 \qquad (6)$$

The subchannel k is composed of subcarriers with index set $\{(kL), (kL+1), (kL+2)... (kL+L-1)\}$, where $k=0,1,2,..., Q-1$. Suppose the kth user is assigned to subchannel k then the complex baseband ZCMT precoded localized OFDMA uplink signal for k^{th} user can be written as follows:-

$$x_n^{(k)} = \frac{1}{\sqrt{N}} \sum_{l=0}^{L-1} (\hat{Y}_l^{(k)} . e^{j2\pi \frac{(kL+l)}{N} n}), \ n = 0,1...N-1 \qquad (7)$$

$\hat{Y}_l^{(k)}$ is modulated signal on subcarrier l for k^{th} user.

The complex passband signal of OFDMA uplink systems after the RRC pulse shaping can be written as follows:-

$$x(t) = e^{j\omega_c t} \sum_{n=0}^{N-1} x_n^{(k)} . r(t-n\breve{T}) \qquad (8)$$

where, ω_c is carrier frequency, $r(t)$ is baseband pulse, $\breve{T} = (\frac{M}{N}).T$ is compressed symbol duration after IFFT and T is symbol duration is seconds. The RRC pulse shaping filter can be defined as follows:-

$$r(t) = \frac{sin\left(\frac{\pi t}{\breve{T}}(1-\alpha)\right) + 4\alpha \frac{t}{\breve{T}}.cos\left(\frac{\pi t}{\breve{T}}(1+\alpha)\right)}{\frac{\pi t}{\breve{T}}.\left(1-\frac{16\alpha^2 t^2}{\breve{T}^2}\right)} \qquad (9)$$

$0 \le \alpha \le 1$, where α is rolloff factor. The PAPR of OFDMA uplink signal in (8) with RRC pulse shaping can be written as follows:-

$$PAPR = \frac{\max_{0 \le t \le N\breve{T}} |x(t)|^2}{\frac{1}{N\breve{T}} \int_0^{N\breve{T}} |x(t)|^2 dt} \qquad (10)$$

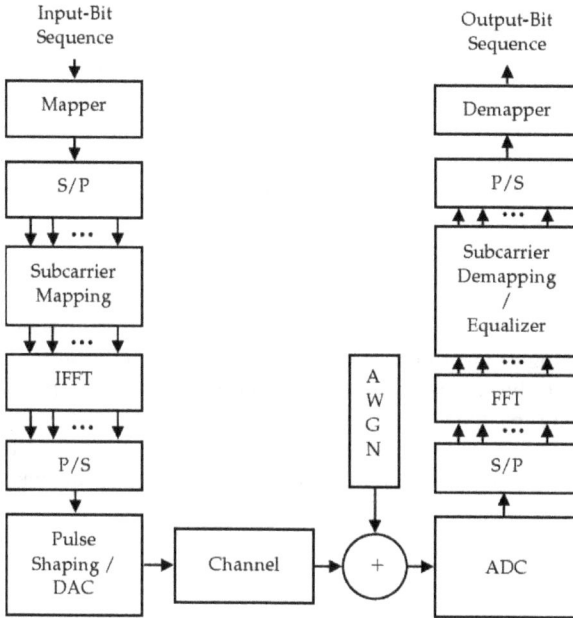

Fig. 4. Random-Interleaved OFDMA uplink system

3. Literature review

A large number of PAPR reduction techniques have been proposed in the literature. Among them, schemes like phase optimization (Nikookar & Lidsheim, 2002), constellation shaping (Kou et al., 2007), selective mapping (SLM) (Lim et al., 2005), nonlinear companding transforms (Jiang et al., 2006), tone reservation (TR), tone injection (TI) (Mourelo., 1999; Yoo et al., 2006), partial transmit sequence (PTS) (Han & Lee, 2004; Müller & Huber, 1997; Cimini & Sollenberger, 2000; Tellambura, 2001), clipping and filtering (Wang & Tellambura, 2005; Li & Cimini, 1998; Nee & Wild, 1998), precoding based techniques (Slimane, 2007; Min & Jeoti, 2007; Baig & Jeoti, 2010a, 2010b, 2010c), precoding based selective mapping (PSLM) techniques (Baig & Jeoti, 2010a, 2010b) and phase modulation transform (Tasi et al., 2006; Thompson et al., 2008) are popular. The precoding based techniques, however, show great promise as they are simple linear techniques to implement without the need of any side $\breve{T} = (\frac{M}{N}).T$ information. Additionally, the precoding based techniques take advantage of frequency variations of the communication channel and offers substantial performance gain in fading multipath channels. In the following sub-section we focus more closely on the PAPR reduction techniques for multicarrier transmission.

3.1 Clipping and filtering techniques

The clipping techniques are simpler and commonly used to reduce the PAPR (Wang & Tellambura, 2005; Li & Cimini, 1998; Nee & Wild, 1998). These techniques apply clipping or nonlinear saturation around the peaks to lower the high PAPR produced by the multicarrier transmitter. It is straightforward to clip the signal parts that are outside the tolerable area. Clipping techniques introduces in-band or out-of-band distortions that can destroy the orthogonality between the subcarriers. Generally, the clipping operation is carried out at the transmitter. On the other hand, the receiver requires to estimate the clipping that has been carried out at the transmitter and to compensate the received OFDM symbol accordingly. Normally, most of the time no more than one clipping happens per OFDM symbol. Hence, the receiver has to approximate the size and the location of the clip. However, it is hard to get related information. After the clipping operation, the filtering operation can noticeably decrease the out-of-band radiation.

Unfortunately, the in-band distortion cannot be reduced by the filtering operation. On the other hand, the clipping can introduce some peak re-growth. So, after the clipping operation and filtering operation the signal may exceed the clipping level at some points. To decrease the peak re-growth, a repeated clipping operation and filtering operation can be carried out to obtain a desirable PAPR at the expense of computational complexity increase. Fig.5 shows represent the clipped edition of the $x^p[m]$, which can be written as follows:-

$$x_c^p[m] = \begin{cases} -A & x^p[m] \le -A \\ x^p[m] & |x^p[m]| < A \\ A & x^p[m] \ge A \end{cases} \tag{11}$$

or

$$x_c^p[m] = \begin{cases} x^p[m] & x^p[m] \le -A \\ \frac{x^p[m]}{x^p[m]}.A & \text{Otherwise} \end{cases} \tag{12}$$

where A is pre-determined clipping level. The equation (12) can be used for both baseband complex-valued signals and passband real-valued signals and the equation (11) can only be used for the passband signals only. The clipping-ratio (CR) normalized by the root-mean-square (RMS) value σ of OFDM signal can be written as follows:-

$$CR = A/\sigma \tag{13}$$

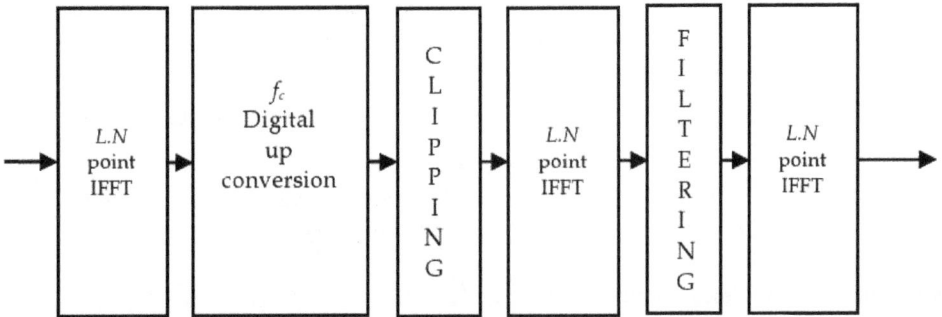

Fig. 5. Block Diagram of OFDM System with Clipping and Filtering (Cho et al., 2011)

3.2 Selective Mapping (SLM)

The SLM is one of the most popular PAPR reduction techniques in the literature (Lim et al., 2005). This technique is based on the phase rotations. In SLM based OFDM (SLM-OFDM) systems, a set of V different data blocks are created at the transmitter representing the identical information and a data block with minimum PAPR is selected for the transmission. Fig.6 shows the general block diagram of the SLM-OFDM system. Every data block is multiplied with the V dissimilar phase sequences, each of length N, $B^{(v)} = [b_{v,0}, b_{v,1}, ..., b_{v,N-1}]^T$, $v= 1, 2...V$, which results in the changed data blocks. Now suppose the altered data block for the vth phase sequence is given by $X^{(v)} = [X_0 b_{v,0}, X_1 b_{v,1}, ..., X_{N-1} b_{v,N-1}]^T$, $v=1, 2... V$. Each X_n^v can be defined as follows:-

$$X_n^v = X_n b_{v,n} \quad , \quad (1 \leq v \leq V) \tag{14}$$

After applying SLM to X, the OFDM signal becomes as follows:-

$$x_n^{(v)} = \frac{1}{\sqrt{N}} \sum_{k=0}^{N-1} X_k^v \cdot e^{j2\pi \frac{n}{N} k} , \text{n=0, 1, 2... N-1} \tag{15}$$

where, $v = 1, 2... V$. Amongst all the tailored data blocks: $x^{(v)}$, $v = 1, 2... V$, the data block with minimum PAPR is selected for the transmission. Side information about the selected phase sequence must be communicated to the receiver which performs the reverse operation in order to recover the actual data block.

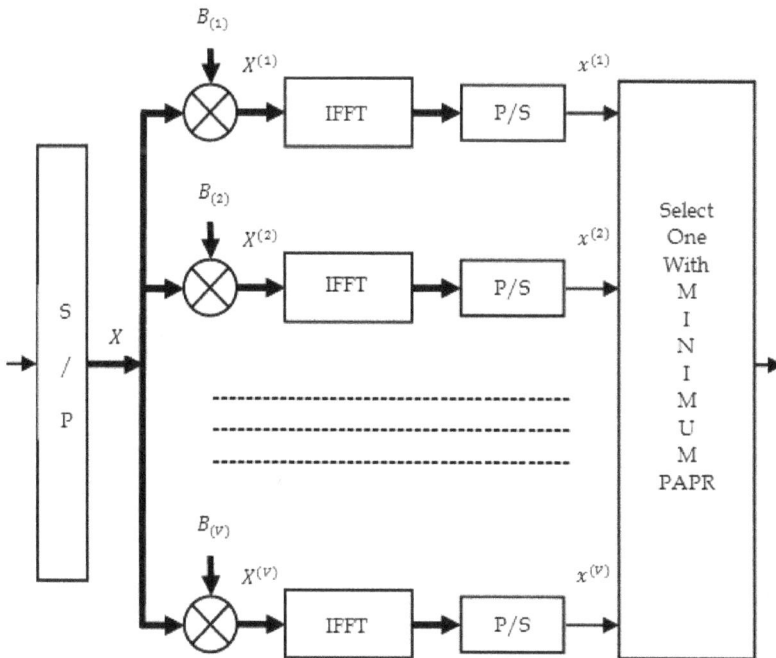

Fig. 6. Block diagram of OFDM system with Selective Mapping (Han & Lee, 2005)

3.3 Partial Transmit Sequence (PTS)

PTS is another very popular PAPR reduction technique (Han & Lee, 2004; Müller & Huber, 1997; Cimini & Sollenberger, 2000; Tellambura, 2001). In this technique, the input data block of N symbols is partitioned into disjoint sub-blocks. In each sub-block, the subcarriers are weighted by a phase factor. The phase factors are chosen in such a way so that the PAPR of the combined signal is reduced. Fig.7 shows the general block diagram of the PTS PAPR reduction technique. In the PTS technique input data block X is partitioned into M disjoint sub-blocks: $X^m = [X^{m,0}, X^{m,1}, ..., X^{m,N-1}]^T$, $m = 1,2,3,...,M$, such that $\sum_{m=1}^{M} X^m = X$ and the sub-blocks are combined to reduce the PAPR in the time-domain. The L-times oversampled time-domain signal of X_m, $m=1,2,3,...,M$ is denoted by: $x^m = [x^{m,0}, x^{m,1}, ..., x^{m,N-1}]^T$, $m=1,2,3,...,M$ is obtained by obtained by taking an IFFT of length NL on X^m concatenated with $(L-1)N$ zeros. These are called PTS. Complex phase factors $b^m = \exp(j\Phi m)$, $m=1,2,3,... M$, are launched to combine the PTSs. The set of phase factors is designated as a vector: $\hat{b} = [\hat{b}^1, \hat{b}^2, ..., \hat{b}^M]^T$. The time domain-signal after combining can be written as follows:-

$$x = \sum_{m=1}^{M} \hat{b}^m . x^m \tag{16}$$

The key idea is to find out the set of phase factors that reduces the PAPR. Generally, to reduce the search complexity, the selection of the phase factors is bounded by a set with a finite number of elements. The set of acceptable phase factors can be written as: $P = \{e^{j2\pi\frac{l}{W}} : l = 0,1,2,...,W-1\}$, where W is the number of permitted phase factors. Additionally, we can set $b^1 = 1$ without any loss of the performance.

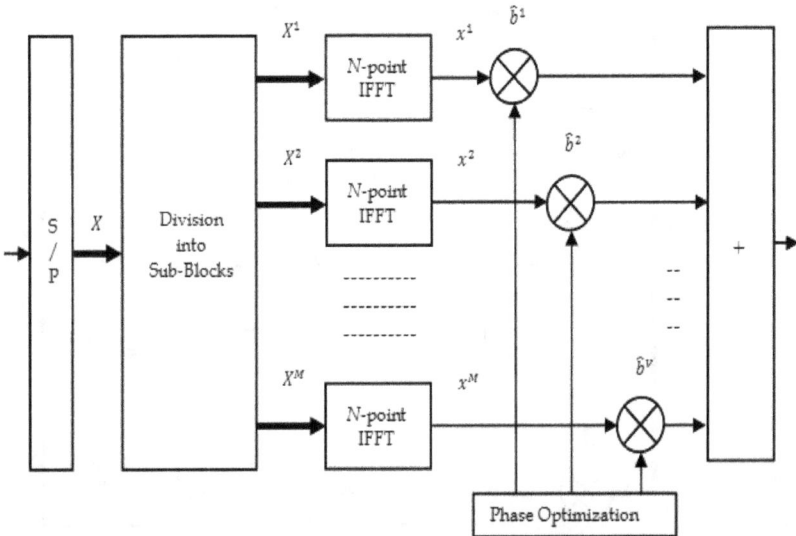

Fig. 7. Block diagram of OFDM system with Partial Transmit Sequence (Müller & Huber, 1997)

Therefore, we should execute a complete search for $(M-1)$ phase factors. So, to find the optimum set of phase factors the W^{M-1} sets of phase factors are searched. If we increase the number of sub-blocks M, the search complexity is increases exponentially. PTS needs M IFFT operations for every data block, and the number of needed side information bits is $[\log_2 W^{M-1}]$. The amount of PAPR reduction is based on the number of sub-blocks M and the number of permitted phase factors W. Subblock partitioning is another factor that may have an effect on the PAPR gain PTS, which is the way of partition of the subcarriers into several disjoint sub-blocks. There are three kinds of sub-block partitioning techniques: interleaved, pseudo-random and adjacent partitioning.

Among them, pseudo-random partitioning has been found to be the best choice for PTS. The PTS technique can work with a random number of subcarriers and any modulation scheme. As mentioned above, the ordinary PTS technique has exponentially increasing search complexity. To lower the search complexity, a range of techniques have been proposed in the literature. Once the PAPR falls below a set threshold, the iterations for updating the set of phase factors must be stopped. Number of techniques has been presented in the literature to reduce the number of iterations. These techniques achieve considerable reduction in search complexity with minor PAPR performance degradation.

Example: The PTS PAPR reduction technique for an OFDM system can be explained with a simple example (Han & Lee, 2005). Here, we take eight subcarriers that are divided into four sub-blocks. The phase factors are selected in $P = \{\pm 1\}$. Fig.8 illustrates the adjacent sub-block partitioning for a data block X of size 8. The original data block X has a PAPR of 6.5 dB. There are 8 ways to mix the sub-blocks with fixed $b^1 = 1$.

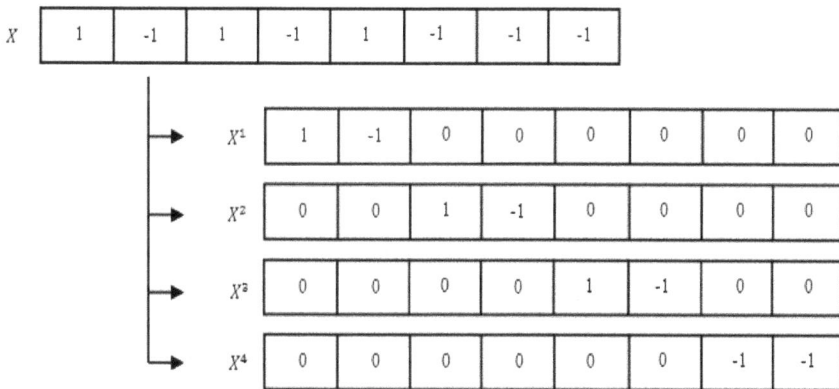

X	1	-1	1	-1	1	-1	-1	-1

X^1	1	-1	0	0	0	0	0	0
X^2	0	0	1	-1	0	0	0	0
X^3	0	0	0	0	1	-1	0	0
X^4	0	0	0	0	0	0	-1	-1

Fig. 8. An example of adjacent subblock partitioning in PTS (Han & Lee, 2005)

Amongst them $[\hat{b}^1, \hat{b}^2, \hat{b}^3, \hat{b}^4]^T = [1, -1, -1, -1]^T$ gets the lower PAPR. The tailored data block will be $X = \sum_{m=1}^{M} \hat{b}^m . X^m = [1,-1,-1,1,-1,1,1,1]$ whose PAPR is 2.2 dB, resulting in a 4.3 dB PAPR gain. In this case, the number of necessary IFFT operations is 4 and the amount of side information is 3 bits. The side information should be transmitted to the receiver for the recovery of original data block. There are many ways to transmit side information; one of the ways is to transmit side information bits with a separate channel other than the data channel. Another ways is to include the side information within the data block but it results in data rate loss.

3.4 Precoding based techniques

Precoding based techniques are simple linear techniques. These techniques can reduce the PAPR up to the PAPR of single carrier systems (Slimane, 2007). Walsh-Hadamard transform (WHT) precoding based techniques, discrete cosine transform (DCT) precoding based techniques, discrete hartley transform (DHT) precoding based techniques are common examples of precoding based PAPR reduction techniques (Slimane, 2007; Min & Jeoti, 2007; Baig & Jeoti, 2010a, 2010b, 2010c).

3.4.1 Walsh-Hadamard Transform (WHT)

WHT is an orthogonal linear transform and can be implemented by a butterfly structure as in FFT. This means that applying WHT does not require the extensive increase of system complexity. The kernel of WHT can be written as follows:-

$$H_1 = [1] \tag{17}$$

$$H_2 = \frac{1}{\sqrt{2}}\begin{bmatrix} 1 & 1 \\ 1 & -1 \end{bmatrix} \tag{18}$$

$$H_{2N} = \frac{1}{\sqrt{2N}}\begin{bmatrix} H_N & H_N \\ H_N & H_N^{-1} \end{bmatrix} \tag{19}$$

where H_N^{-1} denotes the binary complement of H^N.

3.4.2 Discrete Hartley Transform (DHT)

DHT is a linear transform. In DHT N real numbers $x_0, x_1, ..., x_{N-1}$ are transformed in to N real numbers $H_0, H_1, ..., H_{N-1}$. The N-point DHT can be defined as follows:-

$$H_k = \sum_{n=0}^{N-1} x_n \left[cos\left(\frac{2\pi nk}{N}\right) + sin\left(\frac{2\pi nk}{N}\right) \right]$$
$$= \sum_{n=0}^{N-1} x(n). \; cas\left(\frac{2\pi nk}{N}\right) \tag{20}$$

$$p_{m.n} = cas\left(\frac{2\pi mn}{N}\right) \tag{21}$$

Where $cas\theta = cos\theta + sin\theta$ and $k=1, 2, 3... N-1$. The DHT is also invertible transform which allows us to recover the x_n from H_k and inverse can be obtained by simply multiplying DHT of H_k by $\frac{1}{N}$.

3.4.3 Discrete Cosine Transform (DCT)

DCT matrix P of size N-by-N can be created by using equation (22)

$$D_{ij} = \begin{cases} \frac{1}{\sqrt{N}} & i = 0, \quad 0 \leq j \leq N-1 \\ \sqrt{\frac{2}{N}}\cos\frac{\pi(2j+1)i}{2N} & \begin{array}{l} 1 \leq i \leq N-1 \\ 0 \leq j \leq N-1 \end{array} \end{cases} \tag{22}$$

and DCT can be defined as:-

$$X_k = \sum_{n=0}^{N-1} x_n \cdot \cos\left[\frac{\pi}{N}\left(n + \frac{1}{2}\right)k\right] \qquad k=0, 1\dots N\text{-}1 \tag{23}$$

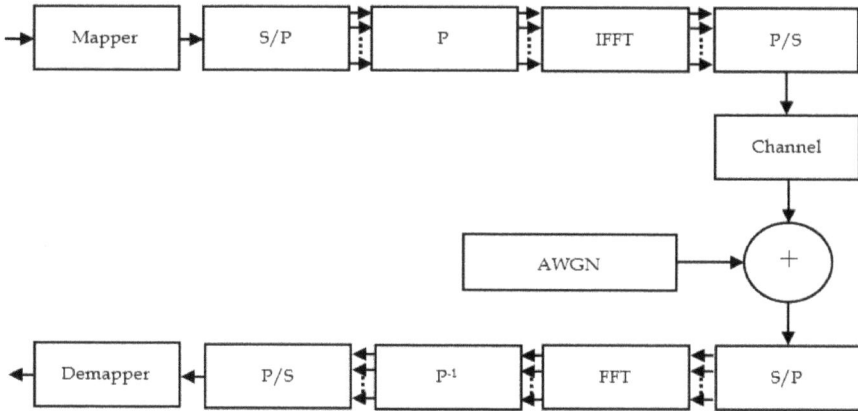

Fig. 9. Block diagram of OFDM system with Precording Techniques (Baig & Jeoti, 2010)

Fig.9 shows the precoding based OFDM system. In these system, the kernel of the WHT/DHT/DCT acts as a precoding matrix P of dimension $N= L{\times}L$ and it is applied to constellations symbols before the IFFT to reduce the correlation among the input sequence. In the precoding based systems baseband modulated data is passed through S/P converter which generates a complex vector of size L that can be written as $X=[X_0, X_1, \dots, X_{L-1}]^T$.Then precoding is applied to this complex vector which transforms this complex vector into new vector of length L that can be written as $Y=PX=[Y_0, Y_1, \dots, Y_{L-1}]^T$, where P is a precoder matrix of size $N=L{\times}L$ and Y_m can be written as follows:-

$$Y_m = \sum_{l=0}^{L-1} p_{m,l} \cdot X_l \qquad m = 0,1,\dots L-1 \tag{24}$$

$P_{m,l}$ means m^{th} row and l^{th} column of precoder matrix. Equation (24) represents the precoded constellations symbols. The complex baseband OFDM signal with N subcarriers can be written as:-

$$x_n = \frac{1}{\sqrt{N}} \sum_{m=0}^{N-1} Y_m \cdot e^{j2\pi\frac{n}{N}m}, \qquad n = 0,1,2\dots N-1 \tag{25}$$

Table 1 summarizes the PAPR reduction techniques presented in the literature review. The clipping techniques have low implementation complexity but on the other hand, the clipping operation may introduce both in-band distortion and out-of-band radiation into the multicarrier signals, which degrades the OFDM system performance including BER and spectral efficiency. SLM and PTS both have high computational complexity. However, the precoding based techniques show great promise as they are simple linear techniques to implement without any complex optimizations.

PAPR Reduction Technique	Implementation Complexity	Distortion	BER Degradation	Bandwidth Expansion	Power Increase	Data Rate Loss
Clipping & Filtering	LOW	YES	YES	NO	NO	NO
Selective Mapping	HIGH	NO	NO	YES	NO	NO
Partial Transmit Sequence	HIGH	NO	NO	YES	NO	YES
Precoding Based Techniques	LOW	NO	NO	NO	NO	NO

Table 1. Comparison of the PAPR Reduction Techniques

The main characteristics of precoding based techniques are, no bandwidth expansion, no power increase, no data rate loss, no BER degradation and distortionless.

4. Proposed PAPR reduction techniques

The random interleaved subcarrier mapping is favourable for the mobile WiMAX because it increases the capacity in the frequency selective fading channels and offers maximum frequency diversity. So, in this section, we present two precoding based random-interleaved OFDMA uplink systems for the PAPR reduction in the mobile WiMAX systems: Zadoff-Chu matrix transform (ZCMT) precoding based random-interleaved OFDMA uplink system and SLM based ZCMT precoded random-interleaved OFDMA uplink system. The PAPR of the proposed system is analyzed with root-raised-cosine (RRC) pulse shaping.

4.1 Zadoff-Chu sequences

Zadoff-Chu sequences are the class of polyphase sequences having optimum correlation properties. These sequences have an ideal periodic autocorrelation, constant magnitude and circular auto-orthogonality. The constant envelope feature of the Zadoff-Chu sequences can greatly alleviate the annoying peak-to-average power (PAPR) problem occurred in orthogonal frequency division multiplexing (OFDM) systems. According to (Chu, 1972; Popovic´, 1997), Zadoff-Chu sequences of length N can be defined as follows:-

$$z_n = \begin{cases} e^{\frac{j2\pi r}{N}\left(\frac{k^2}{2}+qk\right)} & for\ N\ Even \\ e^{\frac{j2\pi r}{N}\left(\frac{k(k+1)}{2}+qk\right)} & for\ N\ Odd \end{cases} \tag{26}$$

Where $k = 0,1,2,\ldots,N-1$, q is any integer, r is any integer relatively prime to N.

4.2 Autocorrelation property

Let us consider the periodic correlation property of Zadoff-Chu sequences with the same prime length. The periodic cross-correlation function can be defined as follows:-

$$\rho(m) = \sum_{n=0}^{N-1} z_n z^*_{(n-m)modN} \\ = \sum_{n=0}^{m-1} z_n z^*_{(n-m+N)} + \sum_{n=m}^{N-1} z_n z^*_{(n-m)} \tag{27}$$

For the sake of simplicity we put $q=0$ and $r=1$ in equation (26), then using equation (26) in the above expression we get:-

$$= \sum_{n=0}^{m-1} e^{\frac{[j\pi n^2]}{N}} . e^{\frac{[-j\pi(n-m+N)^2]}{N}} + \sum_{n=m}^{N-1} e^{\frac{[j\pi n^2]}{N}} . e^{\frac{[-j\pi(n-m)^2]}{N}} \\ = \sum_{n=0}^{m-1} e^{\frac{[j\pi(2mn-2nN-m^2+2mN+N^2)]}{N}} + \sum_{n=m}^{N-1} e^{\frac{[j\pi(n^2-n^2-m^2+2mn)]}{N}} \\ = \sum_{n=0}^{m-1} e^{\frac{[j\pi(2n-m+N)(m-N)]}{N}} + \sum_{n=m}^{N-1} e^{\frac{[j\pi(2mn-m^2)]}{N}} \tag{28}$$

Note that,

$$e^{[\frac{j\pi(2n-m+N)(m-N)}{N}]} = e^{[\frac{j\pi(2mn-2nN-m^2+2mN+N^2)}{N}]} \\ = e^{[\frac{j\pi(2mn-m^2)}{N}]} . e^{[\frac{j\pi(2mN-2nN-N^2)}{N}]} \\ = e^{[\frac{j\pi(2mn-m^2)}{N}]} . e^{j\pi(2m-2n-N)} \tag{29}$$

While N is even $2m-2n-N$ is even. So, the equation (29) can be stated as: $\exp(\frac{j\pi(2mn-m^2)}{N})$.

Hence, we can combine two summations of equation (28) as follows:-

$$\rho(m) = \sum_{n=0}^{N-1} e^{\frac{[j\pi(2mn-m^2)]}{N}} \\ = e^{[\frac{-j\pi m^2}{N}]} . \sum_{n=0}^{N-1} e^{[\frac{j2\pi mn}{N}]} \tag{30}$$

From equation (30), it is obvious that $\rho(m) = 0$, when $m = 0$. Since, m and N are relatively prime to each other, $e^{[\frac{j2\pi mn}{N}]}$ is a primitive N^{th} root of unity. Therefore, $e^{[\frac{j2\pi mn}{N}]}$ is a Nth root of unity but not equal to 1 for the range of m $\{m:1,2,3,\ldots,N\}$ shown in equation (30). Hence, we can employ the theorem as follows:

$$\sum_{n=0}^{N-1} r^n = \begin{cases} N & r = 1 \\ 0 & r \neq 1 \end{cases} \tag{31}$$

where r is N^{th} root of unity, substituting equation (31) into equation (30), we get $|\rho(m)| = 0, \{m : 1,2,3,\ldots,N\}$, At the end, it is concluded that the ideal periodic autocorrelation property of Zadoff-Chu sequences makes it suitable candidate for PAPR reduction in OFDM systems.

4.3 Constant envelope property after IDFT

The Zadoff-Chu sequences have constant amplitude, and its IDFT has also constant amplitude. Additionally, zadoff-chu sequence is also Zadoff-Chu sequence after FFT or IFFT.

4.4 Orthogonality property

The DFT of the Zadoff-Chu sequences equal to the conjugate of the Zadoff-chu sequences as follows:-

$$\mathrm{DFT}(z_n) = (z_n)^* \tag{32}$$

Therefore, the orthogonality in time domain as well as in frequency domain is preserved.

4.5 Zadoff-Chu Matrix Transform (ZCMT)

Zadoff-Chu matrix transform (ZCMT) is used to lower the correlation relationship of the IFFT input sequence. The ZCMT precoding matrix must accomplish the following criteria:-

1. All the elements of the precoding matrix must have the identical magnitude.
2. The magnitude must be equal to: $\dfrac{1}{\sqrt{N}}$.
3. The ZCMT precoding matrix must be non-singular.

The first condition guarantees that each output symbol has the same quantity of information of every input data. The second requirement preserves the power at the precoder output. Finally, the third requirement ensures the recovery of the original data at the receiver. The kernel of the ZCMT is defined in equation (26). For $N = L{\times}L$ and $j = \sqrt{-1}$, the ZCMT kernel Z, of size $N = L{\times}L=L^2$ is obtained by reshaping the Zadoff-Chu sequence row-wise by $k =mL+l$, as hereunder:-

$$
Z = \frac{1}{\sqrt{N}}
\begin{bmatrix}
z_{00} & z_{01} & \cdots & z_{0(L-1)} \\
z_{10} & z_{11} & \cdots & z_{1(L-1)} \\
\vdots & \vdots & \ddots & \vdots \\
z_{(L-1)0} & z_{(L-1)1} & \cdots & z_{(L-1)(L-1)}
\end{bmatrix}
\tag{33}
$$

In other words, the L^2 point long Zadoff-Chu sequence fills the kernel of the matrix transform row-wise.

4.6 Proposed ZCMT precoding based random interleaved OFDMA system

Fig.10 shows a ZCMT precoding based random-interleaved OFDMA uplink system. In this system a precoding matrix Z of dimension $N=L{\times}L$ is applied to constellations symbols before the subcarrier mapping and IFFT to reduce the PAPR. In the ZCMT precoding based random-interleaved OFDMA systems, baseband modulated data is passed through S/P convertor which generates a complex vector of size M that can be written as follows:-

$$X = [X_0, X_1, X_2 \ldots X_{M-1}]^T \tag{34}$$

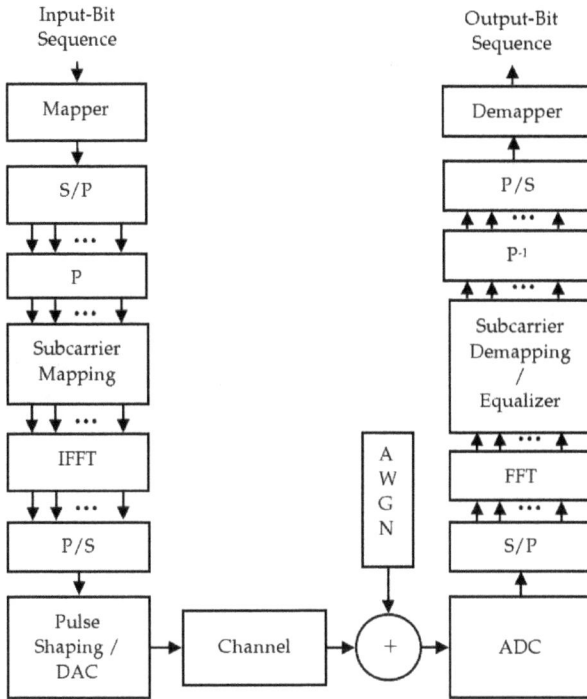

Fig. 10. ZCMT Precording based Random Interleaved OFDMA Uplink System

Then ZCMT precoding is applied to this complex vector which transforms this complex vector into new vector of same length L that can be written as follows:-

$$Y = ZX = [Y_0, Y_1, Y_2 \ldots Y_{M-1}]^T \tag{35}$$

where Z is a precoder matrix of size $N=L \times L$ and Y_m can be written as follows:-

$$Y_l = \sum_{m=0}^{L-1} z_{l,m} \cdot X_m \qquad l = 0,1, \ldots L - 1 \tag{36}$$

$z_{l,m}$ means l^{th} row of precoder matrix. Expanding equation (36), using row wise sequence reshaping $k = mL+l$ in equation (26) we get:-

$$Y_l = \frac{1}{\sqrt{N}} \sum_{m=0}^{L-1} (e^{\frac{j2\pi r}{N}\left(\frac{(mL+l)^2}{2}+q(mL+l)\right)}) \cdot X_m = \frac{1}{\sqrt{N}} \sum_{m=0}^{L-1} (e^{\frac{j2\pi r}{N}\left(\frac{(m^2L^2+l^2+2mlL+2qmL+2lq)}{2}\right)}) \cdot X_m$$
$$= \frac{1}{\sqrt{N}} \sum_{m=0}^{L-1} (e^{\frac{j\pi r m^2 L^2}{N}} \cdot e^{\frac{j\pi r l^2}{N}} \cdot e^{\frac{j2\pi r m l L}{N}} \cdot e^{\frac{j2\pi r q m L}{N}} \cdot e^{\frac{j2\pi r l q}{N}}) \cdot X_m \tag{37}$$

Since, $N=L^2$, then equation (37) can be reduced to

$$= \frac{1}{L} \cdot e^{\frac{j\pi r l^2}{L^2}} \cdot e^{\frac{j2\pi r l q}{L^2}} \sum_{m=0}^{L-1} (e^{\frac{j2\pi r l m l}{L}} \cdot \bar{X}_m) \tag{38}$$

where, $\overline{X}_m = (e^{\frac{j\pi r m^2}{L^2}} . e^{\frac{j2\pi r m l}{L^2}}).X_m$, $m=0,1,2,...,L-1$, $l=0,1,2,...,L-1$. Equation (38) represents the ZCMT precoded constellations symbols. After precoding operation, the subcarrier mapping is performed on these ZCMT precoded constellations symbols in random-interleaved mode. After the subcarrier mapping in random interleaved mode, we get frequency domain samples $\{\hat{Y}_l : l = 0,1,2,...,N-1\}.$. Mathematically, the subcarrier mapping in random interleaved mode can be done as follows:-

$$\hat{Y}_l = \begin{cases} \frac{Y_l}{\hat{Q}} & , l = \hat{Q}.k \quad 0 \le k \le M-1 \\ 0 & otherwise \end{cases} \tag{39}$$

Where $0 \le l \le N-1, N = Q.M, 0 \le \hat{Q} \le Q$, N: System subcarriers, M: User subcarriers (for one user), Q : Subchannels/Users $(Q=N/L)$. The kth subcarrier of each group is assigned to the kth user with index set: $\{(\gamma_{q,1}),(Q+\gamma_{q,2}),....,(L-1)Q+\gamma_{q,L-1})\}$, where $\{(\gamma_{q,1}),(\gamma_{q,2}),...,(\gamma_{q,L-1})\}$ are independent and identically distributed random variables with uniform distribution on $(q=0,1,2,...,Q-1)$. Suppose the kth user is assigned to subchannel k then the complex baseband ZCMT precoded random interleaved OFDMA signal for k^{th} user can be written as:-

$$x_n^{(k)} = \sum_{l=0}^{M-1} \hat{Y}_l^{(k)} . e^{j2\pi \frac{(lQ+\gamma_{q,k})}{N}n}, \quad n = 0,1...N-1 \tag{40}$$

where users index $q =0,1,2,...,Q-1$ and $\hat{Y}_l^{(k)}$ is modulated signal on subcarrier l for k^{th} user. The complex passband signal of ZCMT precoded random-interleaved OFDMA after RRC pulse shaping can be written as follows:-

$$x(t) = e^{j\omega_c t} \sum_{n=0}^{N-1} x_n^{(k)} . r(t - n\breve{T}) \tag{41}$$

where, ω_c is carrier frequency, $r(t)$ is baseband pulse, $\breve{T} = (\frac{M}{N}).T$. is compressed symbol duration after IFFT and T is symbol duration is seconds. The PAPR of the ZCMT precoded random-interleaved OFDMA signal in equation (41) with pulse shaping can be written as follows:-

$$PAPR(dB) = 10log_{10} \frac{\max (|x(t)|^2)}{E\{\max (|x(t)|^2)\}} \tag{42}$$

$E\{.\}$, denote expected value. If the amplitude of all subcarriers are normalized, $E\{max(|x(t)|^2)\} = N$, the equation (42) reduced to:-

$$PAPR(dB) = 10log_{10} \frac{\max (|x(t)|^2)}{N} \tag{43}$$

It should be pointed out that the orthogonality of the symbols after introducing precoding is maintained, as the precoding matrix is cyclic auto-orthogonal (Tasi et al., 2006). The instantaneous power of $x(t)$ can be defined as follows:-

$$p(t) = |x(t)|^2 = x(t) * x^*(t) \tag{44}$$

$$= \frac{1}{N} \sum_i^{N-1} \sum_k^{N-1} x_{zi} x_{zk}^* e^{\{j2\pi(i-k)t\}}$$

$$= \frac{1}{N} [N + 2Re\{\sum_i^{N-2} \sum_{k=i+1}^{N-1} x_{zi} x_{zk}^* e^{\{j2\pi(i-k)t\}}\}] \tag{45}$$

$$= 1 + \frac{2}{N} Re\{\sum_{m=1}^{N-1} e^{(j2\pi t)} \sum_{i=0}^{N-1-m} x_{zi} x_{z(i+m)}^*\}$$

For any complex c,

$$Re(c) \le |c|, \left|\sum c_n\right| \le \left|\sum c_n\right|.$$

That's why,

$$p(t) \le 1 + \frac{2}{N} \sum_{m=1}^{N-1} |\rho(m)| \tag{46}$$

where,

$$\rho(m) = \sum_{i=0}^{N-1-m} x_{zi} x_{z(i+m)}^* , \ m = 0,1,2,\ldots,N\text{-}1$$

is the aperiodic autocorrelation function. It is concluded from equation (46) that if the aperiodic autocorrelation mold of the IFFT input sequence x_z is small ($\rho(m)$ for ≥ 1) then, the peak-power factor of the signal obtained by passing through the OFDM multi-carrier system also can be small (Tellambura, 1997).The peak value of the autocorrelation is the average-power of the input sequence. After that, if the number of subcarriers is not altered, this peak-value completely depends on the input sequence. It means that if the sidelobe of an autocorrelation function of an input sequence has greater value than other input sequences, the former has high correlation property. The IFFT operation can be expressed as multiplying sinusoidal functions to the input sequence, summing and sampling the results. Hence, the high correlation property of the IFFT input causes the sinusoidal functions to be arranged with in-phase form. After summing these in-phase functions, the output might have large peaks.

4.6.1 The effect of Zadoff-Chu Matrix Transform

To verify the contribution of ZCMT, we consider OFDMA system for QPSK modulation. Fig.11 shows that the aperiodic autocorrelation function of randomly generated QPSK sequence with the length 64 is given, which are normalized by the length. Thus the maximum value is 1 which is the average power of the sequence. It is obvious from the Fig.11 that two autocorrelation functions have different sidelobe value. If the sidelobes of autocorrelation have higher values, then the input sequence is highly correlated and its PAPR is high. The high correlation in the input to IFFT causes the subcarriers to align in-phase. After summing these in-phase functions, the output might have high amplitude resulting in higher PAPR. The sidelobe value of the proposed ZCMT is much smaller than the conventional OFDMA systems. Therefore, it is concluded that if we apply ZCMT precoder to the IFFT input sequence, it lower the correlation relationship of the OFDMA input sequence, thus PAPR can be reduced.

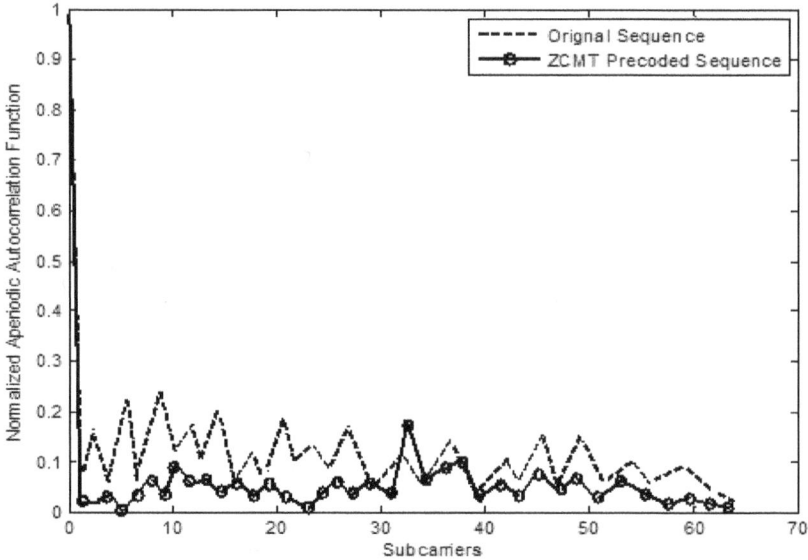

Fig. 11. The normalized autocorrelation function

4.7 Selective mapping to further improve ZCMT precoded random interleaved OFDMA system

Fig.12 shows the block diagram of the proposed SLM based ZCMT precoded random-interleaved OFDMA uplink system. Suppose data stream after S/P conversion is $X=[X_0,X_1,X_2,...,X_{M-1}]^T$, and each data block is multiplied by V dissimilar phase sequences, each length equal to M, $B^{(v)}=[b_{v,0},b_{v,1},...,b_{v,M-1}]^T,(v= 1, 2...V)$, which results in the altered data blocks. Let us denote the altered data block for the vth phase sequence is given by $X^{(v)}=[X_0b_{v,0},X_1b_{v,1},..., X_{N-1}b_{v,M-1}]^T$, where $v =1,2,3,...V$. Then, these altered data blocks are passed through the precoder, which transforms this complex vector into new vector of same length L that can be written as $Y=PX=[Y_0,Y_1,Y_2,..., Y_{L-1}]^T$, where P is a ZCMT precoder matrix of size $N =L\times L$ and Y_m^v can be written as follows:-

$$Y_l^v = \sum_{m=0}^{M-1} z_{m,l}X_m^v \qquad l = 0,1,...L-1$$

(47)

where, $z_{m,l}$ means precoding matrix of mth row and l^{th} column. Equation (47) represents the ZCMT precoded constellations signal. Then the subcarrier mapping of this precoded signal is done in random-interleaved mode. Suppose the kth user is assigned to sub-channel k then the complex baseband SLM based ZCMT precoded OFDMA uplink signal for k^{th} user can be written as follows:-

$$x_n^{(k,v)}=\sum_{l=0}^{M-1} \hat{Y}_l^{(k,v)}.e^{j2\pi\frac{(lQ+\gamma_{q,k})}{N}n}, n = 0,1...N-1$$

(48)

$\hat{Y}_l^{(k,v)}$ is modulated signal on subcarrier m for k^{th} user.

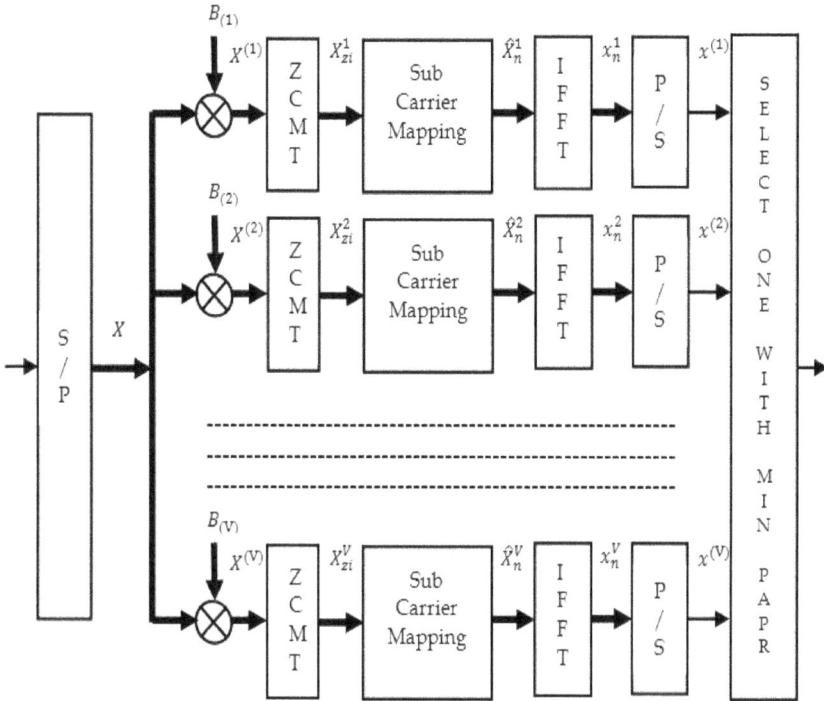

Fig. 12. Block diagram of SLM based ZCMT precoded random-interleaved OFDMA Uplink System

The precoding based SLM technique needs V (dissimilar phase sequences) IFFT operations and the information bits required as side information for each data block is [$\log_2 V$]. Precoding based SLM technique is applicable for any number of subcarriers and all types of modulation techniques. The PAPR reduction for precoding based SLM technique depends on the number of phase sequences V and the output data with lowest PAPR is selected by the transmitter for transmissions. The complex passband signal of random-interleaved OFDMA after RRC pulse shaping can be written as follows:-

$$x(t) = e^{j\omega_c t} \sum_{n=0}^{N-1} x_n^{(k,v)} . r(t - n\breve{T}) \tag{49}$$

where, ω_c is carrier frequency, $r(t)$ is baseband pulse, $\hat{T} = (\frac{M}{N}).T$ is compressed symbol duration after IFFT and T is symbol duration is seconds. The PAPR of the ZCMT precoded SLM based random-interleaved OFDMA uplink signal in (49) with RRC pulse shaping can be calculated by equation (43).

5. Simulation results

Extensive simulations in MATLAB(R) have been carried out to evaluate the performance of the proposed SLM based ZCMT precoded random-interleaved OFDMA uplink system with pulse shaping.

Channel Bandwidth	5MHz
Oversampling Factor	8
User Subcarriers	16
System Subcarriers	512
Precoding	WHT and ZCMT
Modulation	QPSK, 16-QAM, 64-QAM
Pulse Shaping	Root Raised Cosine (RRC)
Typical RRC Roll-Off Factor	$\alpha = 0.22$
Subcarrier Mapping Mode	Random Interleaved
CCDF Clip Rate	10^{-3}

Table 2. System Parameters

To show PAPR analysis of the proposed system, the data is generated randomly then modulated by QPSK, 16-QAM and 64-QAM respectively. We evaluate the PAPR statistically by using complementary cumulative distribution function (CCDF). The CCDF of the PAPR for ZCMT precoded random interleaved OFDMA uplink signal is used to express the probability of exceeding a given threshold $PAPR_0$ (CCDF = Prob (PAPR > $PAPR_0$)). We compared the simulation results of proposed system with WHT precoded random interleaved OFDMA uplink systems and conventional random interleaved OFDMA uplink systems. To show the PAPR analysis of proposed system with pulse shaping in MATLAB® we considered RRC rolloff factor $\alpha = 0.22$ with system subcarriers $N=512$ and user subcarriers $M=16$. All the simulations have been performed on 10^5 random data blocks. Simulation parameters that we use are given in the above Table 2.

Fig.13 shows CCDF comparison of PAPR for the ZCMT precoded random-interleaved OFDMA uplink system and the SLM based ZCMT precoded random-interleaved OFDMA uplink system with the WHT precoded random-interleaved OFDMA uplink systems and the conventional random-interleaved OFDMA uplink systems. At clip rate of 10^{-3}, with user subcarriers $M=16$ and system subcarriers $N=512$, the PAPR is 10 dB, 9.2 dB, 7.4 dB and 5.7 dB respectively, for the conventional random-interleaved OFDMA uplink systems, WHT precoded random-interleaved OFDMA uplink systems, ZCMT precoded random-interleaved OFDMA uplink systems and SLM based ZCMT precoded random-interleaved OFDMA uplink system respectively, using QPSK modulation.

Fig.14 shows CCDF comparison of PAPR for the ZCMT precoded random-interleaved OFDMA uplink system and the SLM based ZCMT precoded random-interleaved OFDMA uplink system with the WHT precoded random-interleaved OFDMA uplink systems and the conventional random-interleaved OFDMA uplink systems. At clip rate of 10^{-3}, with user subcarriers $M=16$ and system subcarriers $N=512$, the PAPR is 9.5 dB, 9.3 dB, 8.2 dB and 6.7

dB respectively, for the conventional random-interleaved OFDMA uplink systems, WHT precoded random-interleaved OFDMA uplink systems, ZCMT precoded random-interleaved OFDMA uplink systems and SLM based ZCMT precoded random-interleaved OFDMA uplink system respectively,, using 16-QAM modulation.

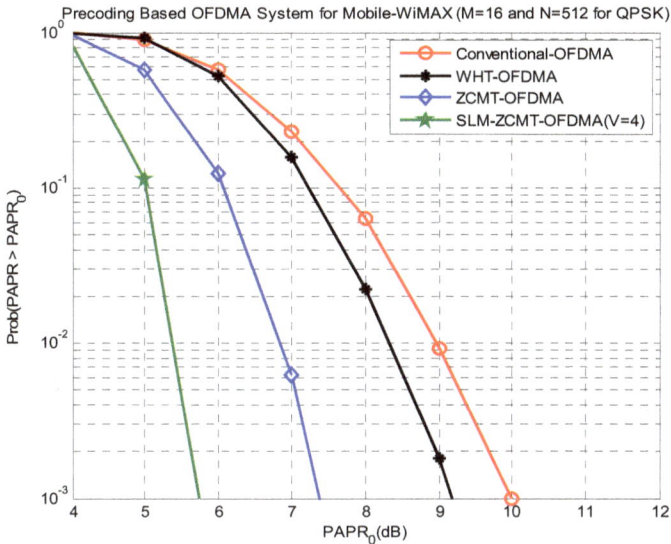

Fig. 13. CCDF Comparison of PAPR of the ZCMT precoded random-interleaved OFDMA uplink system and SLM based ZCMT precoded random-interleaved OFDMA uplink system with the WHT precoded random-interleaved OFDMA uplink system and the conventional random-interleaved OFDMA uplink system respectively, for QPSK modulation.

Fig.15 shows CCDF comparison of PAPR for the ZCMT precoded random-interleaved OFDMA uplink system and the SLM based ZCMT precoded random-interleaved OFDMA uplink system with the WHT precoded random-interleaved OFDMA uplink systems and the conventional random-interleaved OFDMA uplink systems. At clip rate of 10^{-3}, with user subcarriers M=16 and system subcarriers N=512, the PAPR is 9.8 dB, 9.6 dB, 8.7 dB and 6.7 dB respectively, for the conventional random-interleaved OFDMA uplink systems, WHT precoded random-interleaved OFDMA uplink systems, ZCMT precoded random-interleaved OFDMA uplink systems and SLM based ZCMT precoded random-interleaved OFDMA uplink system respectively, using 64-QAM modulation.

Fig.16 shows the SER performance of the ZCMT precoded random-interleaved OFDMA uplink systems; WHT precoded random-interleaved OFDMA uplink systems and the conventional random-interleaved OFDMA uplink systems respectively. The WiMAX Forum recommends using just two out of the six ITU models, which are Pedestrian B and Vehicular A (WiMAX, 2008). So, we use the ITU pedestrian B channel with additive white gaussian noise (AWGN) and MMSE equalization. The parameters for the ITU Pedestrian B channel model can be found in Table 3 (ITU, 1997).

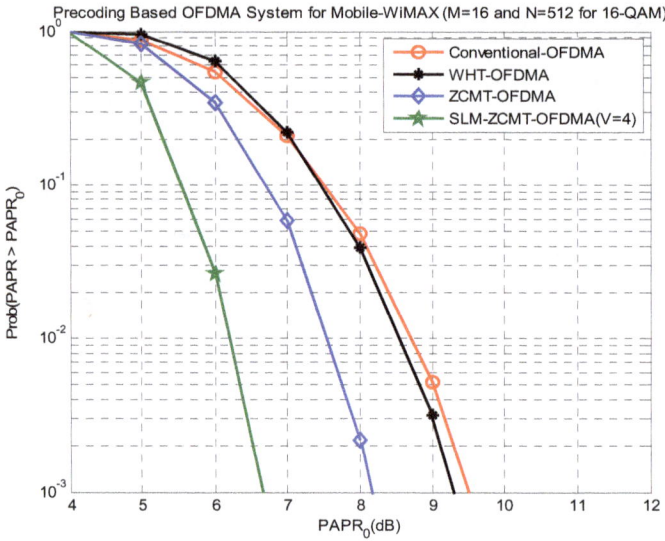

Fig. 14. CCDF Comparison of PAPR of the ZCMT precoded-random interleaved OFDMA uplink system and SLM based ZCMT precoded random-interleaved OFDMA uplink system with the WHT precoded random-interleaved OFDMA uplink system and the conventional random-interleaved OFDMA uplink system respectively, for 16-QAM modulation.

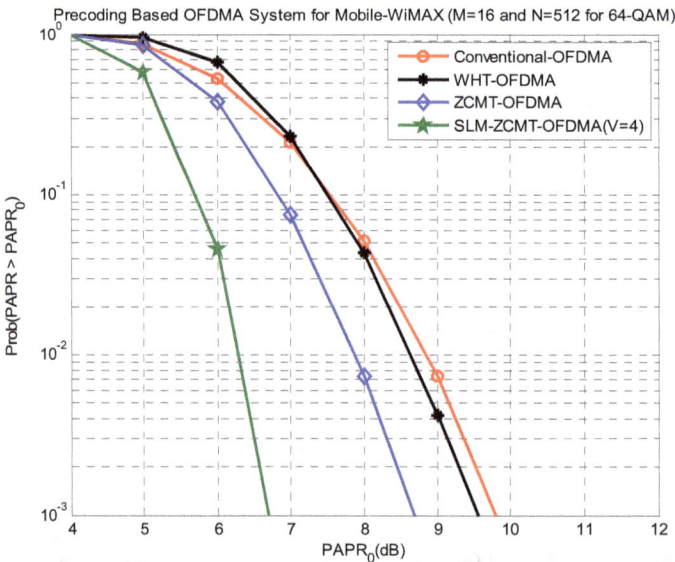

Fig. 15. CCDF Comparison of PAPR of the ZCMT precoded random-interleaved OFDMA uplink system and SLM based ZCMT precoded random-interleaved OFDMA uplink system with the WHT precoded random-interleaved OFDMA uplink system and the conventional random-interleaved OFDMA uplink system respectively, for 64-QAM modulation.

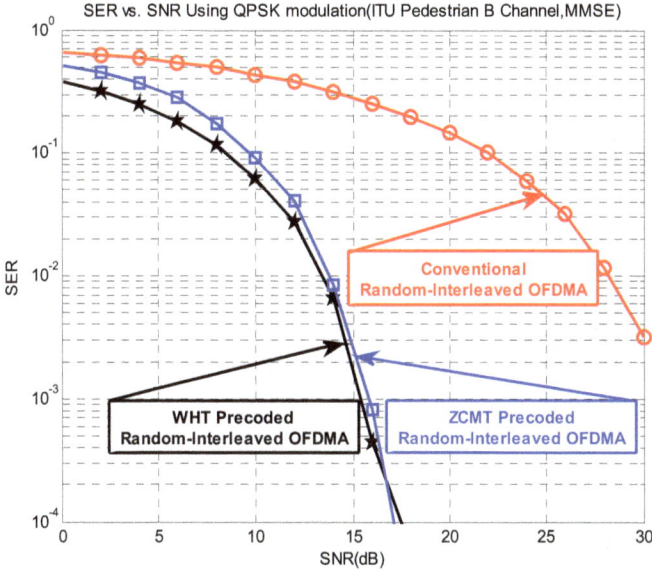

Fig. 16. SER vs. SNR Comparison of the ZCMT precoded Random-Interleaved OFDMA uplink system, the WHT precoded Random- Interleaved OFDMA uplink system and the conventional Random-Interleaved OFDMA uplink system, for sub-band 0 with QPSK modulation.

It is concluded from Fig.16 that the ZCMT precoded random-interleaved OFDMA uplink systems provides approximately same performance as that of the WHT precoded random-interleaved OFDMA systems but a significant SER performance improvement is seen over the conventional random interleaved OFDMA uplink systems for the sub-band 0 using QPSK modulation.

Table 4 summarizes the PAPR of random-interleaved OFDMA uplink systems, WHT random-interleaved OFDMA uplink systems, ZCMT precoded random-interleaved OFDMA uplink systems and SLM based ZCMT precoded random-interleaved OFDMA uplink systems respectively, using QPSK, 16-QAM and 64-QAM. Table 4 concludes that, the ZCMT precoded random-interleaved OFDMA uplink system and the SLM based ZCMT precoded random-interleaved OFDMA uplink system has lower PAPR than the WHT precoded random-interleaved OFDMA uplink systems and conventional random-interleaved OFDMA uplink systems.

Tap No.		1	2	3	4	5	6
ITU Pedestrian B Channel	Relative Delay (ns)	0	200	800	1200	2300	3700
	Average Power (dB)	0.0	-0.9	-4.9	-8.0	-7.8	-23.9
	Doppler Spectrum	Classic	Classic	Classic	Classic	Classic	Classic

Table 3. ITU Pedestrian B channel Parameters

Transmission Scheme	PAPR		
	QPSK	16-QAM	64-QAM
Conventional Random-Interleaved OFDMA	10 dB	9.5 dB	9.8 dB
WHT Random-Interleaved OFDMA	9.2 dB	9.3 dB	9.6 dB
ZCMT Random-Interleaved OFDMA	7.4 dB	8.2 dB	8.7 dB
SLM-ZCMT Random-Interleaved OFDMA	5.7 dB	6.7 dB	6.7 dB

Table 4. At CCDF of 10^{-3}, The PAPR Comparisons of the Conventional Random-Interleaved OFDMA uplink, WHT Random-Interleaved OFDMA uplink, ZCMT Random-Interleaved OFDMA uplink and SLM based ZCMT Random-Interleaved OFDMA uplink respectively, for users subcarriers ($M=16$) and system subcarriers ($N=512$)

6. Conclusion

In this chapter, we present a brief overview of the mobile WiMAX and typical PAPR reduction techniques available in the literature. We also introduce two precoding based systems: ZCMT precoded random-interleaved OFDMA uplink system and SLM based ZCMT precoded random-interleaved OFDMA uplink system, for PAPR reduction in mobile WiMAX systems. Computer simulation shows that the PAPR of the both proposed systems have less PAPR than the WHT precoded random-interleaved OFDMA uplink systems and conventional random-interleaved OFDMA uplink systems. These systems are efficient, signal independent, distortionless and do not require any complex optimizations. Additionally, these systems also take the advantage of the frequency variations of the communication channel and can also offer substantial performance gain in fading multipath channels. Thus, it is concluded that the both proposed uplink systems are more favourable than the WHT precoded random-interleaved OFDMA uplink systems and conventional random-interleaved OFDMA uplink systems for the mobile WiMAX systems.

7. References

Baig, I. & Jeoti, V. (2010). DCT Precoded SLM Technique for PAPR Reduction in OFDM Systems," *The 3rd International Conference on Intelligent and Advanced Systems, Kuala Lumpur, Malaysia*, 2010.

Baig, I. & Jeoti, V. (2010). Novel Precoding Based PAPR Reduction Techniques for Localized-OFDMA Uplink System of LTE-Advanced. *Journal of Telecommunication, Electronic and Computer Engineering (JTEC) Malaysia*, vol.2 no.1, pp 49-58, 2010.

Baig, I. & Jeoti, V. (2010). PAPR Analysis of DHT-Precoded OFDM System for M-QAM. *The 3rd International Conference on Intelligent and Advanced Systems*, Kuala Lumpur, Malaysia, 2010.

Baig, I. & Jeoti, V. (2010). PAPR Reduction in OFDM Systems: Zadoff-Chu Matrix Transform Based Pre/Post-Coding Techniques. *2nd International Conference on Computational Intelligence, Communication Systems and Networks, Liverpool, UK*, 2010.

Cho, Y. S.; Kim, J.; Yang, W. Y. & Kang, C. G. (2011). MIMO-OFDM Wireless Communications with Matlab. John Wiley & Sons, Inc. (Asia) Pte Ltd. ISBN:978-0-470-82561-7.

Chu, D.C. (1972). Polyphase codes with good periodic correlation properties. *IEEE Trans. Inform. Theory*, vol. IT-18, pp. 531-532, 1972.

Cimini, L. J. & Sollenberger, N. R. (2000). Peak-to-Average Power Ratio Reduction of an OFDM Signal Using Partial Transmit Sequences. *IEEE Commun. Lett.*, vol. 4, no. 3, pp. 86–88, Mar. 2000.

Han, S. H. & Lee, J. H. (2004). PAPR Reduction of OFDM Signals Using a Reduced Complexity PTS Technique. *Signal Processing Letters, IEEE*, vol.11, issue.11, pp: 887-890, 2004.

Han, S. H. & Lee, J. H. (2005). An overview of peak-to-average power Ratio reduction a techniques for multicarrier transmission. IEEE Wireless communication, vol.12, issue.2, pp: 56-65, 2005.

ITU, Document. (1997). Rec. ITU-R M.1225-Guidelines for Evaluation of Radio Transmission Technologies for IMT-2000, ITU-R, 1997.

Jayalath, A. & Tellambura, C. (2000). Adaptive PTS Approach for Reduction of Peak-to-Average Power Ratio of OFDM Signal," *Elect. Lett.*, vol. 36, no. 14, pp. 1226–28, July 2000.

Jeoti, V. & Baig, I. (2009). A Novel Zadoff-Chu Precoder Based SLM Technique for PAPR Reduction in OFDM Systems. invited paper, *Proceedings of 2009 IEEE International Conference on Antennas, Propagation and Systems*, Johor, Malaysia, 2009.

Jiang, T.; Yao, W.; Guo, P.; Song, Y. & Qu, D. (2006). Two novel nonlinear Companding schemes with iterative receiver to reduce PAPR in multicarrier modulation systems. *IEEE Trans. Broadcasting*, vol. 52, no. 2, pp. 268–273, 2006.

Knopp, R. & Humblet, P. A. (1995). Information Capacity and Power Control in Single-Cell Multiuser Communications. *IEEE Int. Conf. Communication*, pp. 331-335, 1995.

Kou, Y.; Lu, W. & Antoniou, A. (2007). A new peak-to- average power-ratio reduction algorithm for OFDM systems via constellation extension. *IEEE Trans. Wireless Communications*, vol. 6, no. 5, pp. 1823–1832, 2007.

Li, X. & Cimini, L.J. (1998) Effects of clipping and filtering on the performance of OFDM. IEEE Commun. Letter, 2(20), 131–133, 1998.

Lim, D. W.; No, J. S.; Lim, C. W. & Chung, H.(2005). A new SLM OFDM scheme with low complexity for PAPR reduction. *IEEE Signal Processing Letters*, vol. 12, no. 2, pp. 93–96, 2005.

Min, Y. K. & Jeoti, V. (2007). A Novel Signal Independent Technique for PAPR Reduction in OFDM Systems. *IEEE-ICSCN*, pp. 308-311, 2007.

Mourelo, J. T.(1999). Peak to Average Power Ratio Reduction for Multicarrier Modulation. *PhD thesis, University of Stanford, Stanford*, 1999.

Müller, S. H. & Huber, J. B. (1997). A Novel Peak Power Reduction Scheme for OFDM," *Proc. IEEE PIMRC '97*, Helsinki, Finland, pp. 1090–94, Sept. 1997.

Nee, V. & Wild, A. (1998) Reducing the peak-to-average power ratio of OFDM. IEEE VTC'98, vol.3, pp. 18–21, May 1998.

Nikookar, H. & Lidsheim, K. S. (2002). Random phase updating algorithm for OFDM transmission with low PAPR. *IEEE Trans. Broadcasting*, vol. 48, no. 2, pp. 123–128, 2002.

Popovic´, B. M. (1997). Spreading sequences for multi-carrier CDMA systems. *in IEE Colloquium CDMA Techniques and Applications for Third Generation Mobile Systems, London*, pp. 8/1–8/6, 1997.

Slimane, S. B. (2007). Reducing the peak-to-average power ratio of OFDM signals through precoding. *IEEE Trans. Vehicular Technology*, vol.56, no. 2, pp. 686–695, Mar. 2007.

Tasi, Y.; Zhang, G. & Wang, X. (2006). Orthogonal Polyphase Codes for Constant Envelope OFDM-CDMA System. *IEEE, WCNC*, pp.1396 – 1401, 2006.

Tellambura, C. (1997). Upper bound on peak factor of N-multiple carriers. Electronics Letters, vol.33, pp.1608-1609, Sept.1997.

Tellambura, C. (2001). Improved Phase Factor Computation for the PAR Reduction of an OFDM Signal Using PTS. *IEEE Commun. Lett.*, vol. 5, no. 4, pp. 135–37, Apr. 2001.

Thompson, S. C.; Ahmed, A. U.; Proakis, J. G.; Zeidler, J. R. & Geile, M. J. (2008).Constant envelope OFDM. *IEEE Trans. Communications*, vol. 56, pp. 1300-1312, 2008.

Tse, D. (1997). Optimal Power Allocation over Parallel Gaussian Broadcast Channels. *Proceedings of International Symposium on Information, Ulm Germany*, pp. 27, 1997.

Wang, H. & Chen, B. (2004). Asymptotic distributions and peak power analysis for uplink OFDMA signals. *in Proc. IEEE Acoustics, Speech, and Signal Processing Conference*, vol.4, pp.1085-1088, 2004.

Wang, L. & Tellambura, C. (2005). A Simplified Clipping and Filtering Technique for PAR Reduction in OFDM Systems. *Signal Processing Letters, IEEE*, vol.12, no.6, pp. 453-456, 2005.

WiMAX, Forum. (2008). WiMAX System Evaluation Methodology. Version 2.1, July 2008.

Yoo, S.; Yoon, S.; Kim, S.Y. & Song, I. (2006). A novel PAPR reduction scheme for OFDM systems: Selective mapping of partial tones (SMOPT). *IEEE Trans. Consumer Electronics*, vol. 52, no. 1, pp.40–43, 2006.

12

Peak-to-Average Power Ratio Reduction in Orthogonal Frequency Division Multiplexing Systems

Pooria Varahram and Borhanuddin Mohd Ali
Universiti Putra Malaysia,
Malaysia

1. Introduction

Broadband wireless is a technology that provides connection over the air at high speeds. Orthogonal frequency division multiplexing (OFDM) system has generally been adopted in recent mobile communication systems because of its high spectral efficiency and robustness against intersymbol interference (ISI). However, due to the nature of inverse fast Fourier transform (IFFT) in which the constructive and destructive behaviour could create high peak signal in constructive behaviour while the average can become zero at destructive behaviour, OFDM signals generally become prone to high peak-to-average power ratio (PAPR) problem. In this chapter, we focus on some of the techniques to overcome the PAPR problem (Krongold and Jones, 2003; Bauml, et al. 1996).

The other issue in wireless broadband is how to maximize the power efficiency of the power amplifier. This can be resolved by applying digital predistortion to the power amplifier (PA) (Varahram, et al. 2009). High PAPR signal when transmitted through a nonlinear PA creates spectral broadening and increase the dynamic range requirement of the digital to analog converter (DAC). This results in an increase in the cost of the system and a reduction in efficiency. To address this problem, many techniques for reducing PAPR have been proposed. Some of the most important techniques are clipping (Kwon, et al. 2009), windowing (Van Nee and De Wild, 1998), envelope scaling (Foomooljareon and Fernando, 2002), random phase updating (Nikookar and Lidsheim, 2002), peak reduction carrier (Tan and Wassell, 2003), companding (Hao and Liaw, 2008), coding (Wilkison and Jones, 1995), selected mapping (SLM) (Bauml, et al. 1996), partial transmit sequence (PTS) (Muller and Huber, 1997), DSI-PTS (Varahram et al. 2010), interleaving (Jayalath and Tellambura, 2000), active constellation extension (ACE) (Krongold, et al. 2003), tone injection and tone reservation (Tellado, 2000), dummy signal insertion (DSI) (Ryu, et al. 2004), addition of Guassian signals (Al-Azoo et al. 2008) and etc (Qian, 2005).

Clipping is the simplest technique for PAPR reduction, where the signal at the transmitter is clipped to a desired level without modifying the phase information. In windowing a peak of the signal is multiplied with a part of the frame. This frame can be in Gaussian shape, cosine, Kaiser or Hanning window, respectively. In companding method the OFDM signal is companded before digital to analog conversion. The OFDM signal after IFFT is first companded and quantized and then transmitted through the channel after digital to analog conversion. The receiver first converts the signal into digital format and then expands it. The companding method has application in speech processing where high peaks occur infrequently. In PTS, by partitioning the input signal and applying several IFFT, the optimum phase sequence with lowest PAPR will be selected before being transmitted. This technique results in high complexity. In SLM, a copy of input signal is used to choose the minimum PAPR among the multiple signals. We can conclude that there is always a trade-off in choosing a particular PAPR technique. The trade-off comes in the form of complexity, power amplifier output distortion, cost, side information, PAPR reduction, Bit Error Rate (BER) performance, spectrum efficiency and data rate loss.

2. OFDM signal

In OFDM systems, first a specific number of input data samples are modulated (e.g. PSK or QAM), and by IFFT technique the input samples become orthogonal and will be converted to time domain at the transmitter side. The IFFT is applied to produce orthogonal data subcarriers. In theory, IFFT combines all the input signals (superposition process) to produce each element (signal) of the output OFDM symbol. The time domain complex baseband OFDM signal can be represented as (Han and Lee, 2005):

$$x_n = \frac{1}{\sqrt{N}} \sum_{k=0}^{N-1} X_k e^{j2\pi \frac{n}{N}k}, \quad n = 0,1,2,\ldots\ldots,N-1 \tag{1}$$

where x_n is the n-th signal component in OFDM output symbol, X_k is the k-th data modulated symbol in OFDM frequency domain, and N is the number of subcarrier.

The PAPR of the transmitted OFDM signal can be given by (Cimini and Sollenberger, 2000):

$$PAPR(dB) = \frac{max\left[\left|x_n\right|^2\right]}{E\left[\left|x_n\right|^2\right]} \tag{2}$$

where $E[.]$ is the expectation value operator. The theoretical maximum of PAPR for N number of subcarriers is as follows:

$$PAPR_{max} = 10\log(N) \ dB \tag{3}$$

PAPR is a random variable since it is a function of the input data, while the input data is a random variable. Therefore PAPR can be analyzed by using level crossing rate theorem which calculates the mean number of times that the envelope of a stationary signal crosses a given level. Knowing the amplitude distribution of the OFDM output signals, it is easy to compute the probability that the instantaneous amplitude will lie above a given threshold and the same goes for power. This is performed by calculating the complementary cumulative distribution function (CCDF) for different PAPR values as follows:

$$CCDF = Pr(PAPR > PAPR_0) \tag{4}$$

Here the effect of additive white Gaussian noise (AWGN) on OFDM performance is studied. As OFDM systems use standard digital modulation formats to modulate the subcarriers, PSK and QAM are usually used due to their excellent error resilient properties. The most important block in OFDM is IFFT. IFFT changes the distribution of the signal without altering its average power. The BER or bit error probability P_{be} in an AWGN channel is given by (Han and Lee, 2005):

$$P_{be,MQAM} \approx \frac{4(\sqrt{M}-1)}{k\sqrt{M}}Q\left(\sqrt{\frac{3k}{(M-1)}\cdot\frac{E_b}{N_o}}\right) \tag{5}$$

where M is the modulation order, $k = \log_2(M)$ is the number of bits per symbol, and $Q(.)$ is the Gaussian Q function defined as:

$$Q(y) = erfc(\frac{y}{\sqrt{2}}) \tag{6}$$

In this chapter the performance of BER versus energy per bit to noise power spectral density ratio (E_b/N_o) is analyzed.

3. PAPR reduction techniques

In this section, some of the most important PAPR reduction techniques such as Selected Mapping (SLM), Partial Transmit Sequence (PTS) and Enhanced PTS EPTS) are presented.

3.1 Conventional SLM (C-SLM)

In Conventional SLM (C-SLM) method, OFDM signal is first converted from serial to parallel by means of serial-to-parallel converter. The parallel OFDM signal is then multiplied by several phase sequences that are created offline and stored in a matrix. A copy of the OFDM signal is multiplied with a random vector of phase sequence matrix. For each subblock IFFT is performed and its PAPR is calculated to look for the minimum one. The OFDM signal having minimum PAPR is then selected and be transmitted. The main drawbacks of this technique are the high complexity due to the high number of subblocks and the need to send side information which result in data rate and transmission efficiency degradation, respectively. In Fig. 1, the number of candidate signal or subblocks is given by U, hence $log_2 U$ number of bits is required to be sent as side information.

The other drawback of this method is that by increasing U, higher number of IFFT blocks are required which increase the complexity significantly. Hence, a method with low complexity and high PAPR performance is required.

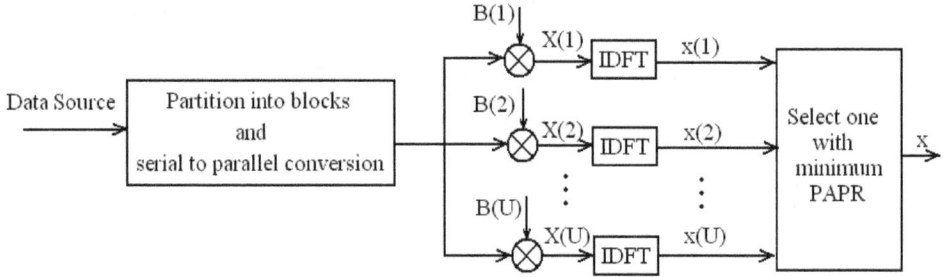

Fig. 1. The block diagram of the C-SLM method.

3.2 Conventional PTS (C-PTS)

To analyze C-PTS let X denotes random input signal in frequency domain with length N. X is partitioned into V disjoint subblocks $X_v=[X_{v,0},X_{v,1},...,X_{v,N-1}]^T$, $v=1,2,...,V$ such that $\sum_{v=1}^{V} X_v = X$ and then these subblocks are combined to minimize the PAPR in time domain.

The Sbblock partitioning is based on interleaving in which the computational complexity is less compared to adjacent and pseudo-random, however it gives the worst PAPR performance among them (Han and Lee, 2005).

By applying the phase rotation factor $b_v = e^{j\phi_v}, v = 1,2,...,V$ to the IFFT of the vth subblock X_v, the time domain signal after combining is obtained as:

$$x'(b) = \sum_{v=1}^{V} b_v x_v \tag{7}$$

where $x'(b) = [x'_0(b), x'_1(b),...x'_{NF-1}(b)]^T$. The objective is to find the optimum signal $x'(b)$ with the lowest PAPR.

Both b and x can be shown in matrix forms as follows:

$$b = \begin{bmatrix} b_1, & b_1 ,...., & b_1 \\ \vdots & \vdots & \vdots \\ b_V, & b_V ,...., & b_V \end{bmatrix}_{V \times N} \tag{8}$$

$$x = \begin{bmatrix} x_{1,0}, x_{1,1},...,x_{1,NF-1} \\ \vdots & \vdots & \vdots \\ x_{V,0}, x_{V,1},...,x_{V,NF-1} \end{bmatrix}_{V \times NF} \tag{9}$$

Fig. 2 shows the block diagram of C-PTS. It should be noted that all the elements of each row of matrix b are of the same values and this is in accordance with the C-PTS method. In order to obtain exact PAPR calculation, at least four times oversampling is necessary (Han and Lee, 2005).

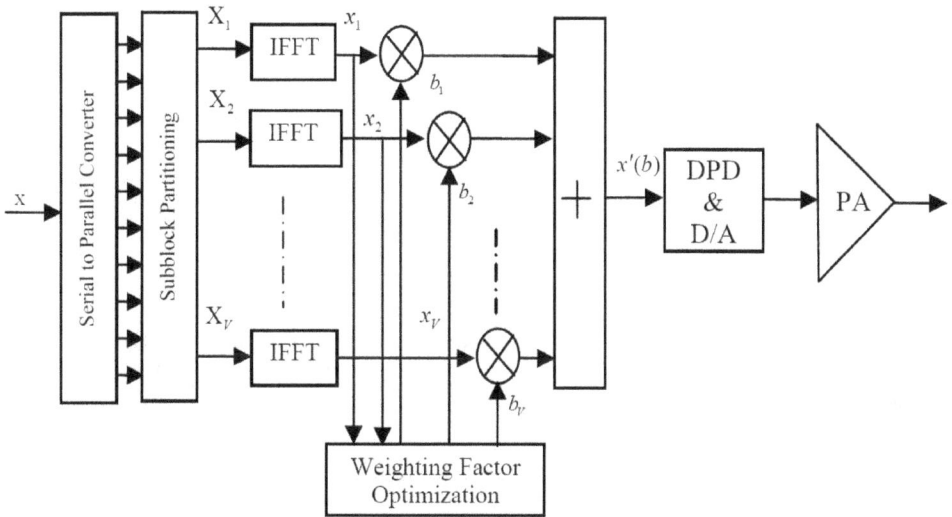

Fig. 2. Block diagram of the C-PTS scheme with Digital predistortion and power amplifier in series

This process is performed by choosing the optimization parameter $\tilde{b}$ which satisfies the following condition:

$$\tilde{b} = arg\,min(\max_{0 \le k \le NF-1} \left| \sum_{v=1}^{V} b_v x_v \right|) \qquad (10)$$

where V is the number of subblocks partitioning and F is the oversampling factor. After obtaining the optimum $\tilde{b}$, the signal is transmitted.

For finding the optimum $\tilde{b}$, we should perform exhaustive search for $(V-1)$ phase factors since one phase factor can remain fixed, $b_1=1$. Hence to find the optimum phase factor, W^{V-1} iteration should be performed, where W is the number of allowed phase factors.

3.3 Enhanced PTS (EPTS)

In order to decrease the complexity of C-PTS, a new phase sequence is generated. The block diagram of the enhanced partial transmit sequence (EPTS) scheme is shown in Fig. 3.

This new phase sequence is based on the generation of N random values of {1 -1 j -j} if the allowed phase factors is $W=4$. The phase sequence matrix can be given by:

$$
\hat{b} =
\begin{bmatrix}
b_{1,1} & ,...,& b_{1,N} \\
\vdots & \vdots & \vdots \\
b_{V,1} & ,...,& b_{2,N} \\
b_{V+1,1} & ,...,& b_{V+1,N} \\
\vdots & \vdots & \vdots \\
b_{P,1} & ,...,& b_{P,N}
\end{bmatrix}_{[P\times N]}
\tag{11}
$$

where P is the number of iterations that should be set in accordance with the number of iterations of the C-PTS and N is the number of samples (IFFT length) and V is the number of subblock partitioning. The value of P is given as follows:

$$
P = DW^{V-1} \quad , \quad D = 1,2,...,D_N
\tag{12}
$$

where D is the coefficient that can be specified based on the PAPR reduction and complexity requirement and D_N is specified by the user. The value of P explicitly depends on the number of subblocks V, if the number of allowed phase factor remains constant.

There is a tradeoff for choosing the value of D. higher D leads to higher PAPR reduction but at the expense of higher complexity; while lower D results in smaller PAPR reduction but with less complexity. For example if $W=2$ and $V=4$, then in C-PTS there are 8 iterations and hence $P=8D$. If $D=2$, then $P=16$ and both methods have the same number of iterations. But when $D=1$, then number of iterations to find the optimum phase factor will be reduced to 4 and this will result in complexity reduction. The main advantage of this method over C-PTS is the reduction of complexity while at the same time maintaining the same PAPR performance. In the case of C-PTS, each row of the matrix $\hat{b}$ contains same phase sequence while each column is periodical with period V, whereas in the proposed method each element of matrix $\hat{b}$ has different random values. The other formats that matrix in (11) can be expressed are as follows:

$$
\hat{b} =
\begin{bmatrix}
\overbrace{b_{1,1},...,b_{1,N/P}}^{P} & ,..., & b_{1,1},...,b_{1,N/P} \\
\vdots & \vdots & \vdots \\
b_{V,1},...,b_{V,N/P} & ,..., & b_{V,1},...,b_{V,N/P} \\
b_{V+1,1},...,b_{V+1,N/P}, & ..., & b_{V+1,1},...,b_{V+1,N/P} \\
\vdots & \vdots & \vdots \\
b_{P,1},...,b_{P,N/P} & ,..., & b_{P,1},...,b_{P,N/P}
\end{bmatrix}_{[P\times N]}
\tag{13}
$$

$$
\hat{b} =
\begin{bmatrix}
\overbrace{b_{1,1},...,b_{1,1}}^{P}, & \overbrace{b_{1,2},...,b_{1,2}}^{P} & ,..., & \overbrace{b_{1,N/P},...,b_{1,N/P}}^{P} \\
\vdots & \vdots & & \vdots \\
b_{V,1},...,b_{V,1}, & b_{V,2},...,b_{V,N/P}, & ..., & b_{V,N/P} \\
b_{V+1,1},...,b_{V+1,1} & ,..., & & b_{V+1,N/P},...,b_{V+1,N/P} \\
\vdots & \vdots & & \vdots \\
b_{P,1},...,b_{P,1} & ,..., & & b_{P,N/P},...,b_{P,N/P}
\end{bmatrix}_{[P\times N]}
\tag{14}
$$

where (13) and (14) are the interleaved and adjacent phase sequences matrix, respectively.

As an example take the case of $N=256$, and the number of allowed phase factor and subblock partitioning are $W=4$ and $V=4$ respectively. With C-PTS there are $W^{M-1}=64$ possible iterations, whereas for the proposed method, in the case of $D=2$, the phase sequence is a matrix of [128x256] elements according to (11). In this case 64 iterations are required for finding the optimum phase sequence, because each two rows of the matrix in (11) multiply point-wise with the time domain input signal x_v with length [2x256].

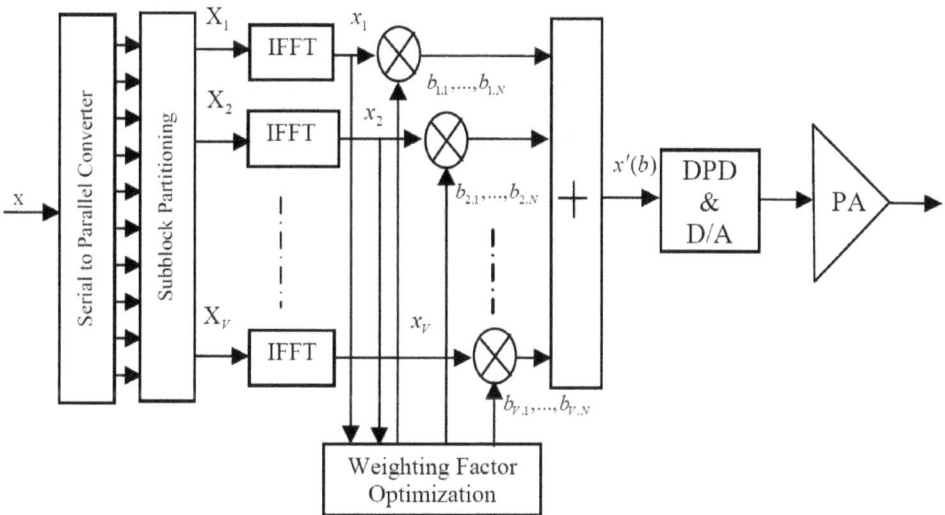

Fig. 3. The block diagram of enhanced PTS

The reduction of subblocks to 2 is because it gives almost the same PAPR reduction as C-PTS with $V=4$. It should be noted that if $D=1$ then the complexity increases while if $D>2$ then the PAPR reduction is less.

Therefore the algorithm can be expressed as follows:

Step 1: Generate the input data stream and map it to the M-QAM modulation.

Step 2: Construct a matrix of random phase sequence with dimension of [PxN].

Step 3: Point-wise multiply signal x_v with the new phase sequence.

Step 4: Find the optimum phase sequence after P iterations to minimize the PAPR.

3.3.1 Numerical analysis

In order to evaluate and compare the performance of the PAPR methods with C-PTS, simulations have been performed. In all the simulations, we employed QPSK modulation with IFFT length of $N=512$, and oversampling factor $F=4$. To obtain the complementary cumulative distribution function (CCDF), 40000 random OFDM symbols are generated.

Fig. 4 shows the CCDF of three different types of phase sequences interleaved, adjacent and random for $D=2$. From this figure, PAPR reduction with random phase sequence outperforms the other types and hence this type of phase sequence is applied in the following simulations.

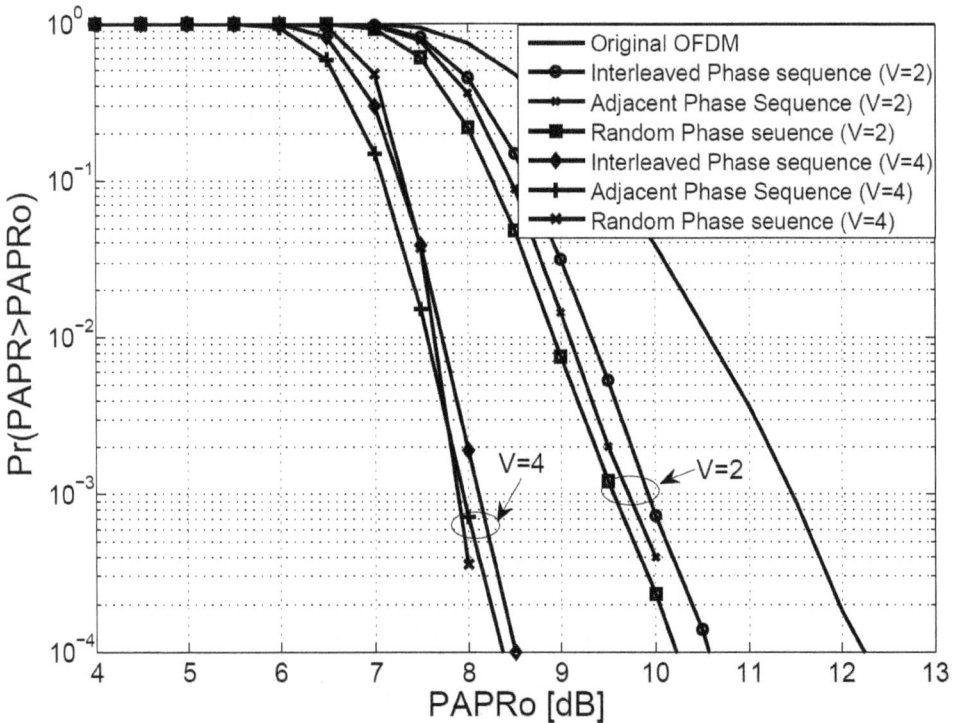

Fig. 4. CCDF of PAPR of the proposed method for different phase sequence when $D=2$

Fig. 5 shows the CCDF comparison of the PAPR of the C-PTS and EPTS for $V=2$ and 4. It is clear that the proposed EPTS shows better PAPR performance compared to C-PTS where almost 0.3 dB reduction is achieved with EPTS.

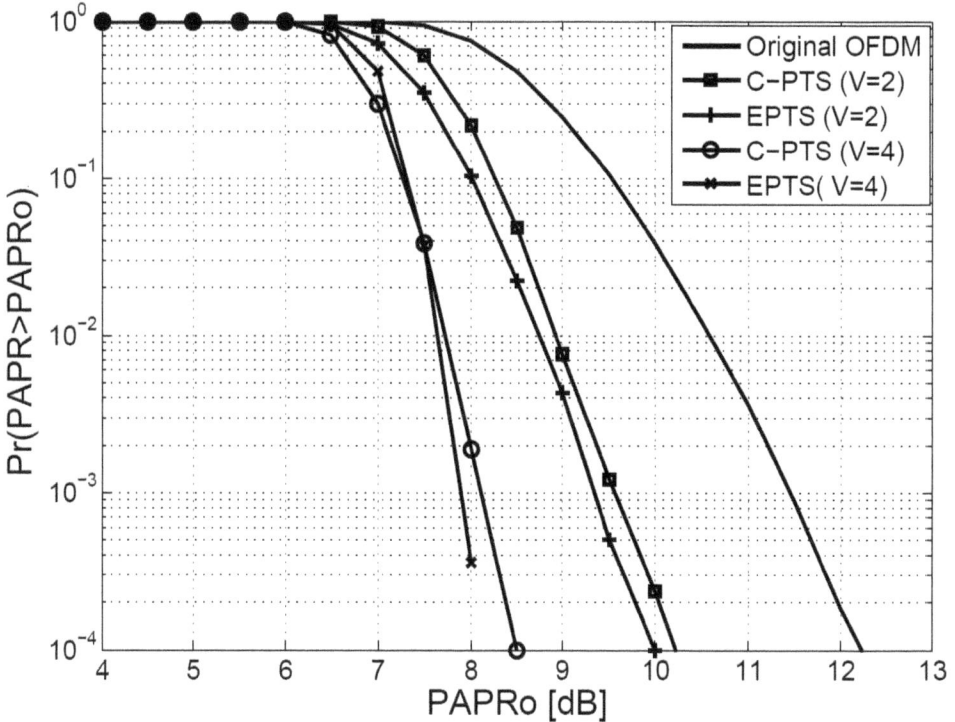

Fig. 5. CCDF comparison of PAPR of the proposed EPTS and C-PTS

3.4 Dummy Sequence Insertion (DSI)

The DSI method reduces PAPR by increasing the average power of the signal. Here, after converting the input data stream into parallel through the serial to parallel converter a, dummy sequence is inserted in the input signal. Therefore, the average value in Equation (2) is increased and the PAPR is subsequently reduced (Ryu, et al. 2004). IEEE 802.16d standard, specifies that the data frame of OFDM signal is allocated with 256 subcarriers which is composed of 192 data subcarriers, 1 zero DC subcarrier, 8 pilot subcarriers, and 55 guard subcarriers. Therefore, the dummy sequence can be inserted within the slot of 55 guard subcarriers without degradation of user data. However, if added dummies are more than 55, the length of the data and the bandwidth required, will be increased. This will degrade the Transmission Efficiency (TE) which is defined as:

$$TE = \frac{K}{K+L} \times 100\% \tag{15}$$

where K is the number of the subcarriers and L is the number of dummy sequence. In this chapter we apply a different DSI method from the one in (Ryu, et al. 2004), where the TE is always 100%.

3.5 Dummy Sequence Insertion with Partial Transmit Sequence (DSI-PTS)

The block diagram of this technique is shown in Fig. 6. A complex valued dummy signals are first generated and then added to the vector of data subcarriers. The new vector in frequency domain is then constructed from K-data and L-dummy subcarriers, respectively. L can be any number less than K. The new vector S is given by:

$$S = \left[X_k, W_l \right] \qquad (16)$$

where $X_k = [X_{k,0}, X_{k,1}, ..., X_{k,N-L-1}], k = 1, 2, ..., K$ is the data subcarrier vector and $W_l = [W_{l,0}, W_{l,1}, ..., W_{l,L-1}], l = 1, 2, ..., L$ is the dummy signals vector.

After generation of the optimum OFDM signal then the PAPR is checked with the acceptable threshold that was pre-defined before. If the PAPR value is less than the threshold then the OFDM signal will be transmitted otherwise the dummy sequence is generated again as depicted with the feedback in Fig. 6. This process is one iteration. The number of iterations can be increased to achieve the desired PAPR ($PAPR_{th}$) reduction but the processing time will also increase likewise and causes the system performance to drop.

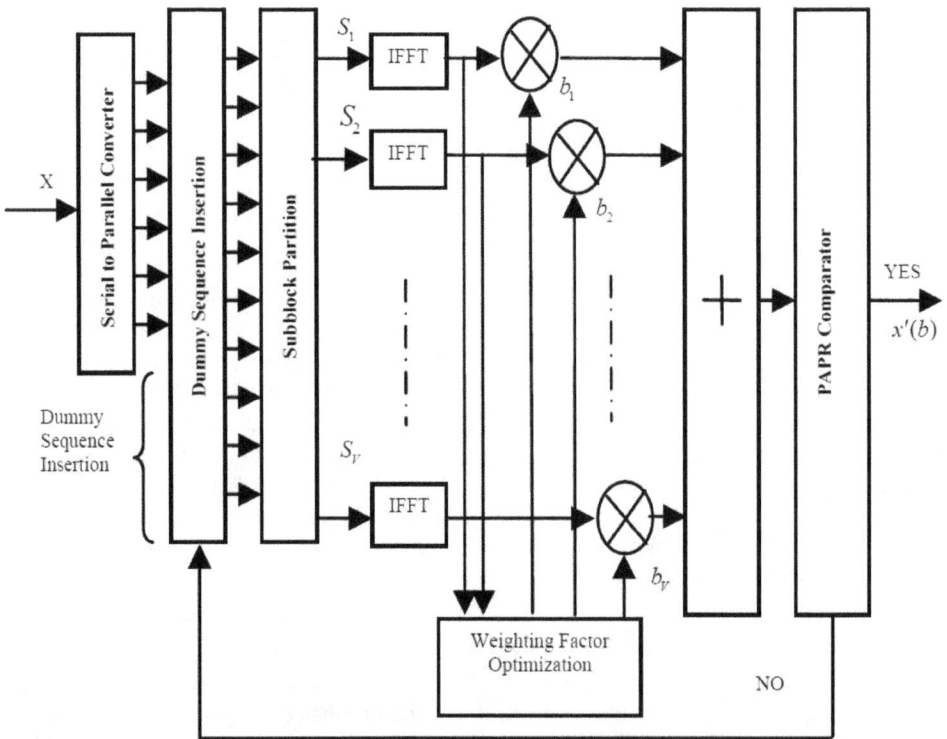

Fig. 6. Block diagram of DSI-PTS technique

As for the DSI-PTS method, consider L as the number of dummy sequence which later will be shown to be $L \leq 55$ and N is the IFFT length which is 256 in the case of fixed WiMAX that includes 192 data carriers, 8 pilots and 55 zero padding and 1 dc subcarrier. Here complementary sequence is applied for the DSI (Ryu, et al. 2004).

From the block diagram in Fig. 6, X is the input signal stream with length N after which the dummy sequence is added. The dummy sequence can be replaced with zeros in data sample. This makes the IFFT length remain unchanged and decoding of the samples in receiver becomes simpler. Then the signal is partitioned into V disjoint blocks

$$S_v = [S_1, S_2, ..., S_V]$$

such that

$$\sum_{v=1}^{V} S_v = S$$

and then these subblocks are combined to minimize the PAPR in time domain. In time domain the signal s_v is oversampled F times which is obtained by taking an IFFT of length FN on signal X_v concatenated with $(F-1)N$ zeros. After partitioning the signal and performing the IFFT for each part, then the phase factors $b_v = e^{j\phi_v}, v = 1, 2, ..., V$ are used to optimize the S_v. In time domain the OFDM signal can be expressed as:

$$s'(b) = \sum_{v=1}^{V} b_v s_v \tag{17}$$

where $s'(b) = [s'_0(b), s'_1(b), ... s'_{NF-1}(b)]^T$. The objective is to find the optimum signal $s'(b)$ with the lowest PAPR. Notice that here $N = K + L$ which means that there is no change in the length of the input signal after the addition of dummy sequence. The subblock partition type here is based on interleaving which is the best choice for PTS OFDM in terms of computational complexity reduction as compared to adjacent and pseudo-random method, however it gives the least PAPR reduction among them.

Then, the process is continued by choosing the optimization parameter $\tilde{b}$ with the following condition:

$$\tilde{b} = arg\,min(\max_{0 \leq k \leq NF-1} \left| \sum_{v=1}^{V} b_v s_{v,k} \right|) \tag{18}$$

After finding the optimum $\tilde{b}$ then the optimum signal $s'(b)$ is transmitted to the next block. Then the PAPR of $s'(b)$ is checked whether it lies in the range of the PAPR threshold ($PAPR_{th}$). After this additional task, the signal is transmitted otherwise it is returned to the DSI block to generate the dummy sequence again. This process will continue until the PAPR is less than the $PAPR_{th}$.

Fig. 7 shows the CCDF curves of conventional PTS and DSI-PTS techniques. We assume here that the number of dummy sequence insertion (L) is 55 which bears no significant effect on the transmission efficiency ($TE = 100\%$). These results are obtained after 10 iteration (I). It can be observed that the PAPR reduction of our proposed PTS scheme outperforms the conventional PTS scheme with an improvement by 2 and 1 dB respectively at $CCDF = 0.01\%$, when $V = 2,4$ respectively. Even though this reduction seems minor the complexity according to Table 1 is reduced significantly.

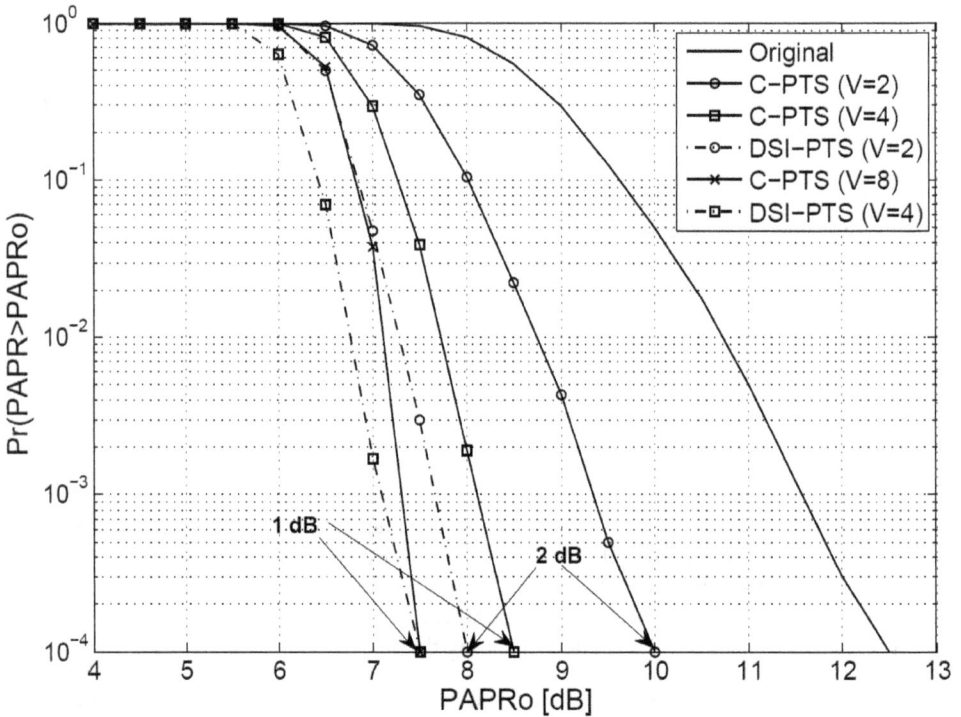

Fig. 7. CCDF of the PAPR of conventional PTS and DSI-PTS technique (L=55, I=10).

Fig. 8 shows the result for different length of dummy sequence. As discussed earlier the maximum length of dummy sequence that can be applied is 55 and this figure shows that with this length the reduction obtained is slightly better than when is 30. It is observed that the reductions of PAPR at $CCDF = 0.01\%$ are 1 dB, 1.5 dB and 2 dB for dummy length of 5, 30 and 55 respectively.

Fig. 9 shows the effect of different iteration number on the PAPR performance. From this figure maximum PAPR reduction is achieved which is 7 dB at $CCDF = 0.01\%$ at 100 iterations with $L = 55$. But increasing the number of iterations will reduce the data rate.

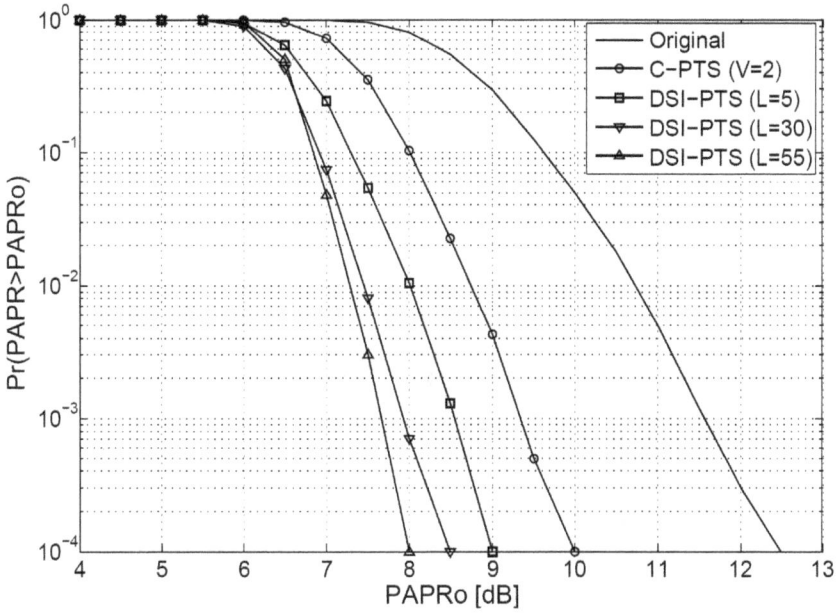

Fig. 8. CCDF of PAPR of DSI-PTS technique for different length of dummy sequence when I=10.

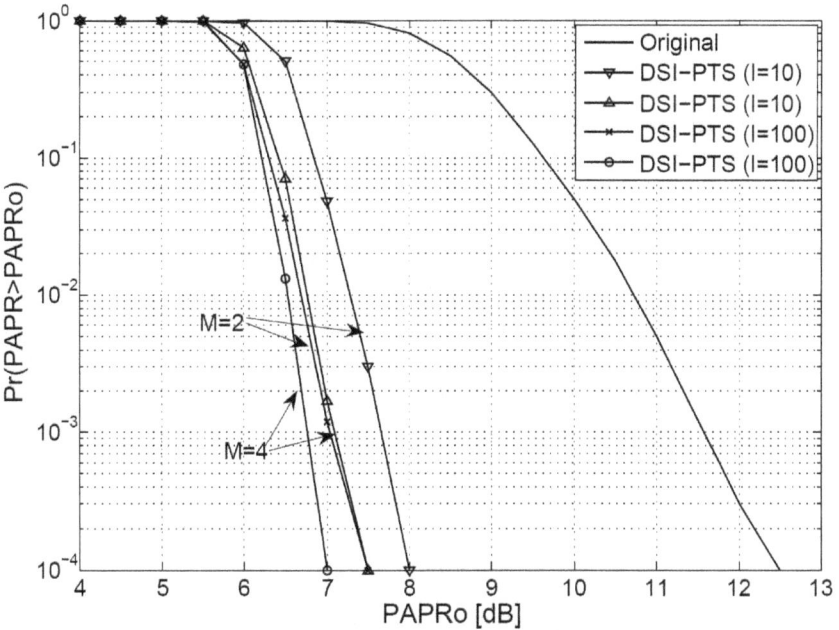

Fig. 9. CCDF of PAPR of DSI-PTS technique for different number of iterations when the L=55

There is about 0.5 dB improvement in PAPR reduction when the number of iteration is 100 compared to 10 iteration for both cases of $V = 2,4$ as shown in Fig. 9.

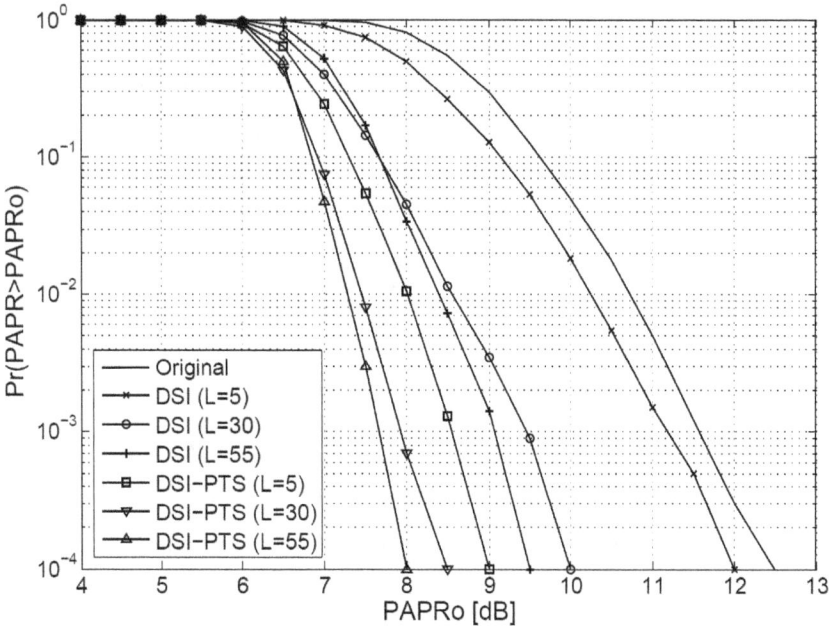

Fig. 10. CCDF of PAPR of DSI-PTS technique compared to DSI when the number of iterations is 10 and $V=2$.

Fig. 10 demonstrates the PAPR reduction capacity in DSI and DSI-PTS techniques. It should also be highlighted on that. The DSI-PTS technique offers about 1.5 dB further reduction in PAPR compared to DSI when the number of dummy sequence $L = 55$ and $V = 2$.

3.6 Dummy Sequence Insertion with Enhanced Partial Transmit Sequence (DSI-EPTS)

The block diagram of this technique is shown in Fig. 11. Here as in DSI described previously, the complex valued dummy signals are first generated and then added to the vector of data subcarriers. The new vector in the frequency domain is then constructed from K-data and L-dummy subcarriers, respectively. L can be any number less than K. The new vector U is given by:

$$U = \left[X_k, W_l \right] \tag{19}$$

where $X_k = [X_{k,0}, X_{k,1}, ..., X_{k,N-L-1}]$, $k = 1,2,...,K$ is the data subcarrier vector and $W_l = [W_{l,0}, W_{l,1}, ..., W_{l,L-1}]$, $l = 1,2,...,L$ is the dummy signals vector.

After generation of the optimum OFDM signal, PAPR is checked with the acceptable threshold that has been predefined before. If the PAPR value is less than the threshold then the OFDM signal will be transmitted otherwise the dummy sequence is generated again as shown by the feedback loop in Fig. 11. This process is one iteration. The number of iterations can be increased to achieve the desired PAPR ($PAPR_{th}$) reduction but the processing time will also increase likewise and cause the system performance to drop. From the block diagram in Fig. 11, X is the input signal with length N. After that dummy sequence is added which causes an increase in the IFFT length.

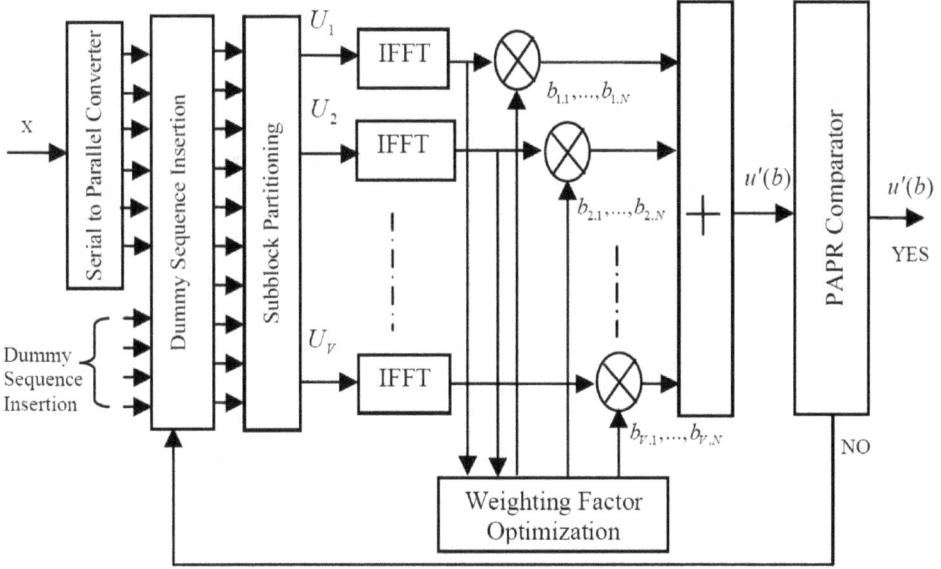

Fig. 11. Block diagram of the proposed DSI-EPTS scheme

The same procedure similar to the one discussed in section 3.5 for DSI-PTS scheme is performed here except the phase sequence is taken from the EPTS scheme discussed earlier in section 3.5.

3.6.1 Computational complexity

The total complexity of the C-PTS with oversampling factor $F=1$, is given by (Baxley and Zhou, 2007):

$$T_{C-PTS} = 3VN / 2 \log N + 2VW^{V-1}N \tag{20}$$

Whereas for the Enhanced PTS this value is:

$$T_{EPTS} = 3 / 4VN \log N + PVN \tag{21}$$

where P is the number of iterations and V is the number of subblocks.

In (Varahram, et al. 2010), the complexity is calculated only for IFFT section, but here we require the total complexity. Hence the total complexity for the DSI-PTS method is given by:

$$T_{DSI-PTS} = 3 \, / \, 4VN \log N + 2VNW^{V-1} + QL \tag{22}$$

The total complexity of DSI-EPTS is given by;

$$T_{DSI-EPTS} = 3 \, / \, 4VN \log N + PVN + QL \tag{23}$$

where Q is the number of iterations for the DSI loop.

It can be observed that (22) and (23) consist of two parts; the first part is actually the complexity of the IFFT itself and the second part is the complexity of the searching algorithm. Most of the papers did not consider the second part which causes wrong calculation of the complexity. It should be noted that the number of IFFT in (24) and (25) is halved which basically is concluded from the simulation results. From the simulation results given in the following section the PAPR performance of the proposed method when the number of IFFT is half of the C-PTS, is almost the same. This is shown for different number of subblocks which proves that in the DSI-EPTS the number of IFFT is halved compared to the C-PTS but gives the same PAPR performance.

	No. of Subblocks	C-PTS	DSI-PTS	CCRR
Total Complexity	V=4	60416	46760	22.6%
	V=8	1103872	1076392	2.4%

Table 1. Computational Complexity of the DSI-PTS and the conventional PTS when N=512 and W=2, Q=3, L=56

	No. of Subblocks	C-PTS	DSI-EPTS		CCRR (%)	
			D=1	D=2	D=1	D=2
Total Complexity	V=4	60416	30376	46760	49.7	22.6
	V=8	1103872	552104	1076392	49.6	2.7

Table 2. Computational Complexity of the DSI-EPTS and the conventional PTS when N=512 and W=2, Q=3, L=56

The computational complexity reduction ratio (CCRR) of the proposed technique over the C-PTS is defined as (Baxley and Zhou, 2007):

$$CCRR = (1 - \frac{Complexity \, of \, the \, DSI - EPTS}{Complexity \, of \, the \, C - PTS}) \times 100\% \tag{24}$$

Table 1 presents the computational complexity of C-PTS and DSI-PTS, for N=512 and W=2.

Table 2 presents the computational complexity of C-PTS and proposed DSI-EPTS, for the same value of N and W, while D is the coefficient that can be specified based on the PAPR reduction and complexity according to equation (12).

It is clear from Table 2, that CCRR is improved for both V=4 and V=8. It should be noted that when D increases, the complexity reduction becomes less while PAPR performance improves, as shown in the simulations.

3.6.2 Side information

The other important factor in studying the PAPR reduction method is the side information which has to be transmitted to the receiver to extract the original signal. One method is that the side information can be transmitted in a separate channel but this comes at the expense of spectrum efficiency degradation.

The number of required side information bits in C-PTS is

$$\left\lfloor log_2 W^{V-1} \right\rfloor$$

where W is the number of allowed phase factors and the sign $\lfloor \ \rfloor$ indicates the floor of y. In DSI-EPTS, the side information can be allocated in the dummy signals and therefore does not have impact on spectrum efficinecy and data rate loss; however, the only drawback of this method is that, because of the increase in the phase sequence matrix, higher memory space is required.

3.6.3 System performance

In C-PTS, even though an OFDM signal does not experience distortion the signal after power amplifier could exhibit distortions if PAPR is higher than the expected value. In this case the power amplifier should back off which degrades the efficiency of the system. In DSI-EPTS, the addition of dummy sequences causes the transmission efficiency to change as follows:

$$TE = \frac{K}{K+L} \times 100[\%] \tag{25}$$

where K is the length of subcarriers and L is the length of dummy sequences. In actual applications where the cost of the system is the main issue, the other block also have to be considered, the digital predistortion (DPD) (Varahram and Atlasbaf, 2005), (Varahram, et al. 2005). By applying DPD technique, it is possible to increase the linearity of the power amplifier and as a result, higher peak signals can be transmitted by the power amplifier and the performance of the PAPR can be improved. This also increases the efficiency of the power amplifiers and decreases the cost of the system.

Fig. 12 shows the CCDF comparison of PAPR of DSI-EPTS with C-PTS. It is clear that by applying the DSI-EPTS when D=2, the PAPR performance is more superior over that of C-PTS for both V=4 and V=8 respectively. But PAPR reduction when D=1 is almost the same as C-PTS for V=4 and V=8 respectively.

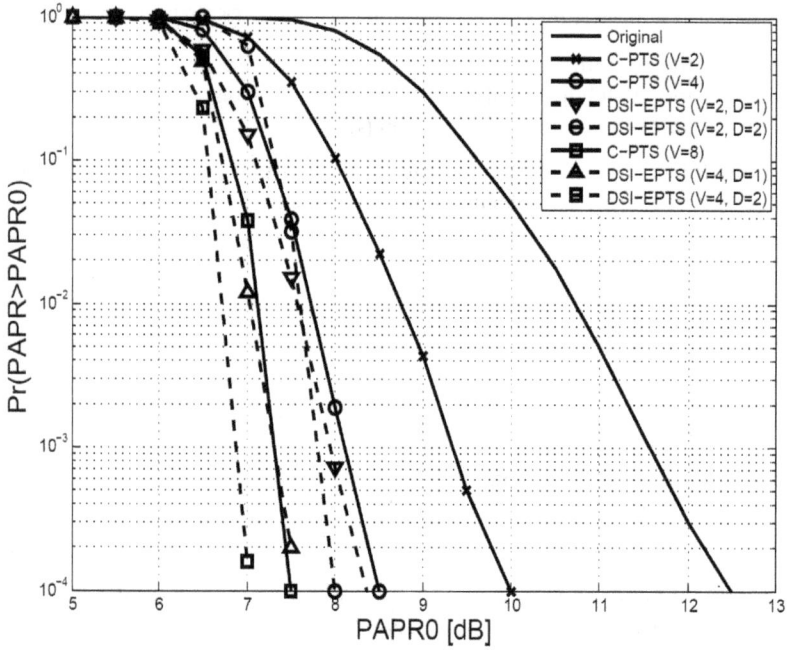

Fig. 12. CCDF comparison of PAPR of the DSI-EPTS and C-PTS

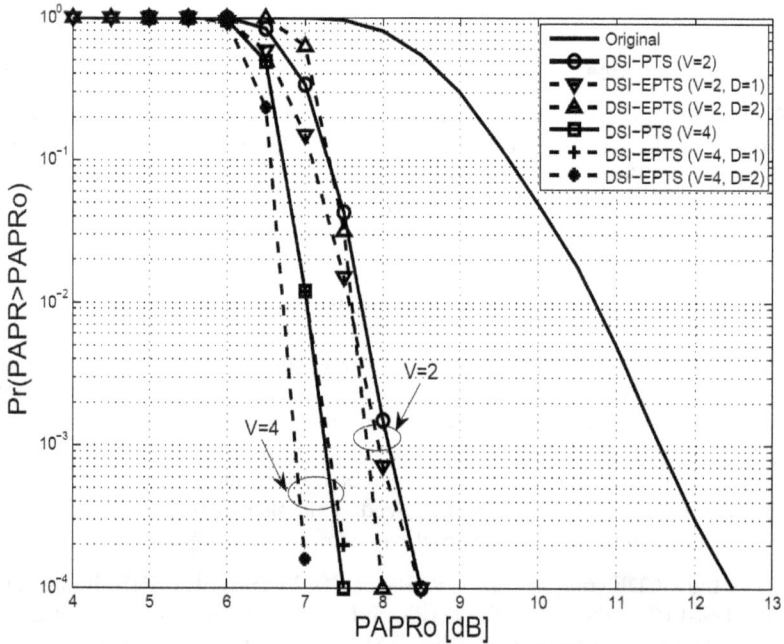

Fig. 13. CCDF comparison of PAPR of the DSI-EPTS and DSI-PTS when L=56

The highest PAPR reduction is achieved when D=2 and V=4. From table 2, the complexity reduction is minimum when D=2. There is always a trade off between PAPR reduction performance and complexity reduction.

Fig. 13 shows the CCDF comparison of PAPR of the DSI-EPTS and DSI-PTS when L=56. The results are shown for V=2 and V=4. The CCDF results show that PAPR of the DSI-EPTS outperforms DSI-PTS for both V=2 and V=4 respectively.

Fig. 14 shows a comparison of Bit Error Rate (BER) performance of the conventional PTS and the proposed EPTS and DSI-EPTS method in Additive White Gaussian Noise (AWGN) channels. The length of dummy sequence and iterations is L=56. From this figure, we can see that the BER is slightly increased with DSI-EPTS method compared to conventional PTS, but PAPR is much improved according to the result of Fig. 13. The performance of the system shows improvement at the cost of BER.

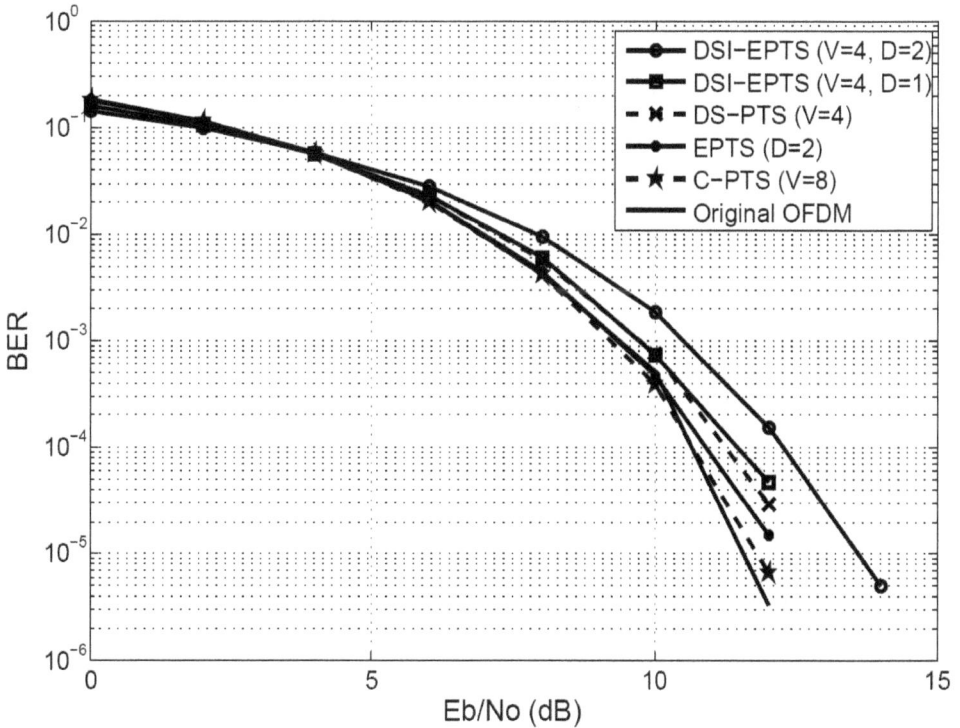

Fig. 14. Comparison of BER performance of the conventional PTS and DSI-PTS technique in AWGN channels.

4. Conclusion

In this chapter we have studied and discussed several PAPR redcution techniques. Their advantages and disadvantages have been analyzed and by performing the simulation results, the PAPR performance of those techniques have been compared. Also the complexity of each technique has been computed and finally compared. These PAPR techniques is ideal for the latest wireless communications systems such as WiMAX and long term evolution (LTE).

5. Acknowledgment

This work was supported by Universiti Putra Malaysia under the Research University Grant Scheme (No. 0501090724RU).

6. References

Al-Azoo W. F., Ali B. M, Khatun S., Bilfagih S. M, and Noordin N. K., "Addition of Gaussian random signals for peak to average power ration reduction in OFDM systems," *IEEE ICCCE'08 International conference*, PP. 1344-1347, 2008.

Bauml R. W., Fischer R. F. H. and Huber J. B., "Reducing the peak-to-average power ratio of multicarrier modulationby selected mapping," *Electron. Lett.*, vol. 32, pp. 2056-2057, 1996.

Bauml v, Fischer R. F. H. and Huber J. B., "Reducing the peak-to-average power ratio of multicarrier modulationby selected mapping," *Electron. Lett.*, vol. 32, pp. 2056-2057, 1996.

Baxley R. J. and Zhou T., "Comparing slected mapping and partial transmit sequence for PAPR reduction", *IEEE Trans. Broadcast.*, vol. 53, no. 4, pp.797-803, December 2007.

Cimini L. J. and Sollenberger N. R., "Peak-to-average power ratio reduction of an OFDM signal using partial transmit sequences," *IEEE Commun. Lett.*, vol. 4, no. 3, pp. 86–88, March. 2000.

Han S. H. and Lee J. H., "An overview of peak-to-average power ratio reduction techniques for multicarrier transmission," *IEEE Wireless Commun.*, vol. 12, no. 2, pp. 56-65, Apr. 2005.

Hao M. and Liaw C., "A companding technique for PAPR reduction of OFDM systems," *IEICE Trans. Commun.*, vol. E91-B, pp. 935-938, 2008.

Foomooljareon P. and Fernando W. A. C., "Input sequence envelope scaling in PAPR reduction of OFDM," in 2002.

Jayalath v and Tellambura C.. (2000, Reducing the peak-to-average power ratio of orthogonal frequencydivision multiplexing signal through bit or symbol interleaving. *Electron. Lett. 36(13)*, pp. 1161-1163.

Krongold B. S. and Jones D. L., "PAR reduction in OFDM via active constellation extension," *IEEE Trans. Broadcast.*, vol. 49, 2003.

Kwon U., Kim D. and Im G., "Amplitude clipping and iterative reconstruction of MIMO-OFDM signals with optimum equalization," *IEEE Transactions on Wireless Communications*, vol. 8, pp. 268-277, 2009.

Mohammady S., Varahram P., Sidek R. M., Hamidon M. N. and Sulaiman N., "Efficiency improvement in microwave power amplifiers by using Complex Gain Predistortion technique," *IEICE Electronics Express (ELEX)*, vol. 7, no. 23, pp.1721-1727, Oct. 2010.

Muller S. H. and Huber J. B., "A novel peak power reduction scheme for OFDM," in *Personal, Indoor and Mobile Radio Communications, 1997.'Waves of the Year 2000'. PIMRC'97., the 8th IEEE International Symposium on,* 1997.

Nikookar H. and Lidsheim K. S., "Random phase updating algorithm for OFDM transmission with low PAPR," *IEEE Trans. Broadcast.,* vol. 48, pp. 123-128, 2002.

Qian H., "Power Efficiency Improvements for Wireless Transmissions," *Power Efficiency Improvements for Wireless Transmissions,* 2005.

Ryu H. G., Lee J.E. and Park J.S., "Dummy sequence insertion (DSI) for PAPR reduction in the OFDM communication system," *IEEE Trans. Consumer Electr.,* vol. 50, no. 1, pp. 89-94, Feb. 2004

Tellado J., *Multicarrier Modulation with Low PAR: Applications to DSL and Wireless.* Kluwer Academic Pub, 2000.

Tan C. E. and Wassell I. J., "Data bearing peak reduction carriers for OFDM systems," *Information, Communications and Signal Processing, 2003 and the Fourth Pacific Rim Conference on Multimedia. Proceedings of the 2003 Joint Conference of the Fourth International Conference on,* vol. 2; 2, pp. 854-858 vol.2, 2003.

Van Nee R. and De Wild A., "Reducing the peak-to-average power ratio of OFDM," in *48th IEEE Vehicular Technology Conference, 1998. VTC 98,* 1998.

Varahram P., Mohammady S., Hamidon M. N., Sidek R. M. and Khatun S., "Digital Predistortion Technique for Compensating Memory Effects of Power Amplifiers in Wideband Applications", *Journal of Electrical Engineering,* vol. 60, no. 3, 2009.

Varahram P., Mohammady S., Hamidon M. N., Sidek R. M. and Khatun S., "Power amplifiers linearization based on digital predistortion with memory effects used in CDMA applicationsm," *18th European Conference on Circuit Theory and Design ECCTD,* pp. 488 – 491, 2007.

Varahram P., Atlasbaf Z., "Adaptive digital predistortion for high power amplifiers with memory effects," *Asia-Pacific Microwave Conference Proceedings, APMC.* 3(1606627), 2005.

Varahram P., Azzo W. A., Ali B. M., "A Low Complexity Partial Transmit Sequence Scheme by Use of Dummy Signals for PAPR Reduction in OFDM Systems," *IEEE Trans. Consumer Electron,* vol. 56, no. 4, pp. 2416-2420, Nov. 2010.

Varahram P., Atlasbaf Z., Heydarian N., "Adaptive digital predistortion for power amplifiers used in CDMA applications," *Asia-Pacific Conference on Applied Electromagnetics, APACE Proceedings,* (1607810):215-21, 2005.

Vijayarangan V. and Sukanesh D., "AN OVERVIEW OF TECHNIQUES FOR REDUCING PEAK TO AVERAGE POWER RATIO AND ITS SELECTION CRITERIA FOR ORTHOGONAL FREQUENCY DIVISION MULTIPLEXING RADIO SYSTEMS," pp. 25, 2009.

Wilkison T. A. and Jones A. E., "Minimazation of the peak to mean envelope power ratio of multicarrier transmission schemes by block coding," in *IEEE Vehicular Technology Conference*, 1995.

13

Hybrid ARQ Utilizing Lower Rate Retransmission over MIMO Wireless Systems

Cheng-Ming Chen and Pang-An Ting
Information and Communication Laboratories,
Industrial Technology Research Institute (ITRI), Hsinchu,
Taiwan

1. Introduction

Hybrid automatic repeat request (Hybrid ARQ or HARQ), an extension of ARQ that incorporates forward error correction coding (FEC), is a retransmission scheme with error-control method employed in current communications systems. In standard ARQ, redundant bits are added to data to be transmitted using an error-detecting code, e.g., cyclic redundancy check (CRC). The contribution of HARQ is its efficient utilization of the available resources and the provision of reliable services in latest-generation systems.

This chapter focuses on wireless systems using HARQ with emphasis on the multiple-input multiple-output (MIMO) paradigm. In this chapter, the architecture of MIMO transceivers that are based on bit-interleaved coded modulation (BICM) and it employs HARQ is described. MIMO system is an attractive technique that can enhance the spectral efficiency through spatial multiplexing (SM) [Foschini 1995]. However, many wireless environments may suffer from ill-conditioned channels or multipath fadings, which degrade the system performance. The aim of this chapter is to find an efficient MIMO scheme for retransmission to combat the ill-ranked channel and enhancing the reliability.

In IEEE 802.16e [WiMAX 2007] standard, Space Time Coding (STC) subpacket combining, which retransmits with a different MIMO format, has been introduced. One possible way to combine the initial signal and the retransmitted signal is the utilization of Maximum Ratio Combining (MRC) in symbol level based on Virtual Space Time Block Coding (VSTBC) [Gao et. al , 2007]. However, such combining technique works properly only if the channel is quasi-static. In circumstances with high mobility, this technique does not provide satisfactory performance. Another approach is to combine the retransmitted and initial signals using symbol level combining (SLC) before the detector. However, the required buffer in SLC occupies large memory in receivers. Moreover, both SLC receiver architectures are only applicable when the initial and retransmitted symbols are aligned. Thus, they may be impractical for Incremental Redundancy (IR) HARQ [Lin et al. ,1984] and Constellation Rearrangement (CoRe). Although bit level combining (BLC) can solve the problem of huge buffer requirement, using ML-like MIMO detectors, e.g., List Sphere Decoder (LSD) [Damen et al. ,2003], can still increase the complexity significantly.

To reduce buffer size and power consumption, a lower rate MIMO mode in retransmission request, termed as Lower Rate Retransmission (LRR) Scheme [Chen et. al, 2009], is proposed. In this chapter, we define the rate as how much information can be transmitted in single time-frequency resource unit. Two examples of LRR schemes are introduced. For the first scheme, it applies rate-2 SM for initial transmission, and rate-1 STBC [Alamouti ,1998] or Space Frequency Block Coding (SFBC) [Kaiser ,2003] is used for retransmission. For the second scheme, rate-3 or rate-4 SM are leveraged for initial transmission, and a lower rate SM scheme, i.e., rate-2 or rate-3, is employed for retransmission. In order not to decrease the spectral efficiency, only partial coded bits are retransmitted in the proposed schemes.

In LRR, with fewer transmit antennae in retransmission, the acquired transmit power gain could be used to retransmit higher modulation symbols and keep the total retransmitted bits as close to initial transmission as possible. We tabulated this scheme as Lower Rate Retransmission combine with modulation step up (LRRMSU).

The notations of this paper are explained as following: Superscripts T and H indicate matrix transpose and hermitian, respectively. Superscripts $*$ indicates complex conjugate operation. Uppercase boldface denotes a matrix while lowercase boldface denotes a vector. $\mathbf{I}_N$ denotes the $N \times N$ Identity matrix. $[A]_{i,j}$ and $[A]_{:,j}$ represent the element of ith row and jth column of matrix $\mathbf{A}$, and jth column of matrix A, respectively. A circularly symmetric complex Gaussian vector $\mathbf{a}$ with mean $\mathbf{m}$ and covariance matrix $\mathbf{R}$ is denoted as $\mathbf{a} \sim NC [\mathbf{m}, \mathbf{R}]$. Finally, nTX, M_t and M_r refer to the n-th transmission, number of transmit antenna and receive antenna, respectively.

This article is organized as follows. In the next section the architecture of a single-input single-output (SISO) transceiver using BICM and HARQ is presented. The following section narrates conventional MIMO HARQ schemes. The LRR schemes are then elaborated with simpler receiver implementations. The following section contains some discussion of MIMO system design based on the employed HARQ scheme, receiver complexity, and storage requirements. To keep the number of retransmission bits, near to that of the initial transmission a LRRMSU scheme is illustrated to enhance the system throughput. Finally, some concluding remarks are provided.

2. MIMO HARQ scheme in 802.16

Figure 1 shows the block diagram of MIMO HARQ in 802.16e transmitter. It is illustrated that k information bits $\mathbf{b}$, $\{\mathbf{b} \in b^{k \times 1} \mid b \in \{0,1\}\}$ are fit into one forward error control coding (FEC) block. These information bits are regarded as systematic bits AB. The mother code rate R_{MC} of the Convolutional Turbo Code (CTC) in 802.16e is $\frac{1}{3}$. After CTC encoding, the encoded bit length is multiplied by 3, and the output $\mathbf{c}, \{\mathbf{c} \in b^{3k \times 1} \mid b \in \{0,1\}\}$, is consists of systematic bits AB, parity bits Y_1Y_2 and W_1W_2. In this example, the code rate R is $\frac{1}{2}$, hence in the initial transmission, W_1W_2 are punctured in the bit selection block, and the remained encoded bits are $\mathbf{c}_s, \{\mathbf{c}_s \in b^{2k \times 1} \mid b \in \{0,1\}\}$. The bit selection procedure depends on

IR or Chase Combining (CC) HARQ [Chase, 1985] as well as the times of reception of NACK message.

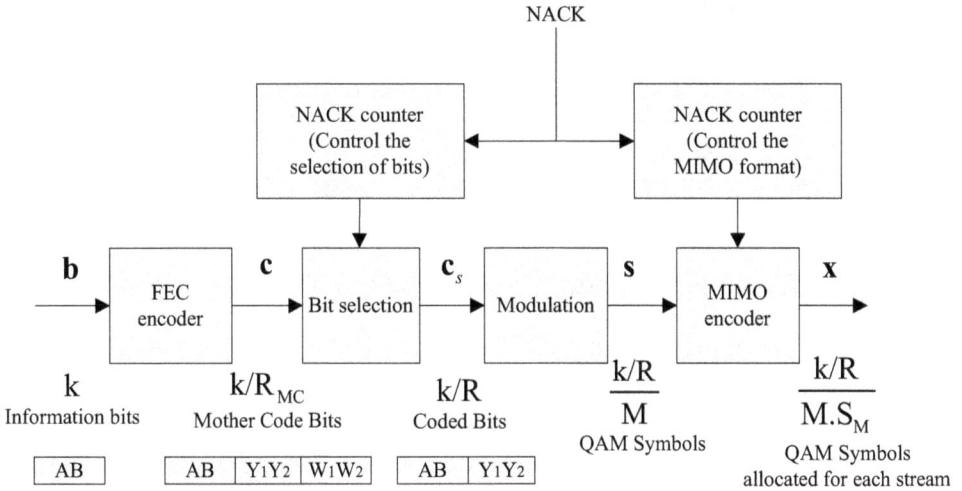

Fig. 1. MIMO HARQ transmitter diagram of 802.16e

Figure 2 gives an example of $R = \dfrac{1}{2}$ transmission with CC and IR modes. In CC mode, the same encoded bits are retransmitted after the request. For IR mode, on the other hand, additional parity check bits are retransmitted. Then, the selected bits, $\mathbf{c}_s$, are modulated into $\overline{M}$-ary QAM symbols,

$$\mathbf{s},\left\{\mathbf{s} \in C^{\frac{2k}{\overline{M}} \times 1}, C \in complex\right\}, \text{ where } M = log_2^{\overline{M}}.$$

It is then followed by a MIMO encoder. The encoder can be either in SM or spatial diversity (S-Div) mode. In SM mode, the MIMO encoder is simply a parser. In S-Div mode, it is Space Time Code (STC) or Space Frequency Code (SFC) encoder. The number of streams of SM and S-Div mode are denoted by S_M and S_D, respectively. Note that a stream is defined as each output of the MIMO encoder. The value of S_M is equivalent to the rate in SM mode. Moreover, in Figure 1, S-Div mode represents rate-1 Alamouti code. Finally, the output of the MIMO encoder is

$$\mathbf{x} \in C^{\frac{2k}{M.S_M} \times S_M.S_D}.$$

The signal of ith MIMO encoder output $\underline{\mathbf{x}}_i = [\mathbf{x}]_{i,:}$ is STC subpacket encoded based on Table 1, and STC subpacket encoding refers to the VSTBC of previous and current transmitted subpacket, as in 802.16e.

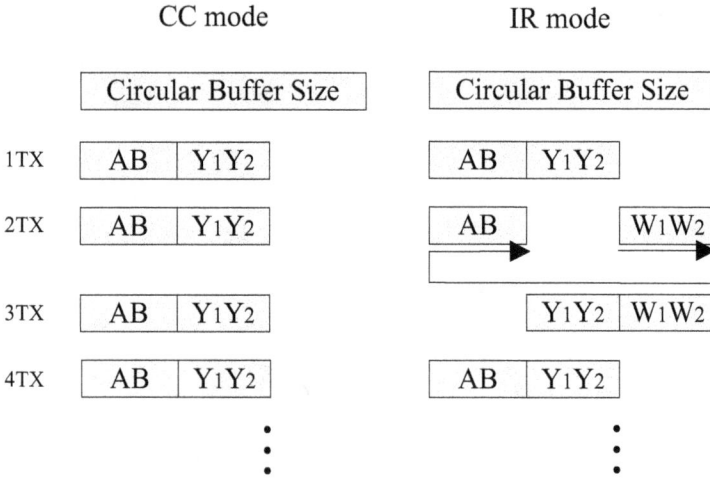

Fig. 2. Different Bit selection manners for CC (left) and IR (right)

reTX / Antenna	Initial TX=1TX	Even TX	Odd TX
$M_t = 2$	$\underline{x}_i = \begin{bmatrix} s_1 \\ s_2 \end{bmatrix}$	$\underline{x}_i = \begin{bmatrix} -s_2^* \\ s_1^* \end{bmatrix}$	$\underline{x}_i = \begin{bmatrix} s_1 \\ s_2 \end{bmatrix}$
$M_t = 3$	$\underline{x}_i = \begin{bmatrix} s_1 \\ s_2 \\ s_3 \end{bmatrix}$	$\underline{x}_i = \begin{bmatrix} -s_2^* \\ s_1^* \\ s_3^* \end{bmatrix}$	$\underline{x}_i = \begin{bmatrix} s_1 \\ s_2 \\ s_3 \end{bmatrix}$
$M_t = 4$	$\underline{x}_i = \begin{bmatrix} s_1 \\ s_2 \\ s_3 \\ s_4 \end{bmatrix}$	$\underline{x}_i = \begin{bmatrix} -s_2^* \\ s_1^* \\ -s_4^* \\ s_3^* \end{bmatrix}$	$\underline{x}_i = \begin{bmatrix} s_1 \\ s_2 \\ s_3 \\ s_4 \end{bmatrix}$

Table 1. STC subpacket encoding

3. Lower Rate Retransmission (LRR) scheme

In practice, due to the propagation mechanisms, MIMO system may be suffered from high spatial correlation, which degrades the system capacity. If there is no feedback information regarding the channel rank, it is always a good approach to retransmit with a lower rate MIMO mode to provide a robust transmission. For instance, a rate-1 SFBC or STBC is recommended for retransmission for initial transmission in rate-2 SM mode. For rate-3 and rate-4 initial transmission in SM mode, on the other hand, rate-2 and rate-3 SM mode are recommended for retransmission. A list of possible MIMO mode selection for LRR is shown in Table 2. Since an open loop system is being considered, thus the transmitter does not possess channel state information (CSI). Note that LRR can be leveraged in association with a stream-to-antenna mapping technique such as precoding or antenna selection.

reTX Antenna	Initial TX=1TX	Request TX
$M_t = 2$	rate-2 SM	rate-1 SFBC/STBC
$M_t = 3$	rate-3 SM	rate-2 SM
$M_t = 4$	rate-4 SM	rate-3 SM

Table 2. MIMO mode selection in LRR

In order to maintain the same spectral efficiency as the conventional HARQ schemes, fewer bits are encoded by LRR MIMO encoder. Although the number of retransmitted bit is reduced, the reliability is improved. Since only a portion of bits is retransmitted, the bit selection should be modified. An example of retransmissions with coding rate $R = \frac{1}{2}$ in CC and IR modes are shown as Figure 3.

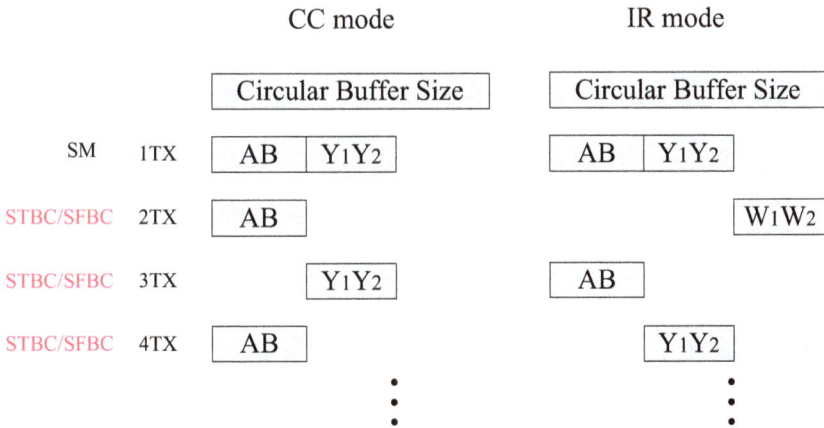

Fig. 3. Different Bit selection methods for CC (left) and IR (right) with LRR scheme.

4. Receiver architectures

In this section, four types of MIMO receivers are elaborated. Firstly, we illustrate the combining methods for both VSTBC and STBC/SFBC in symbol and bit levels. Then, two SM detection algorithms in BLC are described: Soft Linear Minimum Mean Square Error (LMMSE)[Lee & Sundberg, 2007] algorithm and LSD algorithm. All algorithms use soft decision information generated from CSI. Finally, a complexity analysis is carried out in a rate-2 MIMO HARQ scheme.

4.1 VSTBC-MRC with SLC and STBC/SFBC-MRC with BLC

A flat-fading MIMO system can be expressed as:

$$y = \underline{H}Px + n = Hx + n, \tag{1}$$

where H is the $M_r \times M_t$ MIMO channel matrix of an OFDM subcarrier with unitary power complex Gaussian elements, x is the $S_M \times 1$ transmitted signal vector with unit total transmission power per subcarrier, and y is the $M_r \times 1$ received signal vector. Moreover, P is the $M_t \times S_M$ precoding matrix, and $n \sim N_C\left[0, N_o.I_{M_r}\right]$ is the $M_r \times 1$ noise vector with complex Gaussian elements. Without loss of generality, a MIMO scenario with $M_t = M_r = 2$ and identity matrix P is described.

The signal in the odd transmission (as defined in Table 1) is:

$$x_{odd} = \begin{bmatrix} x_0 \\ x_1 \end{bmatrix} \tag{2}$$

And the signal in the even transmission is:

$$x_{even} = \begin{bmatrix} -x_1^* \\ x_0^* \end{bmatrix} \tag{3}$$

Here, subscripts *odd* and *even* indicate different subpacket information at odd and even retransmission. Moreover, the initial transmission is started with the first odd transmission. Hence, a pair of received signal in subpackets can be formed as:

$$y_{odd} = H_{odd}x_{odd} + n_{odd} \tag{4}$$

$$y_{even} = \left(\rho H_{odd} + \Delta H\right)x_{even} + n_{even} \tag{5}$$

Here, we assume $H_{even} = \rho H_{odd} + \Delta H$ and $\Delta H \sim N_C\left[0, \left(1-\rho^2\right)I_{M_r}\right]$ with $0 \le \rho \le 1$ is induced from Doppler effect. Equation 4 and 5 can be further stacked as:

$$\begin{bmatrix} y_{odd} \\ y_{even}^* \end{bmatrix} = \overline{H}x_{odd} + \begin{bmatrix} n_{odd} \\ n_{even}^* \end{bmatrix} \tag{6}$$

Here,

$$\overline{H} = \begin{bmatrix} [H_{odd}]_{:,1} & [H_{odd}]_{:,2} \\ [H_{even}]_{:,2}^* & -[H_{even}]_{:,1}^* \end{bmatrix},$$

and the combined signal vector by MRC is shown in eq. 7.

The MRC for signal operates as following:

$$\begin{bmatrix} \hat{h}_0 \hat{s}_0 \\ \hat{h}_1 \hat{s}_1 \end{bmatrix} = \overline{\mathbf{H}}^H \begin{bmatrix} \mathbf{y}_{odd} \\ \mathbf{y}^*_{even} \end{bmatrix} \tag{7}$$

Where $\hat{h}_0 \hat{s}_0$ and $\hat{h}_1 \hat{s}_1$ are the soft detection symbols of $\hat{s}_0$ and $\hat{s}_1$, respectively. The equivalent channel gain $\hat{h}_0$ and $\hat{h}_1$ can also be obtained by MRC in eq. 8.

The MRC for channel operates as following:

$$\begin{bmatrix} \hat{h}_0 \\ \hat{h}_1 \end{bmatrix} = diag\left(\overline{\mathbf{H}}^H \overline{\mathbf{H}} \right) \tag{8}$$

In equation 7, if the channel is not static for each retransmission, the orthogonality of Alamouti Code is destroyed and the interference is thereby induced. The mathematical description is as follows:

$$\begin{aligned}
\hat{h}_0 \hat{s}_0 = &\left(\left| h_{00}^{odd} \right|^2 + \left| h_{10}^{odd} \right|^2 + \left| h_{01}^{even} \right|^2 + \left| h_{11}^{even} \right|^2 \right) s_0 + \\
&\left\{ (1-\rho^2)\left(h_{00}^{odd*} h_{01}^{odd} + h_{10}^{odd*} h_{11}^{odd} \right) - \right. \\
&\rho\left(h_{00}^{odd*} \Delta h_{01} + h_{01}^{odd} \Delta h_{00}^* + h_{10}^{odd*} \Delta h_{11} + h_{11}^{odd} \Delta h_{10}^* \right) - \\
&\left. \left(\Delta h_{00}^* \Delta h_{01} + \Delta h_{10}^* \Delta h_{11} \right) \right\} s_1 + \\
&\left(h_{00}^{odd*} n_0^{odd} + h_{10}^{odd*} n_1^{odd} + h_{01}^{even} n_0^{even*} + h_{11}^{even} n_1^{even*} \right),
\end{aligned} \tag{9}$$

where the second term is the interference induced from signal $\mathbf{s}_1$. The equivalent SNR of signal $\mathbf{s}_0$ is:

$$SNR = \frac{8}{\left(\left(1-\rho^2\right)^2 \gamma + 2\rho^2\beta + \alpha \right) + 4N_o} \tag{10}$$

with the assumptions of cross correlation

$$E\left\{ \left| h_{ij}^{odd*} h_{mn}^{odd} \right|^2 \right\} = \gamma , \ E\left\{ \left| h_{ij}^{odd*} \Delta h_{mn} \right|^2 \right\} = \beta \ \text{and}$$

$$E\left\{ \left| \Delta h_{ij}^* \Delta h_{mn} \right|^2 \right\} = \alpha, ij \neq mn .$$

The overall receiver architecture is shown in Figure 4. The scheme requires not only symbol level buffer (SLB), but also bit level buffer (BLB), where the BLB is used to store log-likelihood ratios (LLRs). For example, in 3TX, retransmitted packet cannot be combined by MRC with previous symbol values using VSTBC format, hence only previous LLRs is required to be stored in BLB in this retransmission. The SLB and BLB cannot be shared with each other, because both of them are required for 4TX combining.

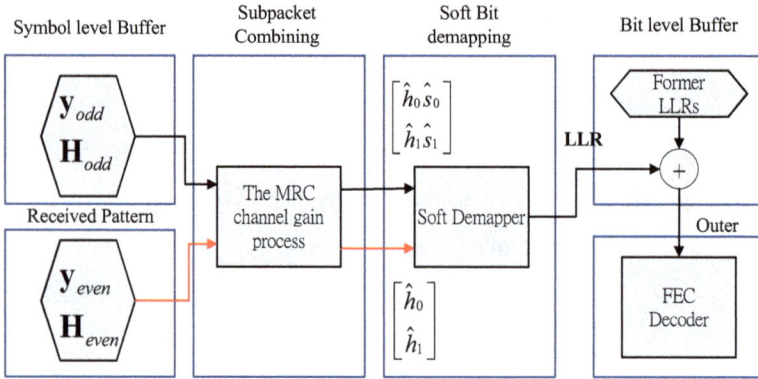

Fig. 4. Subpacket Combining Process

For the proposed LRR scheme with rate-2 SM initial transmission, the scheme is simply tantamount to a SFBC after rate reduction, and it can be detected by MRC method and then combined with previous LLRs in BLC. The SFBC formats of the retransmitted subpacket are expressed as:

$$\mathbf{x}_{sub_0} = \begin{bmatrix} x_0 \\ x_1 \end{bmatrix} \qquad (11)$$

and

$$\mathbf{x}_{sub_1} = \begin{bmatrix} -x_1^* \\ x_0^* \end{bmatrix} \qquad (12)$$

where subscripts sub_0 and sub_1 indicate the subcarrier indices 0 and 1, respectively. The detection is performed for every two consecutive subcarriers and the former detected LLRs is stored in BLB, hence SLB is not required. The overall receiver architecture is shown in Figure 5.

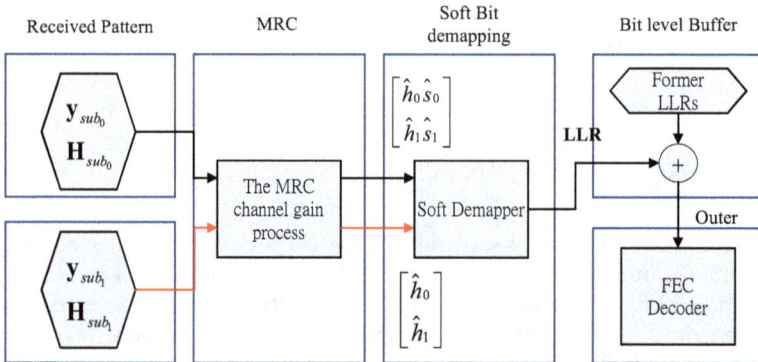

Fig. 5. SFBC MRC with BLC

4.2 Soft LMMSE with BLC

In BLC, there is no restrictions on symbol-alignment for the signals of odd and even retransmissions. Based on the system model in equation 1, and signal vector **x** here is the same as $\mathbf{x}_{odd}$, the detection signal can be expressed as in 13. The LMMSE for signal operates as:

$$\hat{\mathbf{x}} = \left(\mathbf{H}^H\mathbf{H} + S_M N_o\mathbf{I}\right)^{-1}\mathbf{H}^H\mathbf{y} \qquad (13)$$

The equivalent channel gain is the inverse of diagonal terms of MSE matrix. The LMMSE for channel gain operates as:

$$\begin{bmatrix} \dfrac{1}{\hat{h}_0} \\ \dfrac{1}{\hat{h}_1} \end{bmatrix} = diag\left(\left(\mathbf{H}^H\mathbf{H} + S_M N_o\mathbf{I}\right)^{-1}\right), \qquad (14)$$

where $\hat{h}_0$ and $\hat{h}_1$ are the equivalent soft CSI gains and the overall scheme is shown in Figure 6.

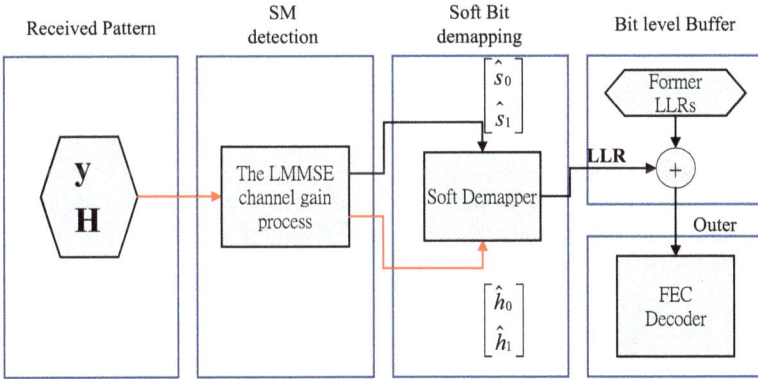

Fig. 6. Soft LMMSE processing with BLC

4.3 LSD with BLC

There are many simplified LLR algorithms for soft values computation for a SM system in the existing literature. Here, we focus on an exhaustive search which results in no penalty in performance. The LLR of the kth bit on the transmitted symbol vector **x** (which contains $S_M \times M$ bits) is:

$$LLR_k \approx \frac{1}{M_t N_o} \times \left(-\min_{k \in k+1}\left\|\mathbf{y} - \mathbf{H}\mathbf{x}_{k,1}\right\|^2 + \min_{k \in k-1}\left\|\mathbf{y} - \mathbf{H}\mathbf{x}_{k,0}\right\|^2\right), \qquad (15)$$

where $\mathbf{x}_{k,1}$ and $\mathbf{x}_{k,0}$ are the symbol vectors with kth bit equals+1 and -1, respectively. The block diagram of LSD with BLC is shown in Figure 7:

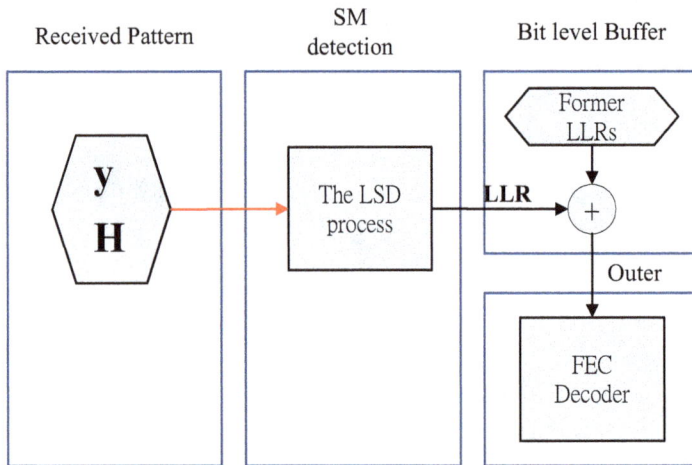

Fig. 7. LSD processing with BLC

4.4 Complexity analysis

An example based on 802.16e Partial Usage of Subchannels (PUSC) format is given here to analyze the complexity of different schemes under our consideration. The resource unit is constructed with one slot with 48 data subcarriers. For sake of simplicity, only the number of complex multiplication operations (CMLs) is considered. A real division operation is assumed to be equivalent to one CML. Besides, the complexity of LSD with BLC is not examined, because its complexity depends on the modulation order and it is undoubtable to be more complicated than the other detectors. The complexity of a MIMO receiver with $M_t = M_r = 2$ is shown as follows.

1.   VSTBC-MRC with SLC:

In this case, there are 48 subcarriers in total to be implemented with the MRC operation, and each of them takes eight CMLs in MRC for signal and eight CMLs (only diagonal elements are needed) in MRC for channel. Hence, it takes 768 CMLs in total.

2.   STBC/SFBC-MRC with BLC:

Every 2-subcarrier pair is to be implemented by one MRC operation in this case, hence the number of required operations is half of VSTBC-MRC with SLC. Thus, the total number of required operations is 384 CMLs.

3.   Soft LMMSE with BLC:

In this scenario, we should consider 48 subcarriers with 2×2 matrix operations in MMSE filtering. Each subcarrier requires eight CMLs to compute $\mathbf{H}^H\mathbf{H}$, eight CMLs for matrix inversion (two CMLs for the determinant and six CMLs for the real divisions), the multiplication of $\left(\mathbf{H}^H\mathbf{H} + S_M N_o \mathbf{I}\right)^{-1}$ with $\mathbf{H}^H$ takes eight CMLs, and finally, $\left(\mathbf{H}^H\mathbf{H} + S_M N_o \mathbf{I}\right)^{-1}\mathbf{H}^H\mathbf{y}$, needs further four CMLs. Hence, there are 1344 CMLs in total.

To recapitulate, our proposed scheme reduces the complexity by about 50% and 70% as compared to VSTBC-MRC with SLC and soft LMMSE with BLC, respectively.

5. Lower Rate Retransmission (LRR) combined with Modulation Step Up (LRRMSU) scheme

There are several pros of the LRR scheme such as 1. a robust retransmission, because the inter-stream interference is reduced, 2. the transmitter side acquires additional transmit power gain for each retransmitted stream, 3. frequency and spatial diversity is gained, because different resource allocation will be automatically guaranteed. However, one possible cons, the total retransmitted bits will be reduced as compare to number of bits in initial transmission, might degrade its performance in high coding rate scenarios. To overcome this deficiency, higher order modulation or called modulation step up is introduced and combined with the LRR scheme in Figure 8. In the initial transmission, the transmission mode operates with 4 transmission antennae with QPSK in each stream. In retransmission, the number of transmission antennae is reduced to 3 but the modulation order is step up to 16 QAM. Therefore, we keep the number of retransmission bits very close to that of traditional scheme.

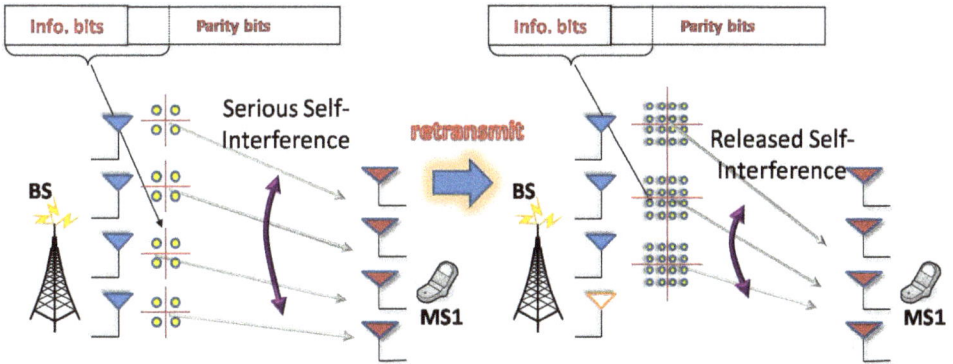

Fig. 8. LRRMSU retransmission scheme

6. Numerical results

In order to verify the superiority of the proposed scheme, the simulation based on a low correlation MIMO model [WiMAX ,2007] with $M_t = M_r = 2$ is undertaken here. In particular, we show two examples of comparison in this paper: *VSTBC in SLC* v.s. *SFBC in BLC* and *SFBC in BLC* v.s. *SM in BLC*. The delay profile of each path is evaluated under ITU-R [ITU-R ,2000] Pedestrian Type-B 3km/hr (PB3) or Vehicular Type-A 60km/hr (VA60). Furthermore, PUSC with 10Mhz bandwidth is assumed and the coding scheme is based on 802.16e CTC. In addition, we postulate that the receiver has perfect CSI. The HARQ round trip interval is 10ms, and the subpacket will be shifted by 3 subchannel length to gain higher diversity in frequency domain. The subpacket size for each coding rate is summarized in Table 3, where N_{EP} is the number of information bits before feeding into FEC encoder. Note that we concentrate on CC mode and the comparison with IR mode is beyond our scope due to page restrictions.

modulation coding scheme	N_{EP}
QPSK 1/2	480
QPSK 3/4	432
16QAM 1/2	480
16QAM 3/4	432
64QAM 1/2	432
64QAM 2/3	384
64QAM 3/4	432
64QAM 5/6	480

Table 3. N_{EP} size of different code rate with CC

Example 1: VSTBC in SLC v.s. SFBC in BLC.

We focus on simulation results of PB3 and VA60. In 1TX, the receiver is a soft MMSE detector for both cases, and in 2TX the VSTBC-MRC or SFBC-MRC is used. Simulation results of 2TX packet error rate (PER) in Figure 9 and 10 show that our proposed scheme has poorer error performance in low mobility. However, the interference terms in equation 9 will impact the performance of VSTBC in SLC scheme as the mobility increases. Hence, the proposed scheme becomes superior in moderate and higher speed scenarios. Nevertheless, a Doppler estimator is generally not available at the receiver; hence conventional 802.16e cannot be guaranteed to be superior at all mobility levels. The PER in 3TX is not evaluated, because the retransmitted subpacket cannot be combined with previous ones in BLC, in 3TX of a 802.16e MIMO-HARQ scheme.

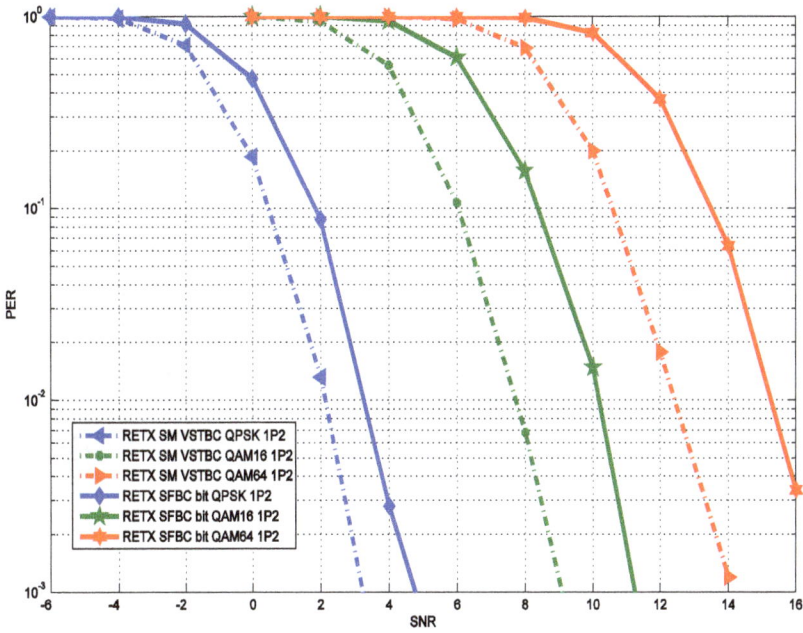

Fig. 9. VSTBC SLC v.s. SFBC BLC in 2TX PB3 CC

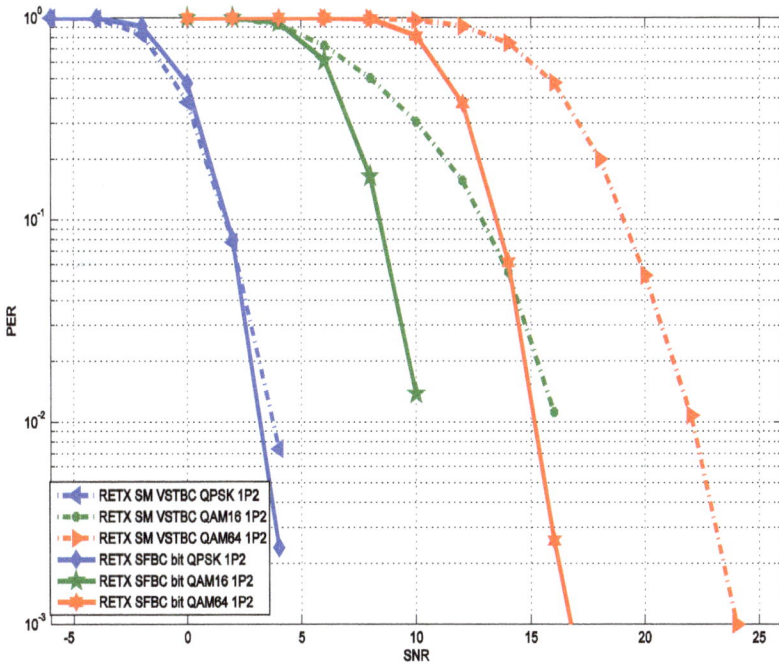

Fig. 10. VSTBC SLC v.s. SFBC BLC in 2TX VA60 CC

Example 2: SFBC in BLC v.s. SM in BLC.

1TX, the SM detection is the same for soft MMSE/LSD for 802.16e and the proposed scheme. For later retransmissions than 1TX, the detection method is soft MMSE/LSD in 802.16e and SFBC-MRC in the proposed scheme. The maximum number of retransmission assumed in our simulation is 3. We first show that the PER results in 11 and 12, and it then follows by

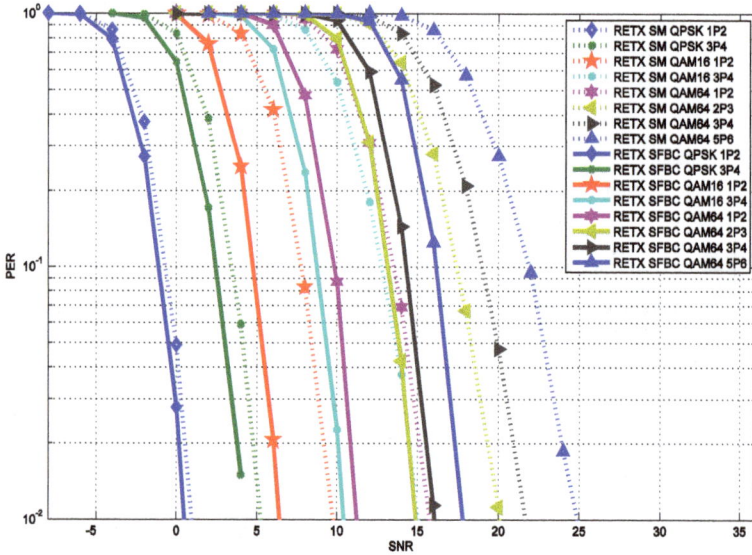

Fig. 11. Soft MMSE BLC v.s. SFBC MRC BLC in 3TX PB3 CC

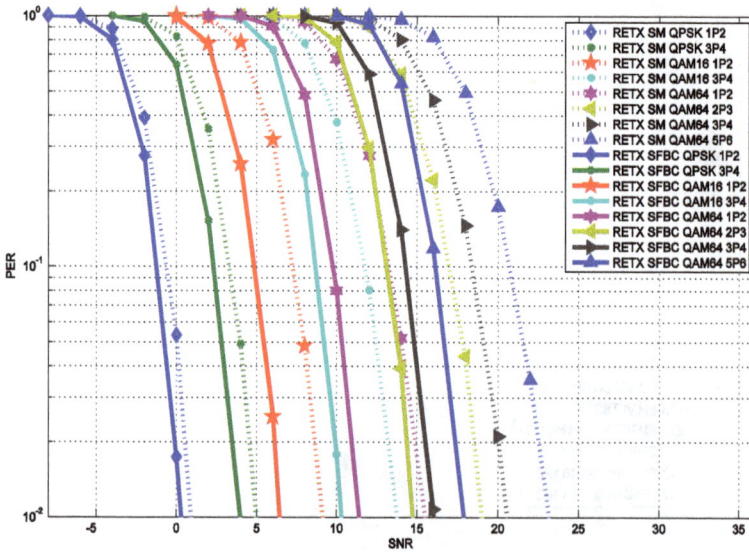

Fig. 12. LSD BLC v.s. SFBC MRC BLC in 3TX PB3 CC

throughput comparison in 13 and 14. Thus, when the channel model is PB3 with low spatial correlation, it is shown that the proposed scheme always outperforms the ones in 802.16e, especially in higher coding rate scenarios. In terms of throughput, the proposed scheme can achieve higher throughput in low signal to noise ratio (SNR) region with the same coding rate. Nonetheless, the throughput curves are similar in high SNR region. The results also show that the proposed scheme is less sensitive to improper link adaptation.

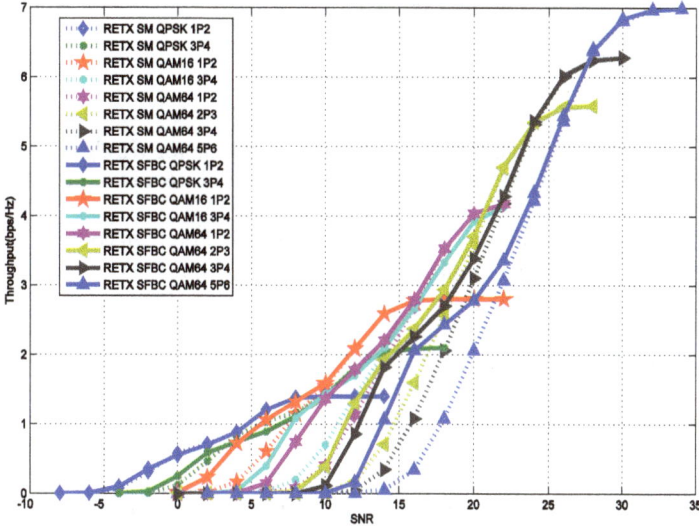

Fig. 13. Soft MMSE BLC v.s. SFBC MRC BLC in PB3 CC

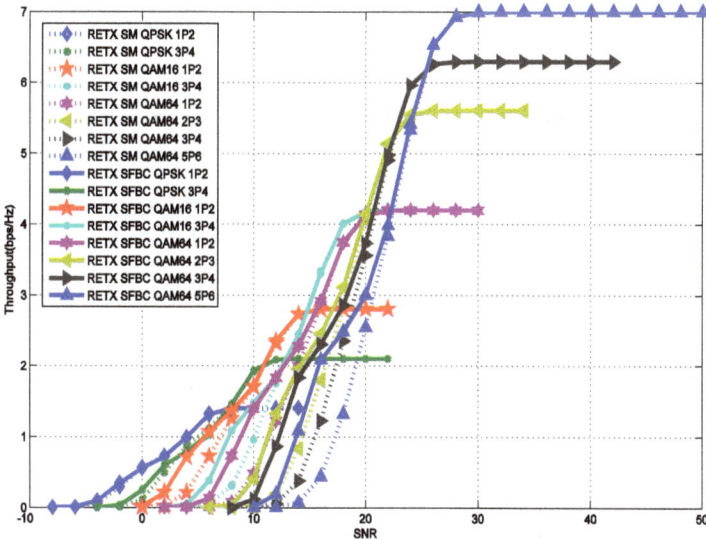

Fig. 14. LSD BLC v.s. SFBC MRC BLC in PB3 CC

7. Conclusions

In this chapter, we proposed a novel scheme to gain better performance, which reduces the receiver complexity (by 50% ~ 70% in 2 $\times$ 2 MIMO scenario) as well as the buffer requirement. From the simulation, it is observable that the proposed scheme can achieve better performance than conventional 802.16e schemes in scenarios with moderate or high mobility. In another set of simulations, we have compared the proposed scheme with the BLC in 802.16e architecture. The results have verified that our scheme is less sensitive to inappropriate link adaptation.

8. References

Alamouti, S. M. (1998). A simple transmit diversity technique for wireless communications, *IEEE Journal on Selected Areas in Communications*, vol. 16, no. 8, pp. 1451– 1458, Oct. 1998.

Chase, D. (1985). Code combining- a maximum-likelihood decoding approach for combining an arbitrary number of noisy packets, *IEEE Transaction Communications*, vol. 33, pp. 385-393, 1985.

Chen, C.M.; Hsu, J.Y.; Kuo, P.H.& Ting, P.A. (2009). MIMO Hybrid-ARQ Utilizing Lower Rate Retransmission over Mobile WiMAX System, IEEE Mobile WiMAX, pp. 129 - 134 , July 2009.

Damen, M. O.; Gamal, H. El & Caire G. (2003). On maximum-likelihood detection and the search for the closest lattice point, *IEEE Transactions on Information Theory*, vol. 49, Issue 10, pp. 2389 - 2402, Oct. 2003.

Foschini, G. (1996). Layered space-time architecture for wireless communication in a fading environment when using multielement antennas, *Bell Labs Technical Journal*, vol. 1, no. 2, pp. 41–59, 1996.

Gao, Y.; Li, G.; Chen, K. & Wu, X. (2007) A Novel HARQ Scheme Utilizing the Iterative Soft-information Feedback in MIMO System *Vehicular Technology Conference*, pp. 423-424, Apr. 2007.

IEEE (2007). IEEE 802.16e-2005, IEEE Standard for Local and Metropolitan Area Networks, Part 16: Air interface for Fixed Broadband Wireless Access Systems, Oct. 2007.

ITU-R (2000). Recommendation ITU-R M.1225 Guidelines for Evaluation of Radio Transmission Technologies for IMT-2000

Jia, M. L.; Kuang J. M. & He, Z. W. (2006). Enhanced HARQ Employing LDPC Coded Constellation Rearrangement with 64QAM. *International Conference on Communication Technology*, pp. 1-4, Nov. 2006.

Kaiser, S. (2003). Space frequency block codes and code division multiplexing in OFDM systems, IEEE *Global Telecommunications Conference*, vol. 4, 1-5, pp. 2360-2364, Dec. 2003.

Lee, I. & Sundberg, C.-E.W. (2007). Reduced-Complexity Receiver Structures for Space–Time Bit-Interleaved Coded Modulation Systems, *IEEE Transactions on Communications*, vol. 55, Issue 1, pp. 142-150, Jan. 2007.

Lin, S.; Costello, Jr., D. J. & Miller, M. J. (1984). Automatic-Repeat Request Error-Control Schemes, *IEEE Communication Magazine*, vol. 22, pp5-17, Dec. 1984.

WiMAX (2007). WiMAX Forum Mobile Radio Conformance Tests (MRCT) Release 1.0 Approved Specification, Revision 2.0.0, Dec. 2007.

Part 3

Mobile WiMAX Techniques and Interconnection with Other Technologies

14

Inter-Domain Handover in WiMAX Networks Using Optimized Fast Mobile IPv6

Seyyed Masoud Seyyedoshohadaei,
Borhanuddin Mohd Ali and Sabira Khatun
Universiti Putra Malaysia (UPM),
Malaysia

1. Introduction

The most attractive feature of WiMAX is arguable the mobility capability that IEEE 802.16e (IEEE, 2004) standard adds to the previous standard. With mobility support, handover has become one of the most important factors that impact the performance of IEEE 802.16e system. Handover is the process of maintaining active sessions of a mobile station when it migrates from current base station to target base station area. Handover occurs when a mobile station changes its point of attachment on the network. However during hard handover, the mobile station cannot receive or send any packet for a short time interval. This is referred to system disruption time because the services are interrupted or handover latency. In WiMAX, when a mobile node or mobile station changes its location, it moves the point of attachment to the network in two different scenarios;

- The mobile station changes its point of attachment between the base stations which reside in the same Access Services Network (ASN) that is called ASN-anchored, intra, micro, or layer 2 handover. In an ASN-anchored handover, the mobile station resides within previous network address (both current and target base stations located in the same IP subnet). In this scenario, the mobile station does not change its IP configuration, only link layer is re-established.
- The mobile station or mobile node changes its point of attachment between the base stations which reside in different ASN (different IP subnets) that is called Connectivity Services Network (CSN)-anchored, inter, macro, or layer 3 handover. In a CSN-anchored handover, in addition to link layer handover a mobile node must perform a new IP configuration to avoid disconnection.

The intra-domain handover procedure requires support from the physical and MAC layers. IEEE 802.16e has its own MAC layer or layer 2 handover algorithm, but a layer 3 handover algorithm is also required to support the Internet Protocol (IP) addressing, for inter-domain handover. A typical protocol in network layer for mobile terminals is Mobile IP include Mobile IPv4 (MIPv4), (IEEE, 2002) and Mobile IPv6 (MIPv6), (IEEE, 2004) that have been standardized by the Internet Engineering Task Force (IETF). There are many problems associated with MIPv4, such as triangular routing, security and limitation of address space which were solved by using MIPv6. But there still remain some other problems, such as long service disruption time (handover latency), signalling overhead and packet loss.

However, MIPv6 does not solve the handover latency problem which is not negligible for real-time applications such as video streaming and Voice over IP (VoIP). Proxy Mobile IPv6 (PMIPv6), Hierarchical Mobile IPv6 (HMIPv6) and Fast Mobile IPv6 (FMIPv6) have been proposed to decrease long handover latency of MIPv6. The MIPv6 Signalling and Handoff Optimization (MIPSHOP) working group has standardized FMIPv6 (IETF, 2005). FMIPv6 is capable decreasing the handover latency and packet loss by mobility detection and creating new address for the target network and receives data through tunnelling in advance. Because of this, FMIPv6 is used as IP layer protocol in WiMAX. However, due to complexity of handover pattern, designing an impressive handover process to support all mobility scenarios with acceptable latency is still a challenge. There have been many proposals on how to effectively coordinate the FMIPv6 handover algorithm in layer 3 with handover algorithm of the IEEE 802.16e system in layer 2. To overcome some of the shortcomings in the proposed proposals an Optimized Fast Handover Scheme (OFHS) is proposed and presented in this chapter.

This chapter is organized as follows. In section 2, the MIPv6, FMIPv6, IEEE 802.16e handover and related works are described. The proposed scheme is explained in section 3. In section 4, a numerical model is developed to evaluate the performance of OFHS compared with that of RFC5270 (IEEE, 2008). T The results and discussion are presented in section 5, and finally, in section 6, conclusions of this chapter are made.

2. Background and previous works

In this section first, some literature that needed to explain proposed method such as mobile IP and the layer 2 handover procedures in IEEE 802.16.e or mobile WiMAX are described. Then some related works are introduced which have focused on how apply FMIPv6 over IEEE 802.16e to support inter-domain handover.

2.1 Background

When a host moved to other subnet, the IP address became incorrect for routing and if hosts used new IP address the connections would be terminated because the new IP address was unknown. Mobile IP mechanism works based on a temporary IP address named Care of Address (CoA). The MIPv4 and MIPv6 have introduced for difference IP addressing. In this work IPv6 has been used for addressing. Therefore, in following sections (2.1.1 and 2.1.2) MIPv6 and Fast MIPv6 are described.

The IEEE 802.16e standard supports mobile user in WiMAX network. It supports only intra-domain handover that movement of the mobile station with in same subnet does not affect the IP address. In section 2.1.3, layer 2 handover procedure that has been defined in IEEE 802.16e explined.

2.1.1 Mobile IPv6

The MIPv6 is a protocol to support inter-domain mobility (in network layer) for IPv6 based network. In MIPv6, the packets that are sent to the mobile node from the correspondent node are intercepted and forwarded by a home agent. The MIPv6 has same functions as MIPv4 that is adapted for MIPv6.

In MIPv6 also, each mobile node has two addresses, a static home address under its home network (HoA), and a care of address (CoA) as the mobile node roams to a foreign network for packet routing. The mobile node can create a CoA from a router advertisement message sent by the new visited network. When the mobile node moves to a foreign network, the mobile node sends Binding Update (BU) messages with its CoA to the home agent in order to update the home agent of its current point of attachment. In this way, mobile node's home agent can always detect coming communication packets to mobile node with home address of mobile node, and dispatching these packets to the mobile nodes' CoA via dynamically created IP tunnels. The signalling and data traffic are all transmitted via a unified IP framework, because, all the MIPv6 signalling messages are formed by extending IP protocols with option headers. However the MIPv6 causes a long latency problem. In order to improve handover performance of MIPv6, IETF introduced some IPv6 mobility protocol solutions such as HMIPv6 and FMIPv6.

2.1.2 Fast mobile IPv6

In MIPv6, the movement detection (based on Router Advertisement in IP-layer) and the address configuration procedures cause a long latency problem. FMIPv6 decreases delay of the movement detection and the address configuration phases of MIPv6. It enables the mobile node to provide the target base station identifier (BSID) and detects upcoming entrance to new subnet. It therefore reduces delay of movement detection. For new address configuration, in the FMIPv6 the mobile node obtains the new associated subnet prefix information in advance, while it is still connecting to the current subnet.

After the mobile node select one of the candidate base stations as target base station according to its policy, it sends the Router Solicitation for Proxy (RtSolPr) to the current access router or previous access router and receives Proxy Router Advertisement (PrRtAdv) messages in return. During exchanges of these messages the mobile node obtains the subnet prefix of the target base station. The current base station configures a new IP address (CoA) based on the subnet prefix of the target base station. After that, the mobile node sends a Fast Binding Update (FBU) message to the previous access router. The purpose of FBU messages is to inform the access router that there is a binding between the current CoA at the current subnet and the new CoA (NCoA) at the target subnet. Then, the Handover Initiation (HI) message is sent to the target or new access router by previous access router. The new access router performs duplicate address detection (DAD) to check validity of NCoA. After DAD procedure the new access router reply with handover acknowledge (HAck) message to the current access router. At this instant, a tunnel between the CoA of and NCoA of mobile node is established. The previous access router sends a fast-binding acknowledgement (FBAck) message to new access router. Fig. 1 illustrates the FMIPv6 procedure for Predictive and Reactive mode. If the mobile node receives the FBAck message in the current subnet before the layer 2 handover is started (there is enough time to exchange required messages to establish tunnel), handover occurs in the predictive mode. Otherwise, if the mobile node is forced to move to the new access router without receiving FBAck, FMIPv6 is in reactive mode.

In the predictive mode, the previous access router first store the tunnelled packets in a buffer. After the mobile node attaches to the new link, mobile node sends a Fast Neighbour

Advertisement (FNA) message to the new access router. Upon reception of an FNA message, the new access router delivers the buffered packets to the mobile station.. In reactive mode, mobile node receives packets from the new access router after the packets are rerouted from previous to new access router.

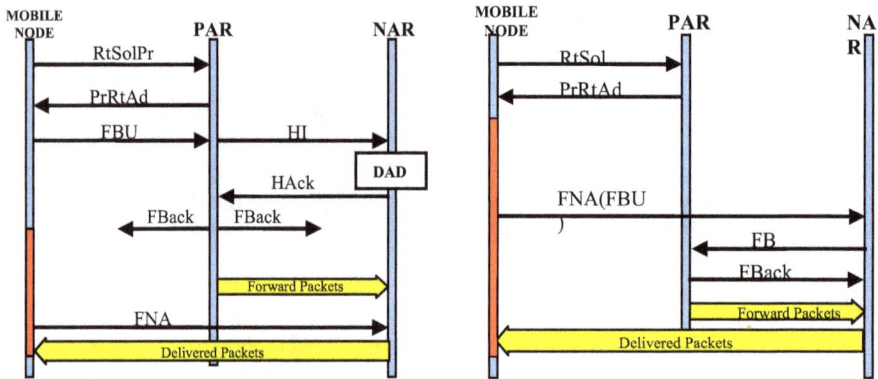

Fig. 1. FMIPv6 Procedure Predictive mode and Reactive Mode

2.1.3 IEEE 802.16e link layer handover

The IEEE 802.16e layer 2 handover procedure can be divided into two steps: handover preparation and handover execution. Fig. 2 illustrates the IEEE 802.16e handover procedure.

The handover preparation can be initiated by either mobile station or base station. During this period, the neighbouring base stations are compared according to its policy. Some metrics such as Quality of Service (QoS) parameters or signal strength are considered to target base station selection. The current base station periodically sends the neighbour advertisement (MOB_NBR-ADV) messages to mobile stations. This message contains information about neighbouring base stations, and the mobile station is capable to select target base stations for a future handover. In order to search for the suitability of neighbouring base stations, mobile station may execute a scanning operation (if necessary). It sends MOB_SCN-REQ to current base station to obtain neighbouring base stations information and the base station reply by MOB_SCN-RSP message. After a mobile station decides to perform handover, it sends a MOB_MSHO-REQ message contain candidate base station identity to the current base station. The current base station negotiates with candidate base stations with exchanges HO-pre-notification and HO-pre-notification-response messages. Then the current base station introduces the recommended base stations by sending an MOB_BSHO-RSP message to mobile station.

The handover execution is started by sending an MOB_HO-IND message from mobile station to the current base station. This message contains selected target base station, and after that packet exchanging between mobile station and current base station is terminate. After IEEE 802.16e network entry process, the mobile station tuned its own parameters to the target base station. The buffered packets are sent to the mobile station from the target base station (it now becomes current base station). If the new base station has a new IP address, a network layer handover mechanism is needed.

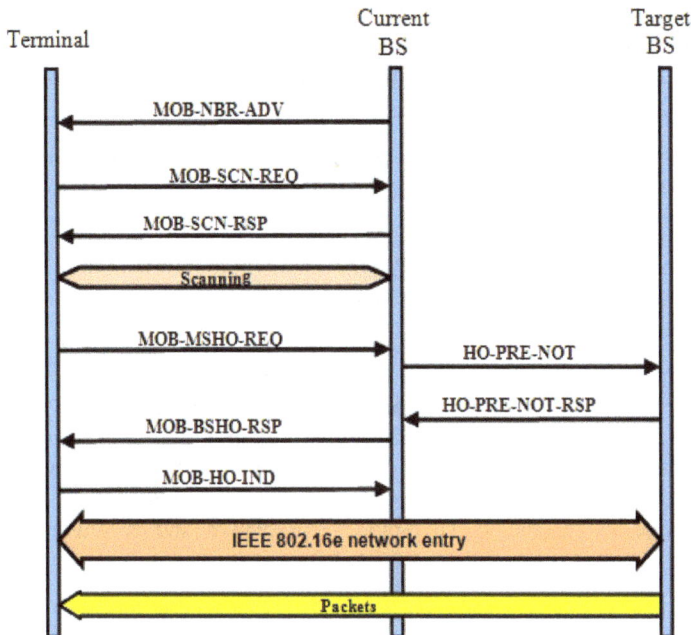

Fig. 2. IEEE 802.16e handover procedure

2.2 Related research works

The reduction of inter-domain handover latency in IEEE 802.16e handover process had been presented in several papers. A link layer optimized scheme that reduces the link-layer handover latency by analyzing and optimizing each step of the procedure is suggested in (Lee, D. et al., 2006). In principle, the overall handover latency does not decrease by simple reduction of the link layer latency. To solve this problem, a cross-layer fast handover scheme for the IEEE 802.16e system is proposed in (Han et al., 2007). It coordinates FMIPv6 with IEEE 802.16e handover procedure to reduce the handover latency. This scheme with a little change is used in RFC5270 (IETF, 2008).

2.2.1 RFC 5270

FH802.16e is a cross layering design for FMIPv6 handover over IEEE 802.16e. One-way signaling is used in the majority of the existing cross layering handovers researches. They usually defined cross layer signals from MAC layer to IP layer. In the Han et al. scheme, two-way signaling between MAC layer and IP layer is defined. This concept helps to achieve faster handover algorithm than previous algorithms. For efficient handovers and reduce the handover latency the authors introduce one command and three events. Same events and command have been proposed in the IEEE 802.21 Media Independent Handover (MIH) (IETF, 2007). They support the interaction between both IP and MAC layers handover procedures. The event are defined as follows:

NEW_CANDIDATE_BS_FOUND: this includes the BSID(s) of candidate base station(s) and is sent by MAC layer to IP layer (FMIPv6) when a new base station(s) is found.

LINK_GOING_DOWN: This is sent by MAC layer to IP layer (FMIPv6) when a mobile node receives an MOB_BSHO-REQ or an MOB_MSHO-RSP message which includes the target BSID. Upon receiving this event, the IP layer of the mobile node performs the handover preparation by sending an FBU message to the current access router.

LINK_SWITCH: This is sent by IP layer (FMIPv6) to MAC layer when the IP layer of a mobile node receives an FBAck message. It caused the mobile node MAC layer start handover execution by sending an MOB_HO-IND message to the current base station.

LINK_UP: This is sent by MAC layer to IP layer (FMIPv6) to inform layer 3 that the network re-entry procedure of IEEE 802.16e is terminated. Upon receiving this event, the IP layer of mobile node sends an FNA message.

The scheme proposed in this article provides RFC5270 and the names of triggers change to: New Link Detected (NLD), Link Handover Impend (LHI), Link Switch (LSW), and Link Up (LUP). Fig. 3 and Fig. 4 show the message sequence diagram of the predictive and reactive FMIPv6 handover initiated by the RFC5270. The handover procedure of RFC5270 consists of two stages: handover preparation and handover execution. Just as FMIPv6 that supports all inter-domain handover scenarios, two modes (predictive and reactive) are defined in RFC5270.

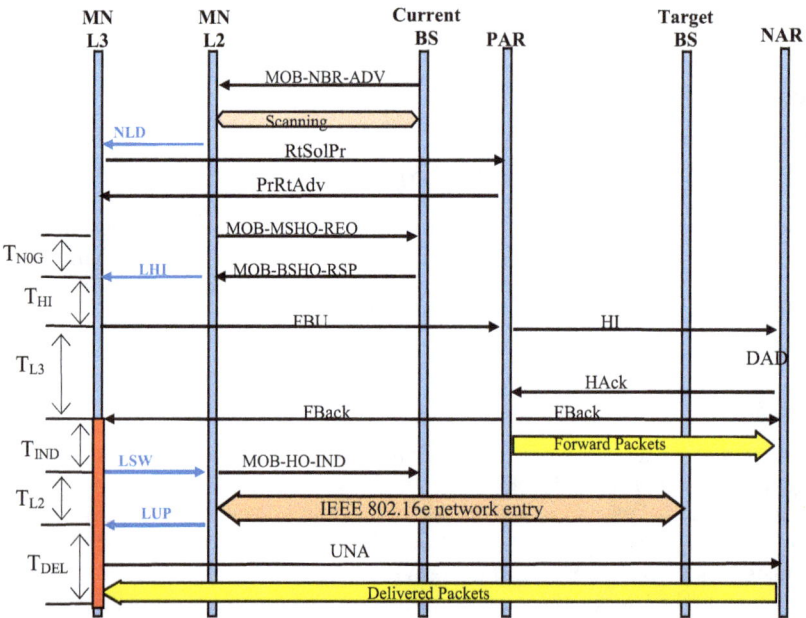

Fig. 3. FMIPv6 over IEEE 802.16e, Predictive Mode

Predictive Mode: Here, the current base station generates and broadcasts a Mobile Neighbor Advertisement (MOB_NBR-ADV) message periodically. It contains the network topology and static link layer information. When the mobile node discovers a new base station in this message, a scanning may be performed to acquire more dynamic parameters for the new base stations. If the newly found base stations are candidates for the target BSs, the NLD event is delivered to its IP layer from the mobile node MAC layer with the found BSIDs. The Router Solicitation Proxy (RtSolPr) message and Proxy Router Advertisement (PrRtAdv) messages are exchanged between the mobile node and previous access router. The terminal initiates handover by sending a Mobile Handover Request (MOB_MSHO-REQ) message to the current base station and receives a Mobile Handover Response (MOB_BSHO-RSP) message in reply with a target base station in it. The current base station may also initiate handover by sending a MOB_BSHO-REQ message to the mobile node.

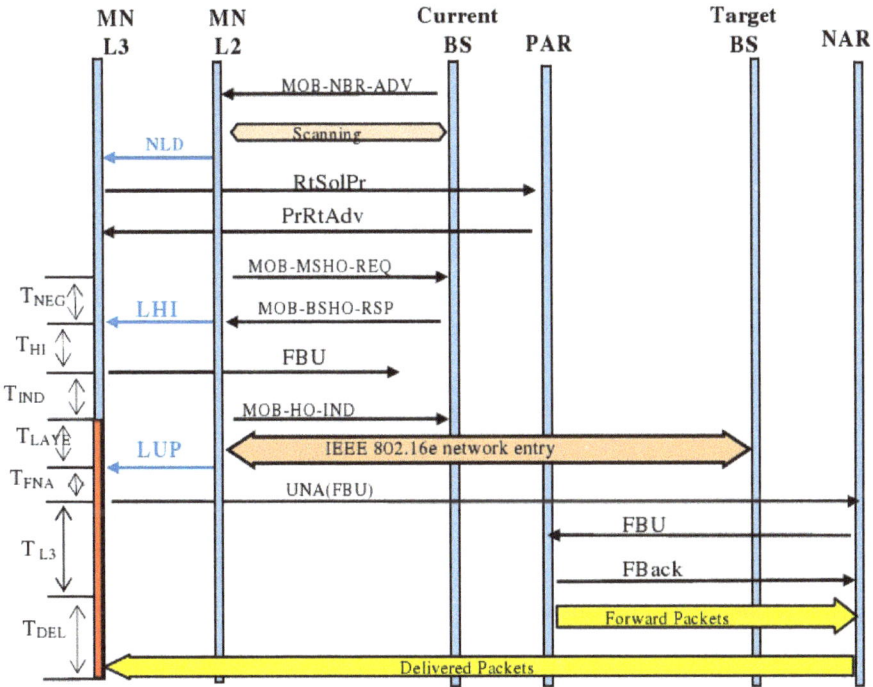

Fig. 4. FMIPv6 over IEEE 802.16e, Reactive Mode

After the mobile node receives MOB_BSHO-REQ or MOB_BSHO-RSP from the base station, the IP layer is triggered by link layer through a LHI to send Fast Binding Update (FBU) to the previous access router. The Handover Indication (HI) and Handover indication Acknowledge (Hack) messages are exchanged between previous and new access routers. The duplicate address detection is performed by new access router (it validates the uniqueness of NCoA in the new subnet, establishes tunnel and sends Fast Biding Acknowledge (FBack) message to the mobile station. Once the tunnel is established, the packets that are destined for the mobile node CoA are forwarded to the NCoA at the new access router through the tunnel. Upon receiving the FBack, the mobile node link layer is signalled by its network layer through a LSW to manage handover by sending a Mobile handover indication (MOB-HO-IND) message to the target base station. This message starts the 802.16e network re-entry process. After re-entry process, the mobile node link layer triggers its network layer with a LUP to send Unsolicited Neighbor Advertisement (UNA) message to the new access router. When the new access router receives the UNA from the mobile node, it delivers the buffered packets to the mobile node.

Reactive Mode: If the mobile node sends the MOB-HO-IND message to the base station before receiving FBack, the mobile station carries out 802.16e network re-entry process without establishing tunnel with selected NAR. At this instant, the mobile node cannot perform predictive mode so it operates in reactive mode as follows. Upon the network entry procedure completion, the link layer of mobile node sends LUP signal to the IP layer. Then the IP layer identifies that it has moved to the target network without receiving the FBack in the previous link. The mobile node sends an UNA to the new access router by using NCoA as a source IP address and sends an FBU to the previous access router. When the new access router receives the UNA and the FBU from the mobile node, it sends the FBack to the previous access router, and the packets that have been forwarded from the previous access router to new access router are delivered to the mobile node (through NCoA) through the new access router.

2.2.2 Cross Layer Handover Scheme (CLHS)

(Chen & Hsieh, 2007) suggested an integrated design of layer 2 and layer 3 called Cross Layer Handover Scheme (CLHS). The main idea of the CLHS is that if the handover procedures of layer 2 and layer 3 can be coincident, the overall overhead of handover will be decreased. In the CLHS, the correlated messages of IEEE 802.16e and FMIPv6 were integrated. The authors show that some FMIPv6 handover information can be exchanged with the messages of IEEE 802.16e. The messages which have the same characteristics during handover procedure are merged. They are described as follows:

FBU-MOB_HO-IND: The original MOB_HO_IND message are modified to include FBU as a new message. There are 6 reserve bits in the MOB_HO_IND message of link layer. One bit of them is used to indicate that the FBU is enabled or disabled. Upon receiving the FBU-MOB_HO-IND message containing FBU bit, the current base station itself (instead of mobile node) sends FBU message to previous access router.

FNA_RNG_REQ: The RNG_REQ message of 802.16e contains 8 reserved bits. They are used to send the information of FNA message of FMIPv6 in reactive mode.

In addition to the two messages, the neighbour advertisement message of layer 3 and the ranging request message of layer 2 were modified and merged. The MOB_NBR_ADV message in IEEE 802.16e and the PrRtAdv message in FMIPv6 have similar functionality. Hence, the CLHS merges these two messages together. The FBack massage of IP layer is combined with the Fast Ranging IE of link layer. Fig. 5 shows message sequence of the CLHS.

2.2.3 Integrated fast handover in IEEE 802.16e (IFH802.16e)

The IFH802.16e proposes a handover scheme for FMIPv6 over the IEEE 802.16e system by integrating FMIPv6 with IEEE 802.16e system. The IFH802.16e used same preparation concept as the previous works. In the IFH802.16e, the previous access router is informed by base station to imitate IP layer handover on behalf of the mobile node.

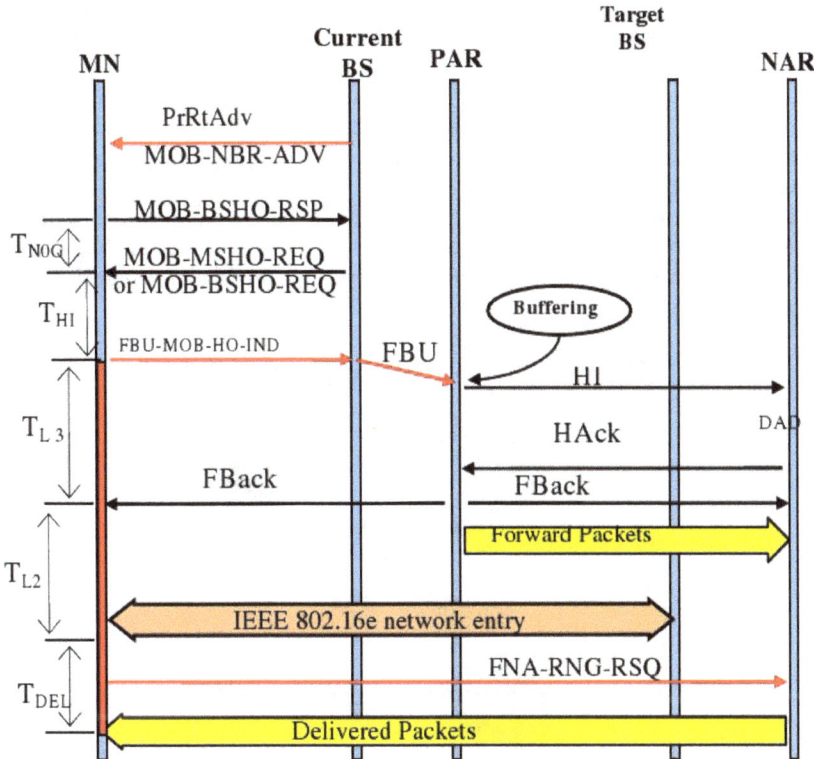

Fig. 5. CLHS procedure

3. Optimized Fast Handover Scheme (OFHS)

In this scheme, pre-established tunnelling mechanism to reduce handover preparation time is used. In addition, a set of messages has been defined to interleave layer 2 and layer 3 procedure. Cross layer design and cross function optimization are used to improve handover performance. The network model is as shown in Fig. 6.

In the OFHS, the serving base station periodically generates and sends the MOB_NBR-ADV message to mobile stations. The MOB_NBR-ADV message of IEEE 802.16e and the PrRtAdv message of FMIPv6 have similar functionality. The information of both messages can be sent through the MOB_NBR-ADV message. Hence, these messages are merged and the PrSolPr message can be eliminated. The mobile station may also perform scanning to obtain link characteristics to evaluate whether to perform handover or otherwise. After the scanning procedure, mobile station selects target base stations among the candidate base stations, based on signal strength, QoS, service price and etc.

If handover is needed, the mobile station sends the MOB_MSHO-REQ message to the possible target base stations that are listed. Then the current base station negotiates with the candidate base stations, and sends the recommended base stations and to mobile station through the MOB_MSHO-RSP message. At the same time the current base station sends the handover notification (HO-NOTIF) message to previous access router. The HO-NOTIF message let the previous access router to start the layer 3 handover. It contains the identities of the recommended base stations and the MAC address of the mobile station. After receiving this message, the previous access router initiates the FMIPv6 handover by sending the handover initiate (HI) message to the next access router associated with target base station. The HI message should contain the NCoA of the mobile station when the stateless address auto-configuration (Thomson et al, 2007) is used. In the OFHS, the NCoA is configured by using the MAC address of the mobile node and the network prefix of new access router. It is performed by previous access router on behalf of the mobile station. The previous access router already knows the network prefix of new access router through some auxiliary protocols (Kwon et al., 2005; Liebsch et al., 2005).

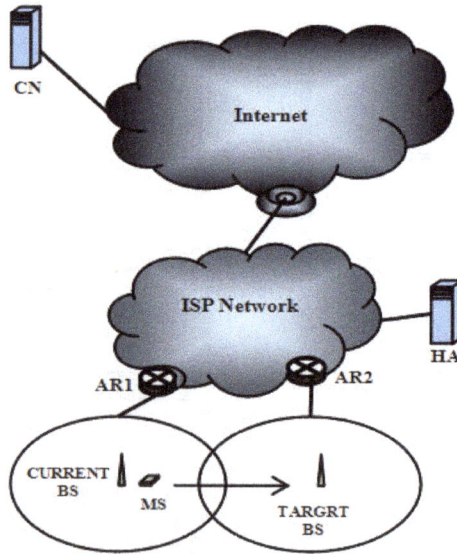

Fig. 6. Network Model

The previous access router exchanges HI and handover acknowledge (HAck) messages with new access router. During this process, a tunnel between the previous and new access routers is set up and the validity of the NCoA is checked with duplicate address detection (DAD). The established tunnel may be more than one based on the recommended base stations. The tunnels are inactive and one of them will be activated only when previous access router receives the handover confirmation (HO-CONFRIM) message that includes the target base station. Once the tunnels are established, previous access router sends an FBack to the mobile station. FBack is applied to inform the status of the configuration of CoA. FBack are sent by the target base station so that the mobile station can be informed that the next CoA is valid. The mobile station can send a MOB-HO-IND message to the target base station according to the policy and then carry out IEEE 802.16e network re-entry process. If the FBack is received by the mobile station before sending MOB-HO-IND message, handover continues in predictive mode. The MOB-HO-IND message contains selected target base station and the MAC address of the mobile station. The current base station notifies the new access router of the target base station by sending the HO-CONFIRM message. The previous access router obtains the exact target base station and related access router by receiving the HO-CONFIRM message. The previous access router starts forwarding the packets destined to the mobile station through one of the tunnels while the other tunnels that are not selected are discarded.

The new access router buffers the packets during the network re-entry procedures. In this scheme layer 3 handover is initiated at the network side while the mobile station performs the layer 2 handover. Because the mobile station is not involved in formulating the NCoA, it should be informed of NCoA. This can be realized by sending the HO-COMPLETE message from target base station to the new access router after the network re-entry procedures of IEEE 802.16e. The target base station sends the REG-RSP message to mobile station and finalizes the network re-entry procedures of IEEE 802.16e and sends HO-COMPLETE message to confirm the layer 3 handover of mobile station. Upon HO-CONFIRM message received by the next access router, it starts delivering the buffered packets to the mobile station. The HO-COMPLETE message is necessary because after the mobile station performed layer 2 handover the NCoA should be notified to the mobile node. The new access router must send the Unsolicited Router Advertisement with Neighbor Advertisement Acknowledgement option to the mobile node. Fig. 7 shows OFHS predictive mode.

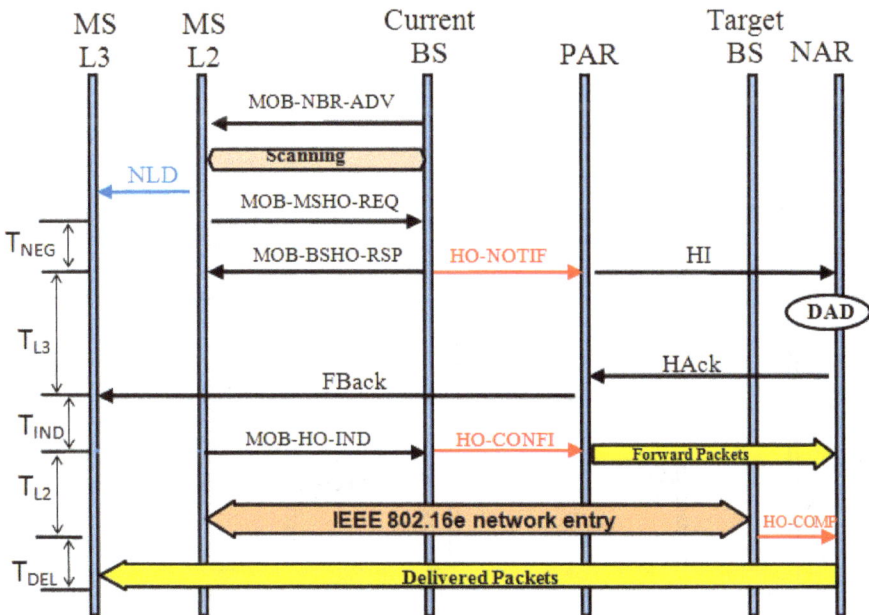

Fig. 7. OFHS Handover Procedure, Predictive Mode

If the mobile node sends MOB-HO-IND message to the current base station before receiving FBack (before establishing tunnel with selected access router), the mobile station starts IEEE 802.16e network re-entry process and the current base station sends HO-CONFIRM message to the previous access router. The previous access router stops sending packets to the mobile node and starts to buffer the packets destined for the mobile station. During the network re-entry procedures of IEEE 802.16e or after that, the previous access router receives the HAck message. There are two scenarios; first, if the previous access router receives HAck messages from the new access router before the end of network re-entry procedures of IEEE 802.16e, the previous access router starts to tunnel the packets destined for the current CoA to the new CoA at the new access router. Then the new access router starts delivering the packets to the mobile station. The previous access router already knows the exact target mobile station and its associated access router, therefore, the previous access router can determine through which tunnel it should start forwarding the packets destined to the mobile station while the other tunnels that are not used will be discarded. The second scenario is that, if the network re-entry procedure of IEEE 802.16e is terminated and the tunnel with selected new access router has not been established yet, the previous access router waits to receive HAck message from the new access router. Upon receiving the HAck message, the previous access router starts to tunnel the packets (destined for the current CoA) to the NCoA at the new access router. Then the new access router starts delivering the tunnelled packets to the mobile node. These two scenarios are called semi-predictive mode defined in OFHS instead of reactive mode defined in the RFC5270. The semi-predictive mode procedure is shown in Fig. 8.

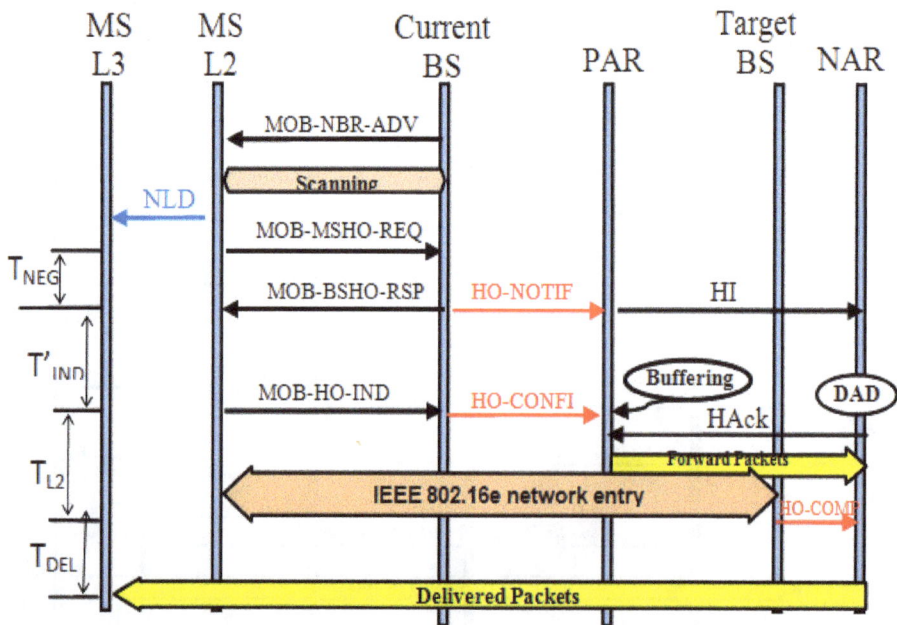

Fig. 8. OFHS Handover Procedure, Semi-Predictive Mode

4. Performance evaluation

In order to evaluate the performance of the proposed method, a numerical model has been developed. In this chapter, the important metrics for evaluating the handover mechanism are total handover procedure time and handover latency respectively. In the evaluation, the OFHS is compared with the RFC5270 as the reference procedure for using FMIPv6 in WiMAX.

To analyze the performance model of the proposed scheme, the duration of each part of the handover procedure are considered. The message interaction is based on the duration of a frame which is an OFDMA type used by IEEE 802.16e air interface. The frame duration is assumed to be at least 1ms and processing time is ignored since it is less than the frame duration. On the other hand, the network nodes message transmission delay is at least a frame long (>1ms). The radio propagation delay is assumed to be smaller than the frame duration, so it is omitted.

4.1 Total handover procedure time

The total handover procedure time (T_{THT}) is defined as the elapsed time between a mobile node sending the MOB_MSHO-REQ message to the current base station and the time the mobile station can receive the first packet through the target access router. $T_{TH-PM-RFC}$ and $T_{TH-PM-OFHS}$ are defined as the total handover time of the predictive mode in RFC5270 and OFHS, respectively. The Equations are defined in term of delay of every routing hop in a wired backbone (T_{HOP}) and frame duration of IEEE 802.16e (T_F). Negotiation between the current base station and the target base station is started by sending MOB-MSHO-REQ. Then the current base station sends handover notification message to target base station and receives handover notification response from it. The procedure is concluded by sending MOB-BS-HO-RSP to current base station. The time lag from the point of sending MOB-MSHO-REQ to receiving MOB-BSHO-RSP or negotiation delay between the current and recommended base stations (T_{NEG}) is given by Equation (1). The time required to perform FMIPv6 in layer 3 from the point of sending FBU to receiving FBack is T_{L3} and the latency of IEEE 802.16e network re-entry procedure is given by T_{L2}. They are expressed in Equations (2) and (3), respectively. $N_{PAR-NAR}$ is the distance between the previous and new access routers in term of number of hops and T_{DAD} is time needed to complete a duplicate address detection procedure. The MAC layer handover time is based on the number of messages exchanged between mobile station and base stations according to the RFC5270. Packet delivery time (T_{DEL}) is the time required from the point of sending the UNA message after IEEE 802.16e handover to receiving the first packet from new access router; this is given by Equation (4).

$$T_{NEG}=4T_{HOP}+2N_{PAR-NAR}\times T_{HOP} \tag{1}$$

$$T_{L3-RFC}=3T_F + 2T_{HOP}+2N_{PAR-NAR} \times T_{HOP} + T_{DAD} \tag{2}$$

$$T_{L2}= 10T_F + 30 \text{ (ms)} \tag{3}$$

$$T_{DEL-RFC}= 3T_F + 2T_{HOP} \tag{4}$$

The elapse time between receiving MOB-BSHO-RSP and starting layer 3 handover is given by T_{HI} (For RFC5270 procedure $T_{HI} = 2T_F$). T_{IND} is elapse time between receiving FBack and sending MOB-HO-IND. To simplify analysis, fixed delay time for T_{IND} is assumed.

The message interaction is based on the duration of a frame, all times expressed as integer number of frame. Therefore, all non-integer times is rounded to the next nearest integer number (this is shown as $[\ ]_F$). In OFHS, T_{NEG}, T_{L2} and T_{DEL} are the same as Equations (1), (2) and (3), and T_{L3} is obtained from Equations (5). Hence, the total handover time of the predictive mode in term of T_F for RFC5270 and OFHS are given by Equation (6) and (7), respectively.

$$T_{L3\text{-}OFHS} = T_F + 2\,T_{HOP} + 2\,N_{PAR\text{-}NAR} \times T_{HOP} + T_{DAD} \tag{5}$$

$$T_{TH\text{-}PM\text{-}RFC} = T_{NEG} + T_{HI} + T_{L3\text{-}RFC} + T_{IND} + T_{L2} + T_{DEL\text{-}RFC} \tag{6}$$

$$= [4T_{HOP} + 2\,N_{PAR\text{-}NAR} \times T_{HOP}]_F + 18T_F + [2T_{HOP}]_F +$$

$$+ [2T_{HOP} + 2N_{PAR\text{-}NAR} \times T_{HOP} + T_{DAD}]_F + T_{IND} + 30(ms)$$

$$T_{TH\text{-}PM\text{-}OFHS} = T_{NEG} + T_{L3\text{-}POR} + T_{IND} + T_{L2} + T_{DEL\text{-}POR} \tag{7}$$

$$= [4T_{HOP} + 2N_{PAR\text{-}NAR} \times T_{HOP}]_F + 12T_F +$$

$$+ [2T_{HOP} + 2N_{PAR\text{-}NAR} \times T_{HOP} + T_{DAD}]_F +$$

$$+ T_{IND} + 30(ms) + [T_{HOP}]_F$$

$T_{TH\text{-}RM\text{-}RFC}$ is the total handover time of the reactive mode of the RFC 5270 and $T_{TH\text{-}SPM\text{-}OFHS}$ as the total handover time of the semi-predictive mode of the OFHS given by Equations (8) and (9) respectively. In reactive mode, after sending FBU, the mobile node does not receive an FBAck from the current access router before the mobile node is forced to move to the target access router. The mobile station must wait for packet rerouting before it can receive any packets from the target access router. T_{FNA} is elapse time between layer 2 handover termination and FNA message, and the time required performing FMIPv6 L3 handover from sending *FBU* to mobile node receiving *FBack* is $T_{L3\text{-}RM}$. In reactive mode and semi-predictive mode T_{IND} has various values depending on location, direction and speed of mobile station. Also, T_{DEL} depends on the number of buffered packets and frame duration.

$$T_{TH\text{-}RM\text{-}RFC} = T_{NEG} + T_{HI} + T_{IND} + T_{L2} + T_{FNA} + T_{L3\text{-}RM} + T_{DEL\text{-}RFC} \tag{8}$$

$$= T_{NEG} + T_{HI} + T_{IND} + T_{L2} + T_{FNA} + T_{L3\text{-}RM} + T_{DEL\text{-}RFC}$$

$$= [2T_{HOP} + 2N_{PAR\text{-}NAR} \times T_{HOP}]_F + T_{IND} + 13T_F + 30(ms) +$$

$$+ [T_{HOP}]_F + [2N_{PAR\text{-}NAR} \times T_{HOP}]_F + [N_{PAR\text{-}NAR} \times T_{HOP}]_F + [T_{HOP}]_F$$

$$T_{TH\text{-}SPM\text{-}OFHS} = T_{NEG} + T_{IND} + T_{L2} + T_{DEL\text{-}PRO} \tag{9}$$

$$= [2T_{HOP} + 2N_{PAR\text{-}NAR} \times T_{HOP}]_F + T_{IND} +$$

$$+ 11T_F + 30(ms) + 2\,[T_{HOP}]_F$$

4.2 Handover latency

Handover latency (T_{HL}) is defined as the elapsed time between a mobile node receiving the last packet through its current access router and the first packet through the target access router. After the previous access router sends the FBAck message to the mobile node, it stops delivering packets to the CoA (sending packets to mobile node). At this time, the current access router re-routes the packets that destined to the CoA to the NCoA in the target access router. Hence, the actual period of handover latency in predictive mode begins when the mobile node receives an FBAck message. In reactive mode, the actual period of the handover latency begins by sending the MOB-HO-IND message. $T_{HL\text{-}PM\text{-}RFC}$ is defined as the handover latency of the predictive mode of the RFC 5270 and $T_{HL\text{-}PM\text{-}OFHS}$ as the handover latency of predictive mode of OFHS given by Equations (10) and (11), respectively.

$$T_{HL\text{-}PM\text{-}RFC} = T_{IND} + T_{L2} + T_{DEL\text{-}RFC} \tag{10}$$

$$= T_{IND} + 14T_F + 30(ms) + [2T_{HOP}]_F$$

$$T_{HL\text{-}PM\text{-}OFHS} = T_{L2} + T_{DEL} = 11T_F + 30(ms) + [T_{HOP}]_F \tag{11}$$

$T_{HL\text{-}RM\text{-}RFC}$ is the handover latency of the reactive mode of the RFC5270 and $T_{HL\text{-}SPM\text{-}OFHS}$ is the total handover latency of the semi-predictive mode of the OFHS given by Equations (12) and (13), respectively.

$$T_{HL\text{-}RM\text{-}RFC} = T_{L2} + T_{FNA} + T_{L3\text{-}RM} + T_{DEL} \tag{12}$$

$$= 11T_F + 30(ms) + [2T_{HOP} + 2 N_{PAR\text{-}NAR} \times T_{HOP}]_F$$

$$+ [N_{PAR\text{-}NAR} \times T_{HOP}]_F + [T_{HOP}]_F$$

$$T_{HL\text{-}SPM\text{-}OFHS} = T_{L2} + T_{DEL} + T'_{IND} \tag{13}$$

$$= 11T_F + 30(ms) + [T_{HOP}]_F + T'_{IND}$$

5. Results and discussion

The parameters of OFHS and RFC570 are compared in this section, based on the previous analysis. Handover parameters are given as in Table 1.

Parameter	Value
T_{HOP}	1 ms
$N_{PAR\text{-}NAR}$	2 hops
T_{DAD}	800 ms
$T_{HI} = T_{IND}$	T_F

Table 1. Network Parameters

Fig. 9 shows total handover time of the RFC5270 and OFHS in term of frame durations for predictive, semi-predictive and reactive modes according to Equations (6) to (9), respectively. Handover latency variation in term of frame duration for all modes of the

RFC5270 and the OFHS are depicted in Fig. 10. The numerical values are obtained from Equations (10) to (13). Fig. 8 and Fig. 9 show that, the delay increases with the frame duration increases. The reason is that the base station replies the received message at the next frame because the current frame resource utilization is scheduled in advance. Additionally the response time is lengthened as the frame duration increases. The OFHS shows better total handover time and handover latency than RFC5270.

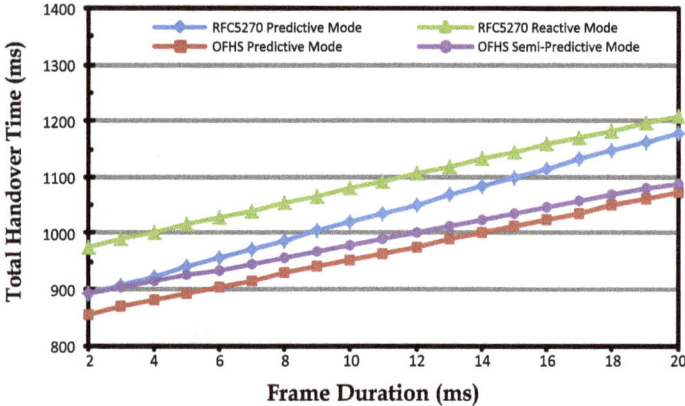

Fig. 9. Total Handover time versus Frame Duration

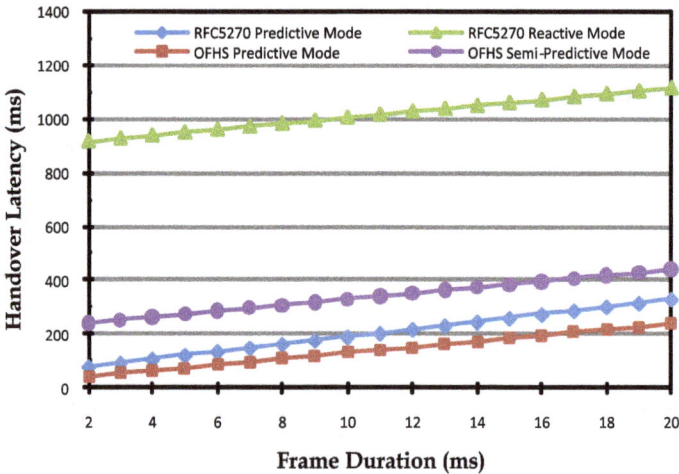

Fig. 10. Handover Latency versus Frame Duration

Usually in IEEE 802.16e, frame duration is considered as 5ms. In Fig. 11 total handover time and handover latency of RFC5270 and OFHS in reactive, predictive and semi-predictive modes for 5ms frame duration are illustrated. When frame duration is 5ms, OFHS decreases total handover time to 47ms for predictive mode, and 90ms for semi-predictive and reactive mode. The OFHS also reduces handover latency to 47ms for predictive mode, and 672ms for semi-predictive and reactive mode compare with RFC5270. The reason is that our scheme needs less number of messages than that of the RFC5270 when performing handover, and pre-established tunnel concept prepare a mechanism to reduce handover time. Also, the additional anticipation time imposed by FMIPv6 that causes the handover execution start earlier than planned is solved. In OFHS, occurrence probability of reactive mode is lower than that of the RFC5270, because earlier handover preparation provides sufficient time for the mobile node to receive FBack and drive predictive mode.

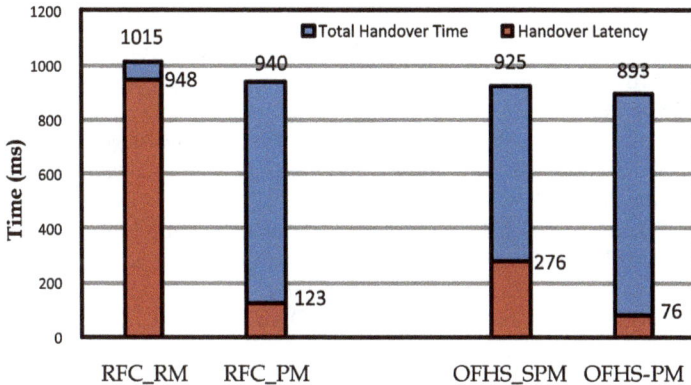

Fig. 11. Handover Latency for Ordinary Frame Duration (5ms)

6. Chapter summary

In this chapter an overview of inter-domain handover in WiMAX networks have been presented. The previous solutions for applying FMIPv6 on IEEE 802.16e have long latency that are not acceptable for real time services such as video streaming and voice over IP. In order to reduce handover latency, an optimized fast IPv6 handover scheme (OFHS) have been proposed. The OFHS combined cross layer design and cross function optimization to achieve lower handover latency. A pre-established multi tunnelling concept and a buffered routers mechanism have used to prepare seamless handover. The Layer 2 handover in 802.16e and layer 3 handover in FMIPv6 procedures are interleaved and the correlated messages for both layers are blended and reconstructed, effectively.

The results show that OFHS reduces handover latency and packet losses, and increase probability of predictive mode that has lower handover latency than reactive mode compared with RFC5270. The OFHS reduces handover latency by 38.2% in predictive mode.

7. References

Chen, Y. and Hsieh, F. (2007). A Cross Layer Design for Handoff in 802.16e Network with IPv6 Mobility, *IEEE Communications Society subject matter experts*, 2007
Han, Y.; Jang, H.; Choi, J.; Park, B. and McNair, J. (2007). A Cross-Layering Design forIPv6 Fast Handover Support in an IEEE 802.16e Wireless MAN, *IEEE Network*, Dec 2007.
IEEE 802.16e (2004). IEEE Standard for Local and Metropolitan Area Networks, part 16: Air Interface for Fixed and Mobile Broadband Wireless Access Systems, *IEEE*

IEEE 802.21 (2009). IEEE Standard for Local and metropolitan area networks- Part 21: Media Independent Handover, *IEEE*

Jang, H. J.; Jee, J. ; Han, Y.H. ; Park, S.D. and Cha., J. (2008). Mobile IPv6 Fast Handovers over IEEE 802.16e Networks, *RFC5270 of Internet Engineering Task Force, Network Working Group*, Mar. 2008.

Johnson, D.; Perkins. C. and Arkko J. (2004). Mobility Support in IPv6, *RFC 3775, Internet Engineering Task Force*

Koodli, R. (2005). Fast Handovers for Mobile IPv6, *RFC 4068, Internet Engineering Task Force, Network Working Group*, July2005

Kwon, D. H.; Kim, Y. S.; Bae, K. J. and Suh, Y. J. (2005). Access router information protocol with FMIPv6 for efficient handovers and their implementations, *Globecom*, pp. 3814-3819, 2005

Lee, D.H.; Kyamakya, K. and Umondi, J.P. (2006). *Fast handover algorithm for IEEE 802.16e Broadband Wireless Access System, In Wireless Pervasive Computing, 1st International symposium*, Jan. 2006

Lee, J.S.; Choi, S.Y. and Eom, Y.I. (2009). Fast Handover Scheme Using Temporary CoA in Mobile WiMAX Systems, *In 11th International Conference in Advanced Communication Technology, ICACT 2009*, pp. 1772–1776 (2009)

Liebsch, M.; Ed.; Singh, A.; Chaskar, H.; Funato, D. and Shim, E. (2005). Candidate Access Router Discovery (CARD), *RFC 4066, Internet Engineering Task Force*, 2005.

Park, J.; Kwon, D.; Suh, Y. (2006). An Integrated Handover Scheme for Fast Mobile IPv6 over IEEE 802.16e Systems, 2006

Perkins, C. (2002). IP Mobility Support for IPv4, *RFC 3344, Network Working Group of Internet Engineering Task Force*, July 2005.

Seyyedoshohadaei, S.M.; Khatun S.; Mohd Ali, B.; Othman, M. and Anwar, F. (2009). An Integrated Scheme to Improve Performance of Fast Mobile IPv6 Handover in IEEE 802.16e Network, *proceeding of MICC 2009 Malaysian International Conference on Communication*, Kuala Lumpur, Malaysia, Des 2009

Seyyedoshohadaei, S.M.; Mohd Ali, B.; Othman, M. and Khatun S. (2011). Network Mobility in IEEE 802.16e Network using Fast Mobile IPv6, *proceeding of ASME International Conference on Communication and Broadband Networking ICCBN 2011*, pp.747-752, Kuala Lumpur, Malaysia, Jun 2011

Thomson, S.; Narten, T. ; and Jinmei, T. (2007). IPv6 Stateless Address Auto configuration, *RFC 4862, Network Working Group of Internet Engineering Task Force*, Sep 2007

15

Interaction and Interconnection Between 802.16e & 802.11s

Tarek Bchini and Mina Ouabiba
ARTIMIA, Malakoff
France

1. Introduction

1.1 Problematic

With the rapid evolution of wireless and mobile networks, and the emergence of several standards that use different technologies, the problem of compatibility between these technologies, or the transition from one technology to the other by the mobile station without interruption of services, becomes a real challenge to face, to ensure a good Quality of Service (QoS) for the client.

In this context, we will analyze Mobile Stream Control Transport Protocol (MSCTP) and IEEE 802.21 technology as two vertical handover mechanisms between two of mobile networks: IEEE 802.11s, and mobile WIMAX. The simulations will be run under Network Simulator 2 (NS2) [1].

1.2 Mobile Wimax (IEEE 802.16e)

The mobile WiMAX (IEEE 802.16e) [2] is a mobile extension of the IEEE 802.16 standard [3].

IEEE 802.16 defines the specifications for radio metropolitan networks or WMAN (Wireless Metropolitan Area Network), offering broadband to achieve a high flow rate and using techniques to cover large areas [3].

The IEEE 802.16e is suitable for any kind of traffic thanks to its flexibility justified by its three MAC layers [2] and its use of IP protocol.

There are two kinds of Handover in the mobile WiMAX: Intra-ASN Handover (layer 2: no change of IP address) and Inter-ASN Handover (layer 3: IP address change) [4] [5]. For Intra-ASN, two mechanisms have been specified: Hard handover for the low speed and Soft handover for the high speed; and for Inter-ASN Handover it defines: Mobile IPv4 or Client-MIPv4, and Proxy-MIPv4 [4].

The architecture of 802.16e is composed of mobile stations (MS), that communicate freely (radio link) with base stations (BS), which act as an intermediates gateways with the terrestrial infrastructure of IP network. The base stations themselves are connected to the network elements called ASN-GW (gateways) which manages their connection with the IP network [4] [5].

The NAP (Network Access Provider) is an entity that provides the infrastructure for radio access to one or more providers of network services. It can control one or more ASN (Access Service Network) which is composed of one or more BS and one or more gateways.

The NSP (Network Service provider) is an entity that provides IP connectivity and network services to subscribers compatible with the level of service it establishes with subscribers. An NSP may also establish roaming agreements with other providers of network services and contractual agreements with third-party providers of application (for example, ASP: Application Service Provider) to provide IP services to subscribers.

A NSP control one or more CSN (Connectivity Service Network) which is the core of the WiMAX network.

The architecture of mobile WiMAX is presented in the figure 1:

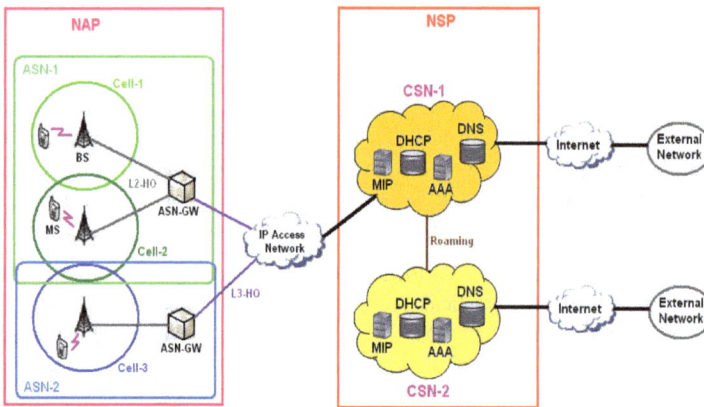

Fig. 1. IEEE 802.16e Architecture [4] [5]

1.3 IEEE 802.11s

IEEE 802.11s [6] [7] is an amendment being developed to the IEEE 802.11 WLAN standard, and aims to implement mobility on Ad-Hoc networks with acceptable debit.

In September 2003, IEEE formed the 802.11s SG which, in July 2004, became the "Extended Service Set (ESS) Mesh Networking" or 802.11s Task Group (TGs), and it is the most advanced group of the 802.11 WG.

The current objective of this TG is to apply mesh technology to WLANs by defining a Wireless Distribution System (WDS) used to build a wireless infrastructure with MAC-layer broadcast/multicast support in addition to the unicast transmissions. The TG should produce a protocol that specifies the installation, configuration, and operation of WLAN mesh. Moreover, the specification should include the extensions in topology formation to make the WLAN mesh self-configure and self-organized, and support for multi-channel, and multi-radio devices. At the MAC layer, a selection path protocol should be incorporated, instead of assigning the routing task to the network layer.

The WLAN Mesh architecture comprises the following IEEE 802.11 based elements:

- Mesh Point (MP) which supports (fully or partially) mesh relay functions, and implement operations such as channel selection, neighbor discovery, and forming and association with neighbors. Additionally, MPs communicate with their neighbors and forward traffic on behalf of other MPs.
- Mesh Access Point (MAP= MP+AP) which is a MP but acts as an AP as well. Therefore, MAPs can operate in a WLAM Mesh or as part of legacy IEEE 802.11 modes.
- Mesh Portal (MPP=MP+Bridge) is another kind of MP that allows the interconnection of multiple WLAN meshes to form a network of mesh networks. Moreover, MPP can function as bridges or gateways to connect to other wired or wireless networks in the DS.
- Simple Station (STA): outside of the WLAN Mesh, connected via Mesh AP.

The architecture of IEEE 802.11s is presented in the figure 2 below:

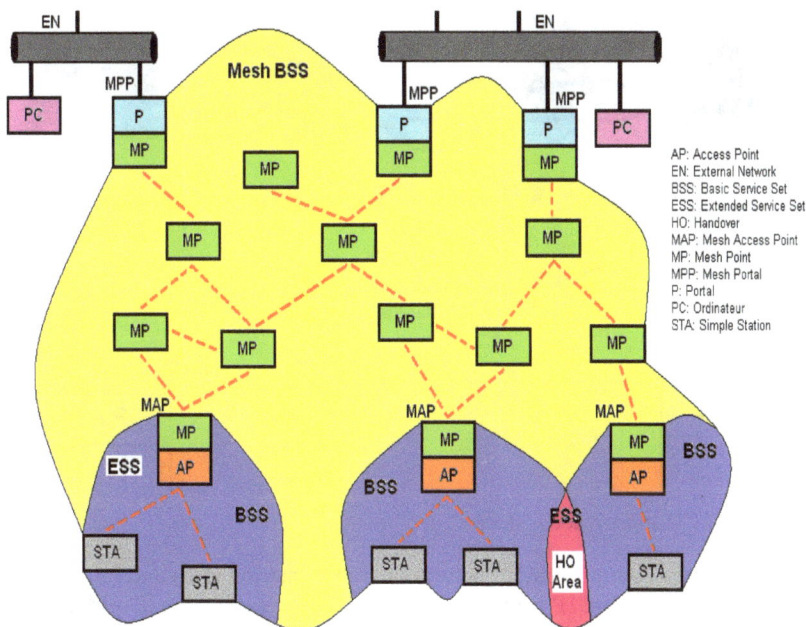

Fig. 2. IEEE 802.11s Architecture

2. Vertical handover mechanisms proposed

In our work, we analyze two vertical handover mechanisms that will be used between the mobile WiMAX and the IEEE 802.11s, and in this section we will present the two proposed mechanisms.

2.1 MSCTP protocol

The transport layer mobility is proposed as an alternative to the network layer mobility to support integrated mobility. The management of mobility in the transport layer is made exclusively by Stream Control Transmission Protocol (SCTP) [8] and its extension: Dynamic Address Reconfiguration (DAR) [9].

SCTP extended with DAR constitute Mobile SCTP (MSCTP) [10] [11] [12].

MSCTP was designed in order to avoid the connection disruptions observed with TCP or UDP during a change of IP address. It is a transport layer protocol similar to Transmission Control Protocol (TCP). It provides point-to-point communication oriented connection between applications running on different hosts. The major difference with TCP is the multi-homing; it allows by multi-homing to manage multiple IP addresses in terminal nodes by conserving the point-to-point connection intact (see figure 3).

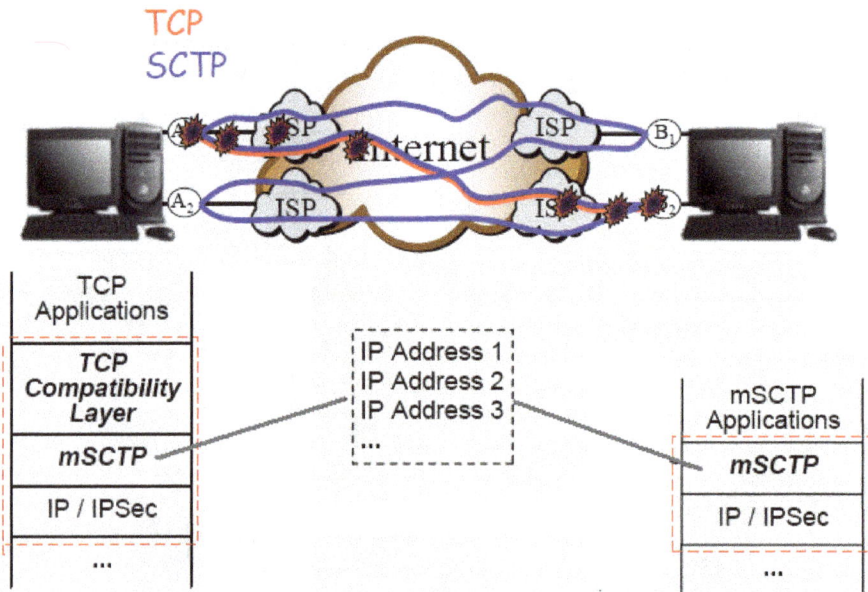

Fig. 3. MSCTP vs TCP and protocol stack

In the beginning of the communication between a mobile station (MS) and its correspondent (CN) implementing both MSCTP protocol; in the MS, there is only one IP address chosen as primary address, and used as destination address for the current transmission. The other IP addresses are used only for retransmissions. The DAR extension allows to MS to add, delete and change IP addresses during a SCTP session, without affecting the connection established, by using address configuration messages.

During the communication with the CN; when the MS changes from its home network to a foreign network passing by handover Area (at the beginning of the coverage area of foreign network), it receives an IP address from the foreign network either by contacting a DHCP, or by automatic configuration of IPv4 address. The MS is now able to establish other link with its CN through this second IP address obtained, and may become accessible via the foreign network. Then it sends its second IP address via the home network to its CN, and the CN will add the new IP address to the association identifying the connection with the MS and sends an ACK to the MS to confirm. After, when the MS begin to leave the coverage area of its home network to the coverage area of the foreign network, it notifies the CN to assign the new IP address as primary IP address, which the CN approves with an ACK.

The new primary IP address is now the second IP address obtained. The CN sends at this moment all the messages to the new IP address of MS via the foreign network. And finally, when the MS leave definitively the coverage area of the home network, it informs the CN to delete the first IP address of the association, which the CN confirms with an ACK [10] [11] [12]. To use MSCTP, the only requirement is that the both endpoints should implement MSCTP protocol.

By applying the MSCTP in the case of vertical handover between 802.16e and 802.11s, the protocol will use the multi-homing technique to open two IP sessions with the BS of the mobile WiMAX network and the AP of the Wireless mesh network to avoid the service interruption during handover.

2.2 IEEE 802.21 or MIH

IEEE 802.21 or Media Independent Handover (MIH) [13] [14] [15] is a recent evolution for all networks, that providing capabilities to detect and initiate handover from one network to another. It designed a new function to control access to the lower layers (Layers 1 and 2). This new function provides new service access points (SAPs) and allows the information to be queried by the upper layers (Layer 3 and higher). Both mobile device and network hardware must implement the standard to work, but everything should remain backward compatible for non-MIH aware devices.

The standard allows simply to provide information that help to the initiation of handover, the selection of the network and the activation of the interface. The execution and the decision of handover is not part of the standard.

In MIH Function (MIHF), there are three services that allow the passage of messages along the stack. The table 1 below compiled from IEEE 802.21, outlines the basic functions of these services [14].

MIH services	Origin	Destination	Use cases
Event	MIHF or lower layer	MIHF or upper layer Remote or local stack	Link up/down/going down, transmissions status
Command	MIHF or upper layer	MIHF or lower layer Remote or local stack	switch links, get status
Information	Upper or lower layer Secure or insecure port	Upper or lower layer Remote or local stack	information elements (IEs), neighbor reports

Table 1. MIH services

The MIH architecture is illustrated in the figure 4 below:

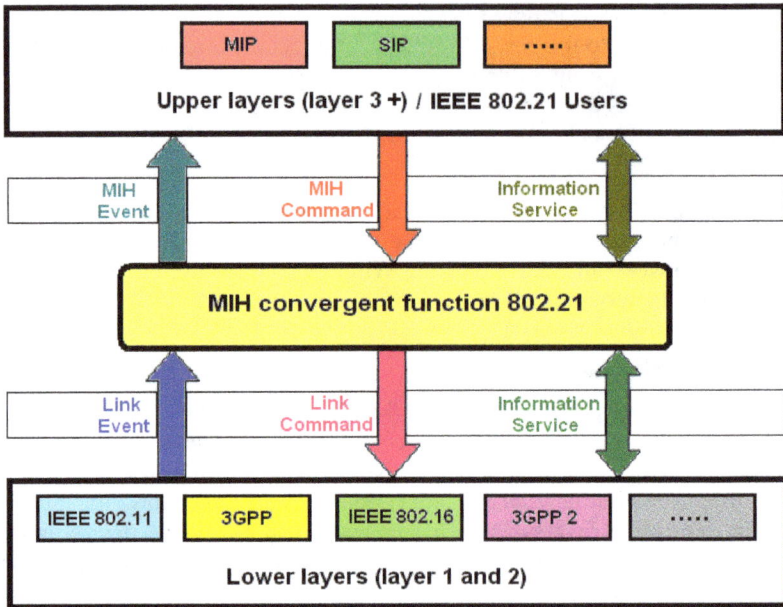

Fig. 4. MIH architecture

3. Interconnection models proposed

3.1 Common simulation model and common scenario

We describe in this section the common interconnection model proposed between IEEE 802.16e and IEEE 802.11s using MSCTP or IEEE 802.21 during the vertical handover; and the MS mobility scenarios between the two networks. We will assume an 802.16e cell with coverage of 1 km radius and an 802.11s cell with coverage of 300 m radius (these choices of radius values are based on the nature of the test environment that is an urban area which is not very dense). And, the two cells have a common area (handover area) with a variable and maximum distance of 180 m between both limits of cells (this choice of the surface of common area between cells is based on the time needed for the handover simulation).

The Base station (BS) of 802.16e network is linked to an ASN-GW that is linked too via IP network to a CSN (WiMAX ISP); the Access Point (AP) of 802.11s network is linked to a router that is linked too via IP network to a WIFI CSN; and the two CSNs are connected together and with the distant servers via Internet network.

We will evaluate the mechanisms through two mobility scenarios for the simulations: the case where the mobile move from 802.16e cell to 802.11s cell, and the opposite case. In the two scenarios, the mobile station (MS) traverses 200 m in 802.16e or 802.11s cell, and traverses in handover area (common area between 802.16e and 802.11s cells) 100 m.

We will propose three MS speeds to see the impact of speed increasing on the handover. So, we will propose: 5 m/s = 18 km/h; 10 m/s = 36 km/h; and 20 m/s = 72 km/h as mobile speeds for the simulations.

3.2 Simulation model based on MSCTP protocol

Based on common model proposed in the section before, with MSCTP protocol, the end users (the MS moving between the two cells and its correspondent node or server) must implement the MSCTP protocol.

The architecture of simulated model proposed for the interconnection between IEEE 802.16e and IEEE 802.11s using MSCTP is illustrated in the figure 5 below:

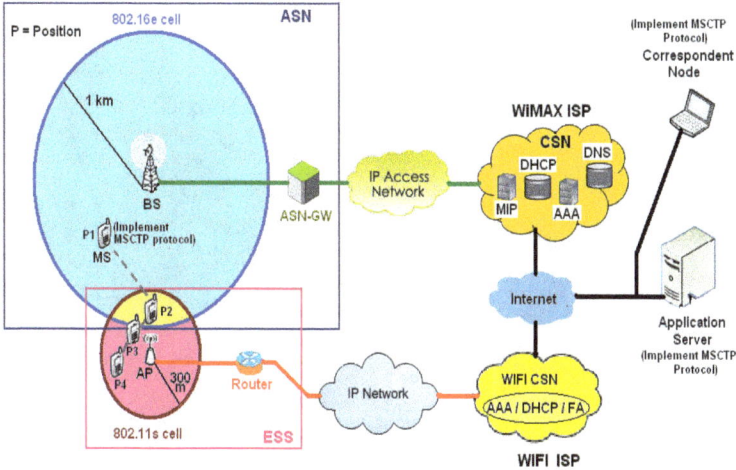

Fig. 5. Interconnection model using MSCTP

The exchange of information between the MS and its correspondent node (CN) during the MS mobility scenario between IEEE 802.16e and IEEE 802.11s using MSCTP is illustrated in the figure 6 below:

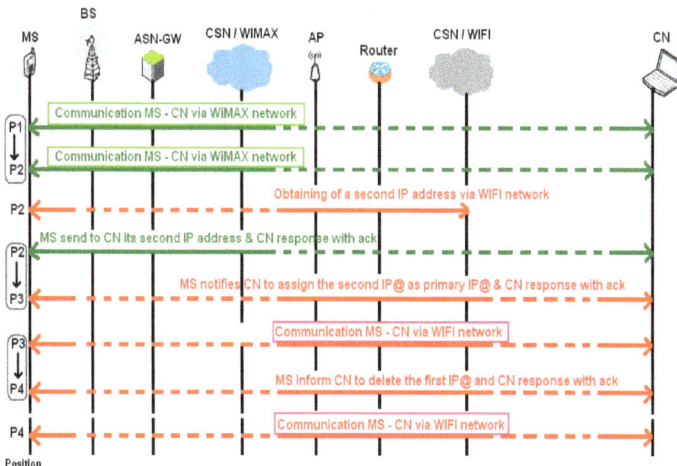

Fig. 6. Messages exchanged during simulation of the protocol MSCTP

3.3 Simulation model based on 802.21 architecture

Based on the common model already described; with IEEE 802.21 architecture, an MIH server must be implemented in the Internet network, more precisely between the WiMAX CSN, and the WIFI CSN; and MIH modules must be implemented in the two CSNs and in the MS.

The architecture of the interconnection model proposed between the two networks using IEEE 802.21 is illustrated in the figure 7 below:

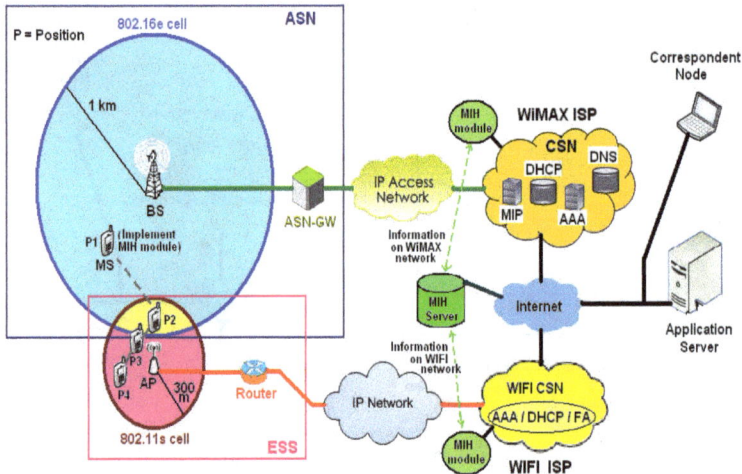

Fig. 7. Interconnection model using IEEE 802.21

The exchange of information between the MS and its correspondent node (CN) during the MS mobility scenario between IEEE 802.16e and IEEE 802.11s using the MIH module is illustrated in the figure 8 below:

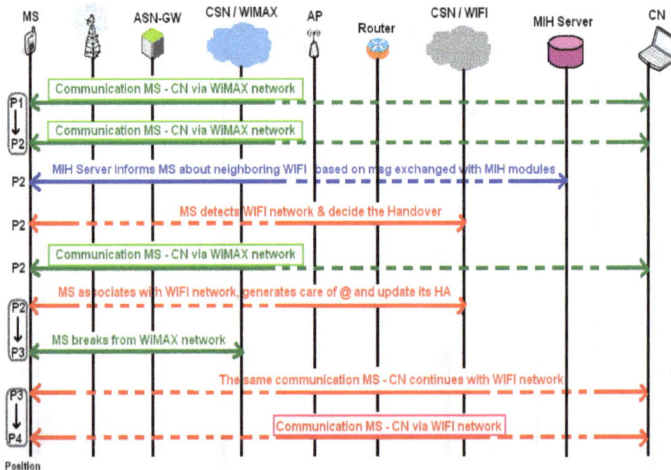

Fig. 8. Messages exchanged during simulation of the protocol MIH

For the simulations, we choose to use Proxy Mobile IP (PMIP) [16] as layer 3 protocol that interacts with lower layers via MIH module.

PMIP is an amelioration of MIP, it introduces a functional entity called Proxy Mobile IP to help MIP traversal across VPN or "NAT and VPN" gateways. The PMIP is in the path between MS and its corresponding HA (Home Agent), and acts as a surrogate MS and HA.

PMIP does not involve a change in the point of attachment address when the user moves, and there is no need for the terminal to implement a client MIP stack.

3.4 Simulation parameters under NS2 simulator

During the simulations and with the two handover mechanisms, we will test three traffics types: the VoIP with a fixed size to 160 byte and a rate of 300 packet/sec, the Data with a fixed size to 640 bytes and a rate of 200 packet/sec, and the video streaming with a fixed size to 1280 bytes and a rate of 100 packet/sec (optimal values usually chosen in NS2).

Under NS2, with MSCTP protocol we use the CBR over MSCTP traffic type; and with IEEE 802.21 architecture, we use the CBR over UDP traffic type.

The duration of one simulation are fixed to 250 seconds, and the results are calculate every 10 seconds in 802.16e or 802.11s cell area, and every 5 seconds during the handover process in common area.

For the two networks: IEEE 802.16s and IEEE 802.11s, the simulation parameters under NS2 are illustrated in the table 2 below:

	IEEE 802.16e	IEEE 802.11s
Transmission Power (Pt_)	15 W	0.2818 W
Receiving Threshold (RXThresh_)	7.59375e-11 W	1.76148e-10 W
Carrier Sending Threshold (CSThresh_)	4.34219e-12 W	3.32874e-11 W
Coverage Radius (Distance D)	1 km	300 m
Radio Propagation Model	Two-Ray Ground [17] $$P_r(d) = \frac{P_t G_t G_r h_t^2 h_r^2}{d^4 L}$$	
Transmit Antenna Gain (Gt_)	1 dB	
Receive Antenna Gain (Gr_)	1 dB	
System Loss (L_)	1 dB	
Transmit Antenna Height (ht_)	1.5 m	
Receive Antenna Height (hr_)	1.5 m	
Modulation	OFDMA	OFDM
Frequency (Freq_)	3.5 Ghz	2.4 Ghz

Table 2. Simulation parameters

3.5 Performance criteria

The performance criteria adopted in our simulations to compare MSCTP and IEEE 802.21 mechanisms in the case of vertical handover between 802.11s and 802.16e networks are: End-to-end delay, packets loss ratio and debit. These parameters are the main criteria of QoS measuring in the networks. To evaluate the QoS degree of these criteria, we will compare simulation results obtained with theoretical thresholds estimated to evaluate the QoS depending on traffic type.

4. Results

4.1 End-to-end delays

The end-to-end delay is a very important parameter to evaluate the QoS for the real time traffic. It is the time needed for a packet to be transmitted across a network from source to destination.

In this section, we will calculate the delays of packets during the simulation time for the three mobile speeds and the three traffic types; with the two vertical handover techniques: MIH architecture and MSCTP protocol; and applying the two scenarios: handover from 802.16e to 802.11s and from 802.11s to 802.16e.

We start by present the results for the VoIP traffic (see figures 9 and 10).

In the two figures 9 and 10, first for all the curves we see that during the handover process, the delays obtained with MSCTP are slightly lower than those obtained with the MIH; and the handover are executed with MIH before that with MSCTP.

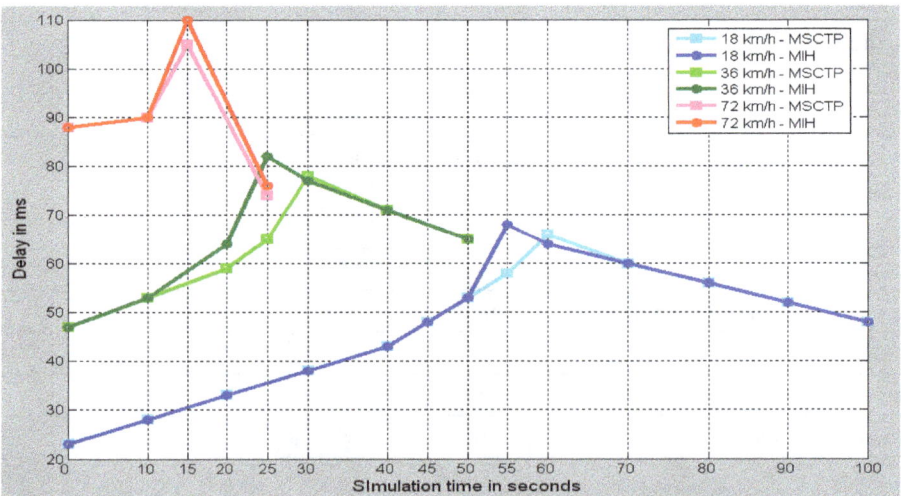

Fig. 9. Delay of HO from 802.11s to 802.16e / VoIP

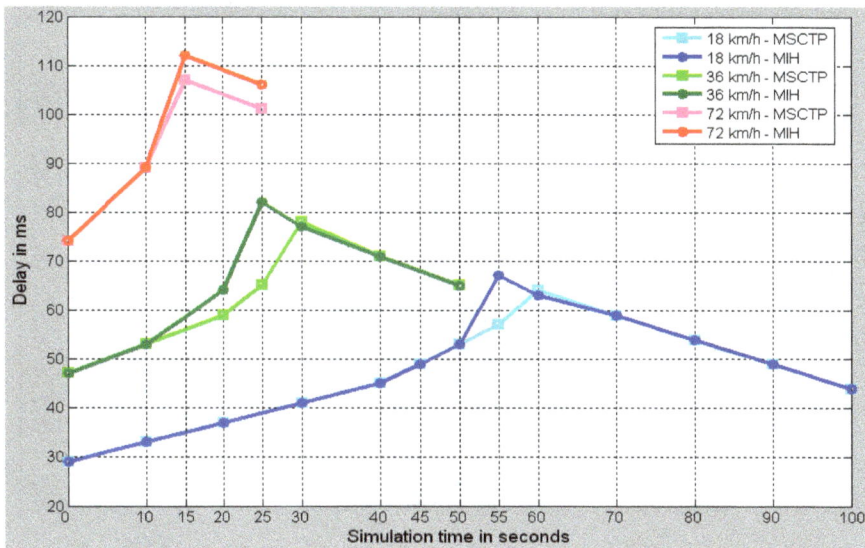

Fig. 10. Delay of HO from 802.16e to 802.11s / VoIP

With the two speeds: 18 and 36 km/h, the QoS level is accepted for the VoIP traffic that needed a minimum delay of 100 ms [18]; but with the speed equal to 72 km/h, the QoS degrades during the handover.

Finally, with a speed of 18 km/h, the QoS is better in 802.11s network; with 36 km/h the QoS is equivalent in the two cells; and with 72 km/h of speed the QoS is better in 802.16e cell.

We present now the results of data traffic (see figures 11 and 12):

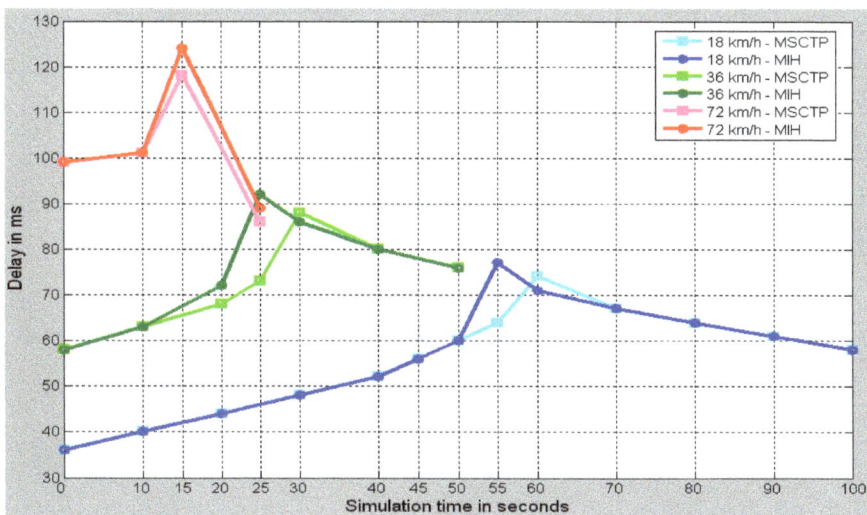

Fig. 11. Delay of HO from 802.11s to 802.16e / Data

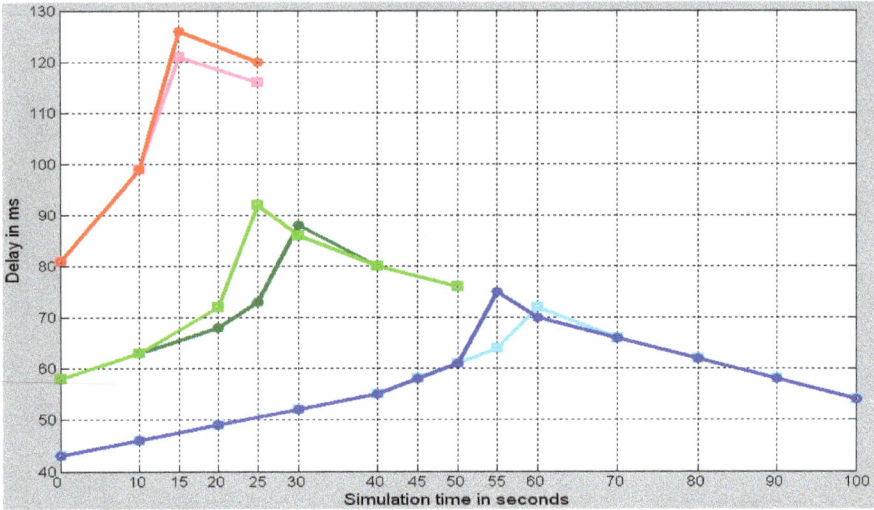

Fig. 12. Delay of HO from 802.16e to 802.11s / Data

With the data traffic, there is no delay constraint, so we can consider the QoS level acceptable for all the cases.

We note that the delays obtained with data are higher than those obtained with the VoIP traffic; for example with MSCTP protocol and with a speed of 18 km/h, the maximum delay obtained of the VoIP traffic is 65 ms versus 73 ms of the data traffic.

We present finally in this section the results of video streaming traffic (see figures 13 & 14):

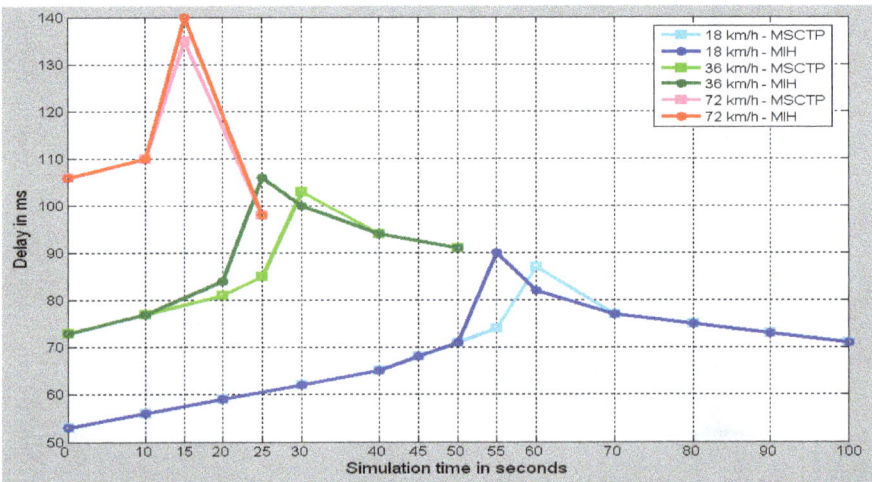

Fig. 13. Delay of HO from 802.11s to 802.16e / Video

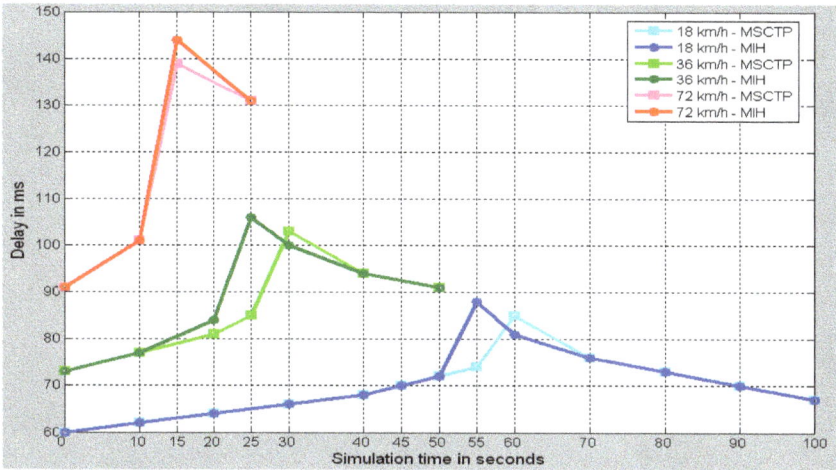

Fig. 14. Delay of HO from 802.16e to 802.11s / Video

With the video traffic, the delay values increase comparing by the VoIP or the data traffic.

With the speed of 18 km/h, the delays not exceed 100 ms; with the speed of 36 km/h the delays exceed slightly 100 ms during the handover; and with the speed of 72 km/h, the delay values exceed largely 100 ms during the handover and in the 802.11s cell.

The delays with MSCTP are slightly lower than those with MIH.

4.2 Packet loss ratio

We calculate in this section the percentage of lost packets with the same cases as those described in the section 4.1. We start with the VoIP traffic (see figures 15 and 16):

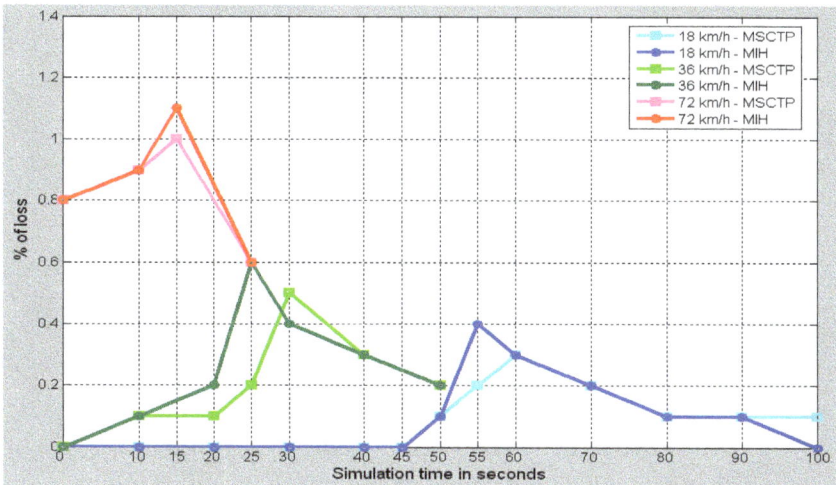

Fig. 15. Percentage of packets lost with HO from 802.11s to 802.16e / VoIP

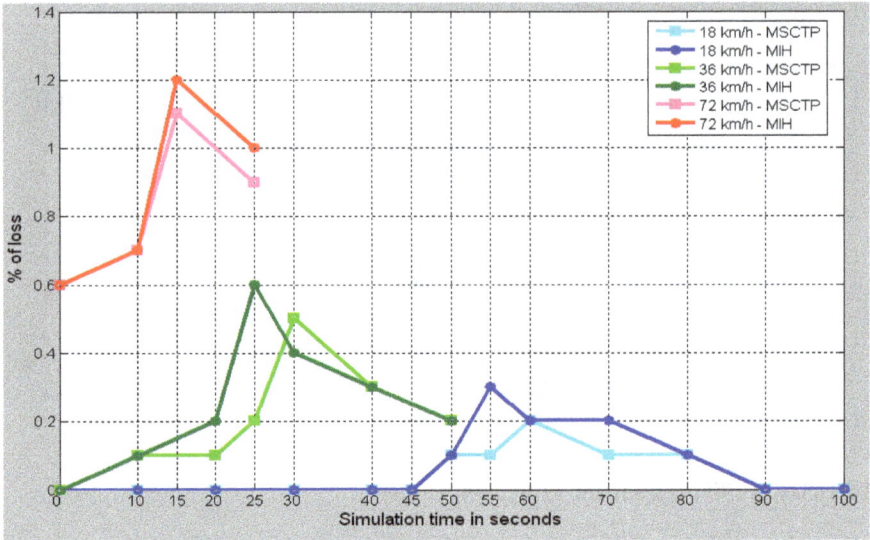

Fig. 16. Percentage of packets lost with HO from 802.16e to 802.11s / VoIP

During the Handover, the only results that converge to the threshold of 1% [18] required by the VoIP traffic] are those corresponding to the speed of 72 km/h. For the two other speeds the QoS level is acceptable.

We present now the results of data traffic (see figures 17 and 18):

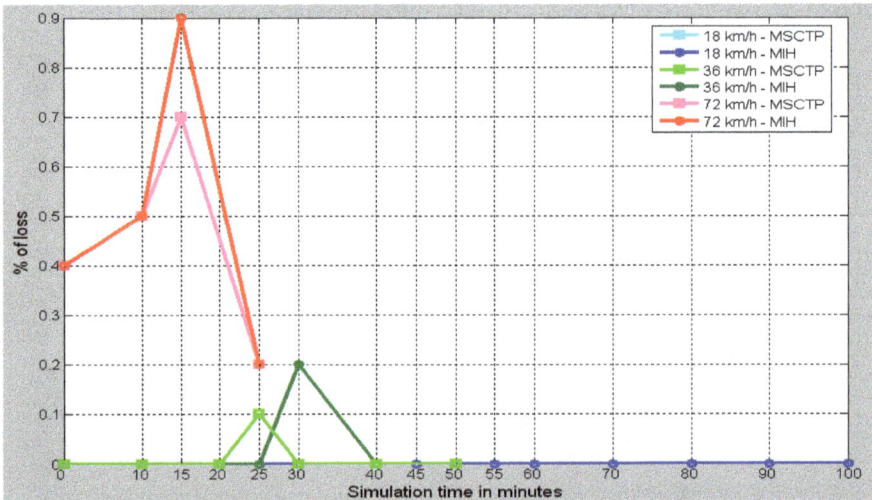

Fig. 17. Percentage of packets lost with HO from 802.11s to 802.11e / Data

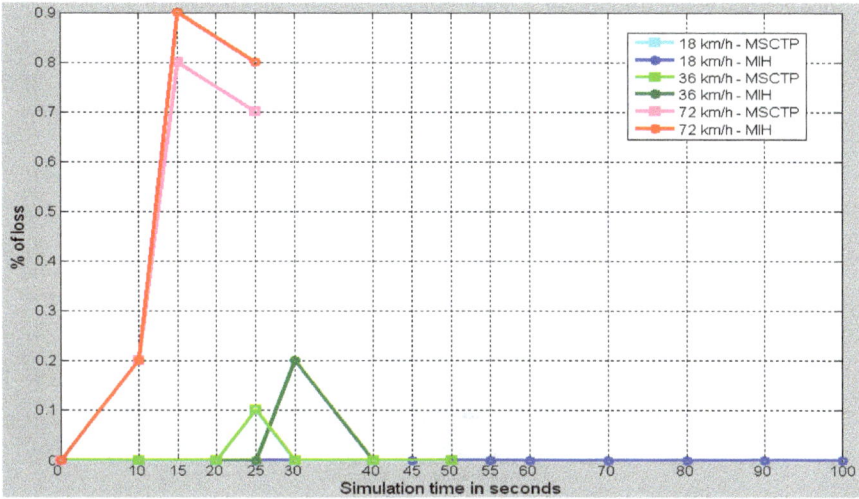

Fig. 18. Percentage of packets lost with HO from 802.16e to 802.11s / Data

With the data traffic, the percentage of loss is zero with a speed of 18 km/h, and it is acceptable with a speed of 36 km/h. It has a maximum of 0.1% with MSCTP and 0.2% with MIH during the Handover.

But with a speed of 72 km/h, the percentage of loss is not acceptable during the handover and in the 802.11s cell.

We pass now to video streaming traffic (see figures 19 and 20):

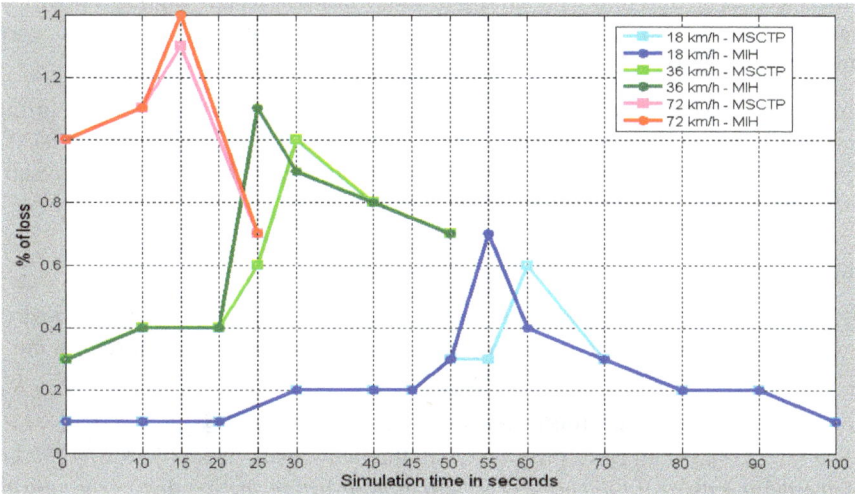

Fig. 19. Percentage of packets lost with HO from 802.11s to 802.16e / Video

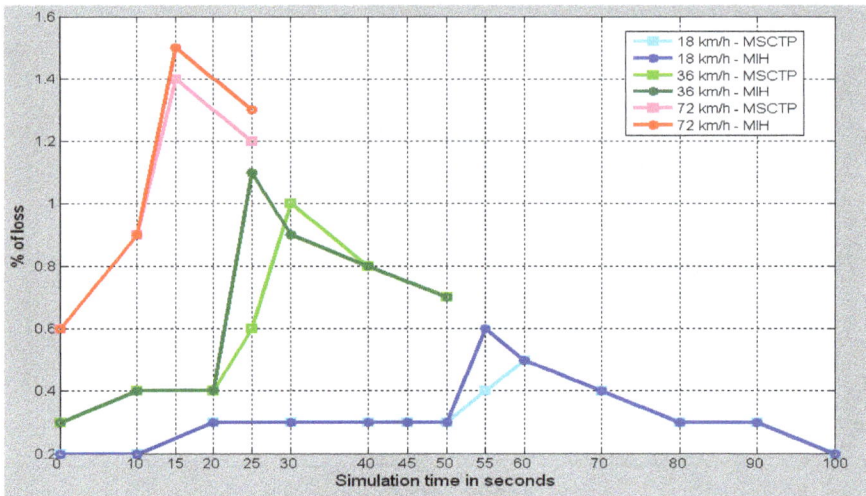

Fig. 20. Percentage of packets lost with HO from 802.16e to 802.11s / Video

With the video traffic, the loss values are higher than those with VoIP or data traffic, and with a speed of 18 km/h the results are acceptable because they not exceed 1% which is a maximum of loss required for video traffic [18].

We note that the case of handover from 802.16e to 802.11s produce results slightly better than those obtained with the opposite case during the handover.

With the speed of 36 km/h, the maximum of the % of loss exceeds slightly 1% with MIH, and is equal to 1% with MSCTP; and the results are equivalent in the two ways of handover.

With a speed of 72 km/h, the results are not acceptable with the two mechanisms during the handover because they exceed largely 1%, and the results of the handover from 802.11s to 802.16e are better than those of the opposite case. And finally with this speed, the results are acceptable in the 802.16e cell but not in the 802.11s cell.

4.3 Debit

Finally, in this section we will evaluate the debit experimented by the mobile during the handover cases already proposed for the simulations in section 4.1. We start with the VoIP traffic (see figures 21 and 22).

Concerning the debit of VoIP traffic and with a minimum required by this type of traffic fixed to 4 kb/s [18]; the all cases proposed with the two speeds: 18 and 36 km/h present an acceptable debit; but with the speed of 72 km/h and during the handover, the results are lower than 4 kb/s, and the debit obtained with MSCTP during the handover is slightly higher than that obtained with MIH.

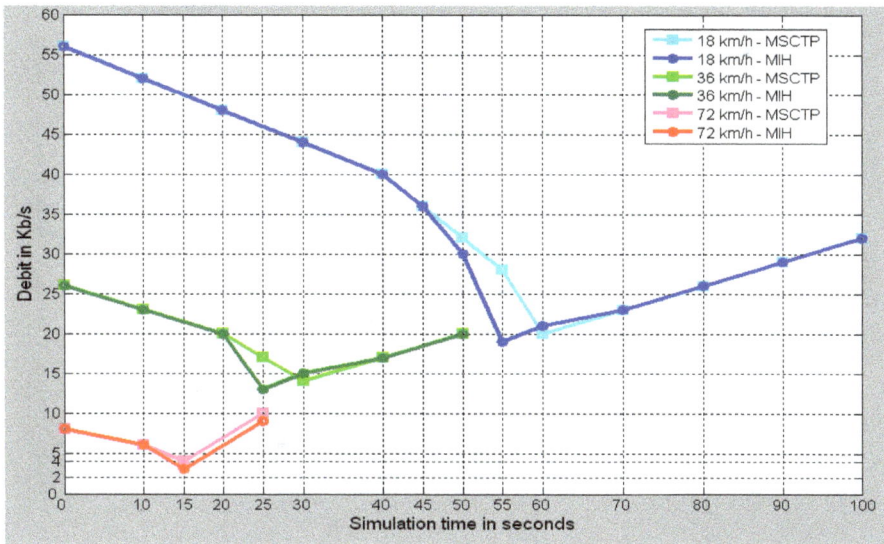

Fig. 21. Debit of HO from 802.11s to 802.16e / VoIP

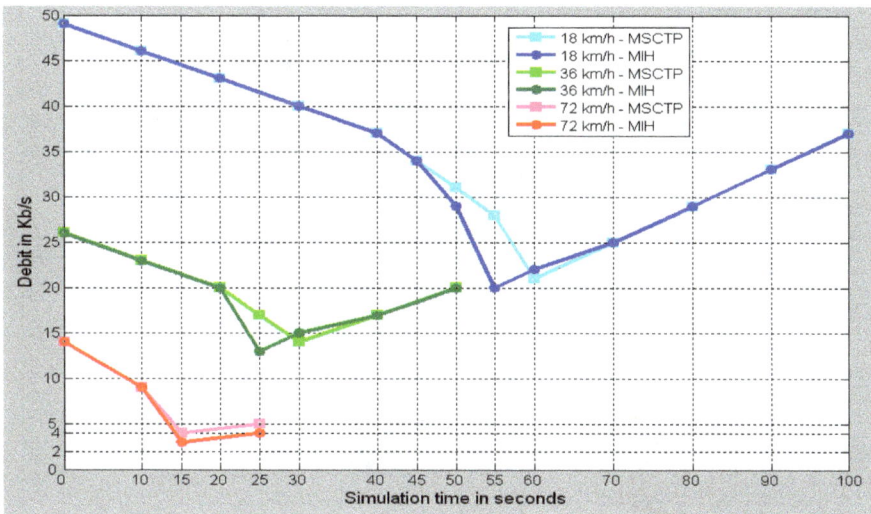

Fig. 22. Debit of HO from 802.16e to 802.11s / VoIP

We present now the results of data traffic (see figures 23 and 24):

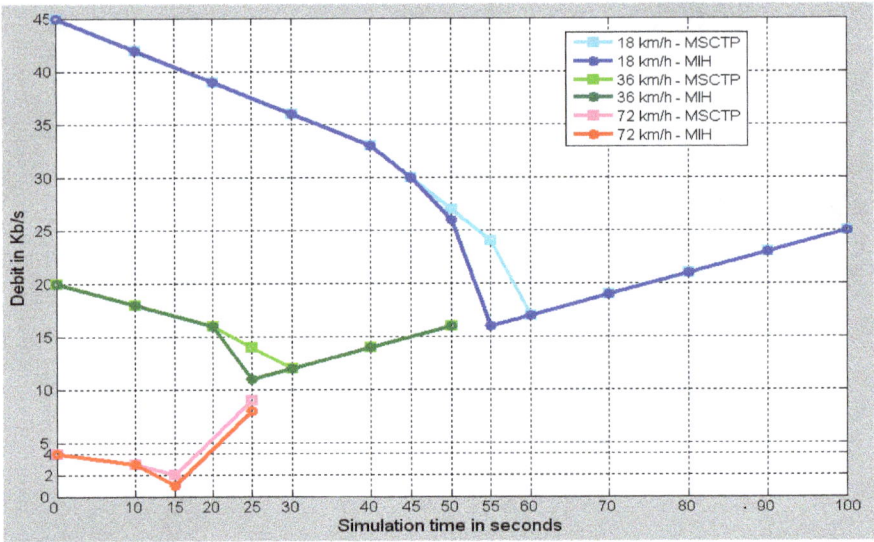

Fig. 23. Debit of HO from 802.11s to 802.16e / Data

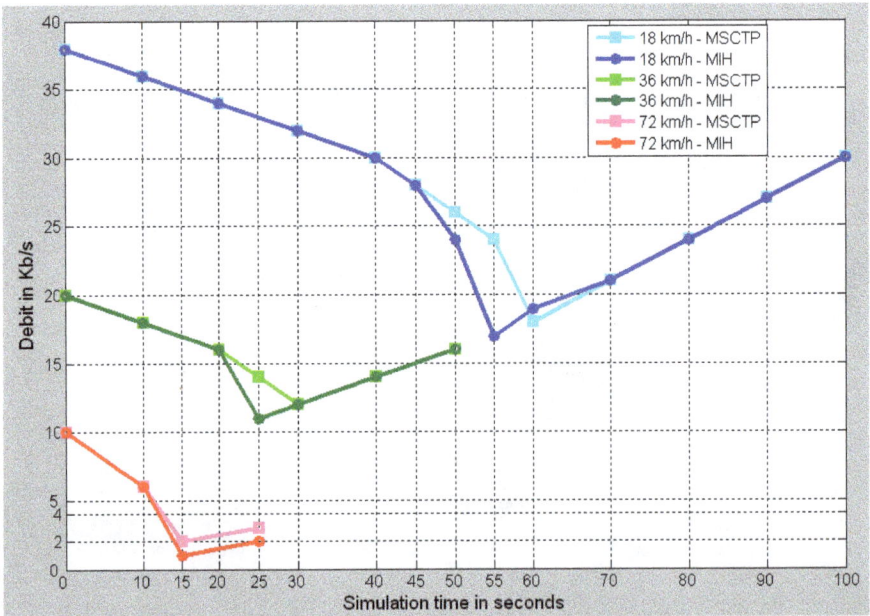

Fig. 24. Debit of HO from 802.16e to 802.11s / Data

With the data traffic, the debit decrease when the speed increase. The debit is better in 802.11s cell when the speed is weak, and it is better in 802.16e when the speed is high.

We present finally the results of video traffic (see figures 25 and 26):

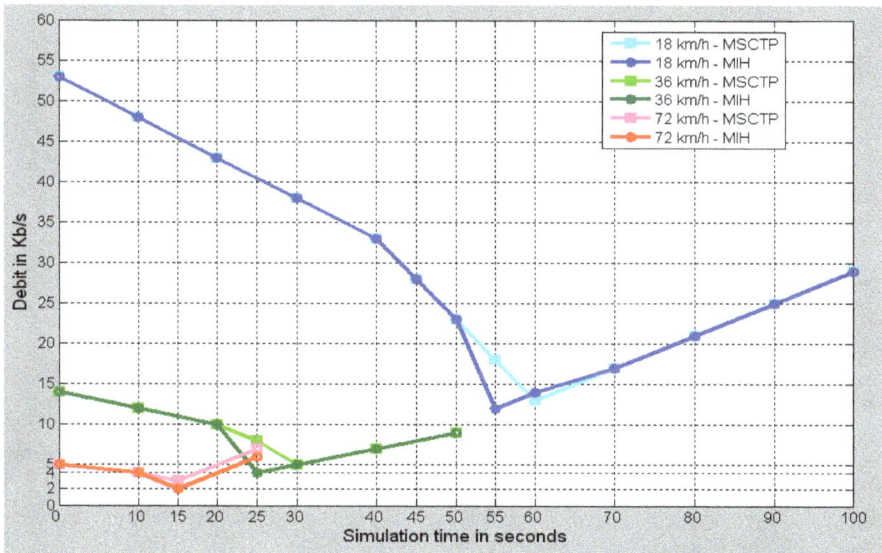

Fig. 25. Debit of HO from 802.11s to 802.16e / Video

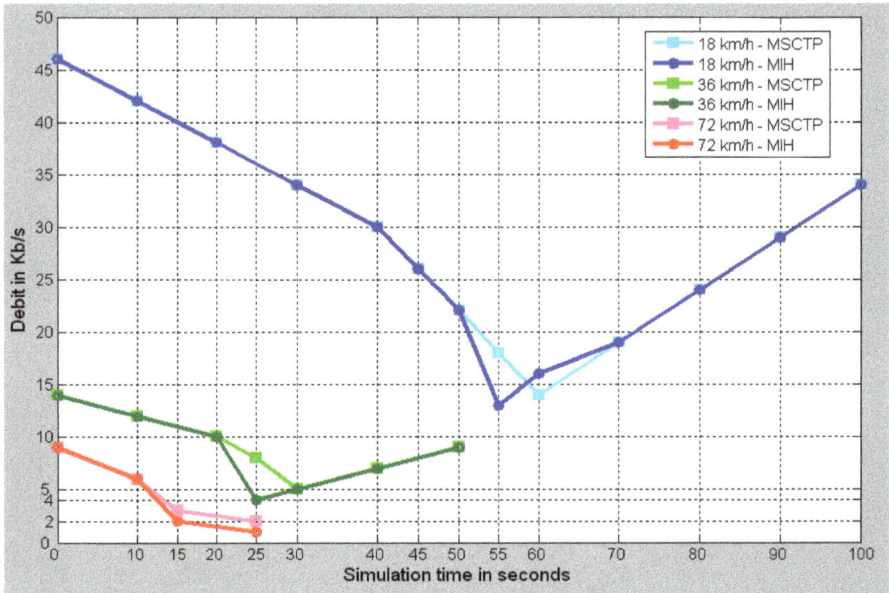

Fig. 26. Debit of HO from 802.16e to 802.11s / Video

With the video traffic, the debit values decrease comparing to the other traffic types.

5. Conclusion

The interoperability and the vertical handover between different networks present currently a real challenge to overcome. The difference of networks operation is the main reason of this problem. And, for pass to the 4G networks, it is important to resolve this problem of interoperability between different networks.

Our work has focused on the interconnection between two wireless radio networks of the IEEE 802 family, and we are concentrating on the QoS aspect for several traffics types especially during the handover process. For doing that, we have proposed two interconnection models based on two recent handover mechanisms, and we have simulated those two models with three mobile speeds and in the both directions of networks.

Observing the results obtained, we can conclude that with a low or medium speed of displacement of a mobile station, the both techniques: IEEE 802.21 and MSCTP present a good solution during the vertical handover. With the two techniques, there are very few interruptions during the vertical handover. But based on details of simulation results, we notice that with MSCTP protocol we obtained a QoS level slightly better than that obtained with MIH architecture.

Also the handover from 802.11s to 802.16e generates results better than the opposite case of handover. But with a high speed, it is the opposite rather because the mobile WIMAX supports better the increasing speeds; and also the results in this case are still not acceptable comparing by QoS level needed for each traffic type.

It should be noted that during all the simulations, the scenarios proposed does not include cell congestion or lack of available resources.

For future work, we will propose interconnection models between networks of different family, we will mix a network world with a telecommunication world, and we will try to propose a handover mechanism adapted to the two entities that we will define.

6. References

[1] The Information Science Institute (ISI), *"The Network Simulator-NS-2"*, http://www.isi.edu/ nsnam/ns/.

[2] EEE Std, "Air Interface for Fixed and Mobile Broadband Wireless Access Systems," IEEE 802.16e, Part 16, February 2006.

[3] IEEE Std, "Air Interface for Fixed Broadband Wireless Access Systems," Local and metropolitan area networks, Part 16, 2004.

[4] arviz Yegani, *"WiMAX Overview,"* White paper, IETF-64 Cisco Systems, 2005.

[5] WiMAX Forum, *"WiMAX End-to-End Network Systems Architecture,"* Draft Stage 2: Architecture Tenets, Reference Model and Reference Points, June 2007.

[6] Steven Conner, Jan Kruys, Kyeongsoo Kim and Juan Carlos Zuniga, *"IEEE 802.11s Tutorial,"* Overview of the Amendment for Wireless Local Area Mesh Networking, IEEE 802 Planary, November 2006.

[7] Guido R. Hiertz, Sebastian Max, Rui Zhao, Dee Denteneer and Lars Berlemann, *"Principles of IEEE 802.11s,"* Computer Communications and Networks, 2007, ICCCN 2007.

[8] RFC 2960, "Stream Control Transmission Protocol," IETF, 2000.

[9] Stewart R., & al., IETF, *"Stream Control Transmission Protocol (SCTP) Dynamic Address Reconfiguration,"* IETF Internet, Draft, draft-ietf-tsvwg-addip-sctp-13.txt, November 2005.

[10] Koh, S., & al., *"mSCTP for Soft Handover in Transport Layer,"* IEEE Communication Letters, Vol. 8, No.3, pp.189-191, March 2004.

[11] Memory graduation, Esteban Zimanyi, "Performance analysis of vertical Handover between UMTS and 802.11 networks," 2005.

[12] Deng Feng, "Seamless Handover between CDMA2000 and 802.11 WLAN using mSCTP," Thesis, 2006.

[13] IEEE 802.21 tutorial, July 2006.

[14] Jared Stein, "Survey of IEEE 802.21 Media Independent Handover Services," April 2006.

[15] V. Gupta, "IEEE 802.21 standard and metropolitan area networks: Media Independent Handover services", Draft P802.21/D05.00, April 2007.

[16] K. Leung, G. Dommety, P. Yegani & K. Chowdhury, *"Mobility Management Using Proxy Mobile IPv4"*, Internet Draft, IETF, 2007.

[17] Information Sciences Institute (ISI), *"NSNAM web pages, 18.2 Two-Ray Ground reflection model,"* http://www.isi.edu/nsnam/ns/doc/node218.html, January 2009.
[18] WiMAX Community, "WiMAX fundamentals, 1.7.3 Quality of Service", June 2007.

Permissions

The contributors of this book come from diverse backgrounds, making this book a truly international effort. This book will bring forth new frontiers with its revolutionizing research information and detailed analysis of the nascent developments around the world.

We would like to thank Prof. Roberto C. Hincapie and Prof. Javier E. Sierra, for lending their expertise to make the book truly unique. They have played a crucial role in the development of this book. Without their invaluable contribution this book wouldn't have been possible. They have made vital efforts to compile up to date information on the varied aspects of this subject to make this book a valuable addition to the collection of many professionals and students.

This book was conceptualized with the vision of imparting up-to-date information and advanced data in this field. To ensure the same, a matchless editorial board was set up. Every individual on the board went through rigorous rounds of assessment to prove their worth. After which they invested a large part of their time researching and compiling the most relevant data for our readers. Conferences and sessions were held from time to time between the editorial board and the contributing authors to present the data in the most comprehensible form. The editorial team has worked tirelessly to provide valuable and valid information to help people across the globe.

Every chapter published in this book has been scrutinized by our experts. Their significance has been extensively debated. The topics covered herein carry significant findings which will fuel the growth of the discipline. They may even be implemented as practical applications or may be referred to as a beginning point for another development. Chapters in this book were first published by InTech; hereby published with permission under the Creative Commons Attribution License or equivalent.

The editorial board has been involved in producing this book since its inception. They have spent rigorous hours researching and exploring the diverse topics which have resulted in the successful publishing of this book. They have passed on their knowledge of decades through this book. To expedite this challenging task, the publisher supported the team at every step. A small team of assistant editors was also appointed to further simplify the editing procedure and attain best results for the readers.

Our editorial team has been hand-picked from every corner of the world. Their multi-ethnicity adds dynamic inputs to the discussions which result in innovative outcomes. These outcomes are then further discussed with the researchers and contributors who give their valuable feedback and opinion regarding the same. The feedback is then collaborated with the researches and they are edited in a comprehensive manner to aid the understanding of the subject.

Apart from the editorial board, the designing team has also invested a significant amount of their time in understanding the subject and creating the most relevant covers. They scrutinized every image to scout for the most suitable representation of the subject and create an appropriate cover for the book.

The publishing team has been involved in this book since its early stages. They were actively engaged in every process, be it collecting the data, connecting with the contributors or procuring relevant information. The team has been an ardent support to the editorial, designing and production team. Their endless efforts to recruit the best for this project, has resulted in the accomplishment of this book. They are a veteran in the field of academics and their pool of knowledge is as vast as their experience in printing. Their expertise and guidance has proved useful at every step. Their uncompromising quality standards have made this book an exceptional effort. Their encouragement from time to time has been an inspiration for everyone.

The publisher and the editorial board hope that this book will prove to be a valuable piece of knowledge for researchers, students, practitioners and scholars across the globe.

List of Contributors

Ismael Gutiérrez and Faouzi Bader
Centre Tecnològic de Telecomunicacions de Catalunya-CTTC, Parc Mediterrani de la Tecnologìa, Castelldefels, Barcelona, Spain

Yu-Jen Chi
Department of Electrical Engineering, National Chiao Tung University, Taiwan

Chien-Wen Chiu
Department of Electric Engineering, National Ilan University, Taiwan

Mohd Faizal Jamlos
School of Computer and Communication Engineering, University of Malaysia Perlis (UniMAP) to University Malaysia Perlis, Kangar, Perlis, Malaysia

Sarawuth Chaimool and Prayoot Akkaraekthalin
Wireless Communication Research Group (WCRG), Electrical Engineering, Faculty of Engineering, King Mongkut's University of Technology North Bangkok, Thailand

Cristian Anghel and Remus Cacoveanu
University Politehnica of Bucharest, Romania

Peter Drotár, Juraj Gazda, Dušan Kocur and Pavol Galajda
Technical University of Kosice, Slovakia

Zdenek Becvar and Pavel Mach
Czech Technical University in Prague, Faculty of Electrical Engineering, Czech Republic

M. Shokair, A. Ebian, and K. H. Awadalla
El-Menoufia University, Egypt

Zhuo Sun, Xu Zhu, Rui Chen, Zhuoyi Chen and Mingli Peng
Beijing University of Posts and Telecommunications, China

Oleksii Strelnitskiy, Oleksandr Strelnitskiy, Oleksandra Dudka, Oleksandr Tsopa and Vladimir Shokalo
Kharkiv National University of Radio Electronics (KhNURE), Kharkiv, Ukraine

Imran Baig and Varun Jeoti
Universiti Teknologi PETRONAS, Malaysia

Pooria Varahram and Borhanuddin Mohd Ali
Universiti Putra Malaysia, Malaysia

Cheng-Ming Chen and Pang-An Ting
Information and Communication Laboratories, Industrial Technology Research Institute (ITRI), Hsinchu, Taiwan

Seyyed Masoud Seyyedoshohadaei, Borhanuddin Mohd Ali and Sabira Khatun
Universiti Putra Malaysia (UPM), Malaysia

Tarek Bchini and Mina Ouabiba
ARTIMIA, Malakoff, France

www.ingramcontent.com/pod-product-compliance
Lightning Source LLC
Chambersburg PA
CBHW070725190326
41458CB00004B/1039